Tues: 6:45 - computers 310 Eng Bldg.

WED: E .I.T. 7p.m.

1

JOHN WILEY & SONS, INC.

New York · London · Sydney

PLASTIC DESIGN of STEEL FRAMES

LYNN S. BEEDLE

Research Professor of Civil Engineering
Fritz Engineering Laboratory
Department of Civil Engineering
Lehigh University

THIRD PRINTING, DECEMBER, 1964

Library of Congress Catalog Card Number: 58–13454

Printed in the United States of America

TO MY FAMILY

Preface

This book presents the principles and methods that are the basis for plastic design and shows how they may be used in the solution of practical building frame design problems. It was written for students of structural engineering—those in colleges and universities on the one hand, and those in engineering practice on the other. As a consequence, the first six chapters present the fundamental concepts and the methods of analysis and design; numerous examples are included for illustration, and problems are provided for the student. The last three chapters are concerned specifically with design.

Theoretically a structure may "fail" in a number of ways. It may reach its limit of usefulness through instability, fatigue, or excessive deflection. Alternatively, if none of these failure modes occur, then the structure will continue to carry load beyond the elastic limit until it reaches its ultimate load, through plastic deformation, and then collapses. Most indeterminate structural steel building frames (and some other structures as well) fall into this latter category. Plastic analysis provides a rational method for basing the design on this most typical mode of failure. In other words, it bases the design on the maximum load the structure will support.

The concept of ductility of structural steel, presented in Chapter 1, forms the basis for the plastic theory of bending which is discussed in Chapter 2. The plastic method of analyzing a structure for ultimate load is given in Chapter 3. Proper attention must be given in design to the effect of shear force, axial force, local and lateral buckling, etc.; further, the beams, columns, and connections must be proportioned to meet the requirements that the method imposes. Emphasis upon these and other design problems commences in Chapter 4, and continues in Chapter 5 ("Connections") and in Chapter 6 ("Deflections"). In order to conserve space it has been necessary frequently to omit detailed derivations, particularly in portions of Chapter 4; in these instances, reference has been made to other sources of information. A summary prepared from a design point of view is given in Chapter 7, together with "general provisions" for materials, fabrication, loads and forces, and the load factor. Design examples are then given in Chapters 8 and 9 to illustrate the principles and methods of the earlier chapters. The calculations in these examples have been made as complete as possible; thus it will be apparent that many obvious short cuts were not used.

What will plastic design mean? Since the structure will look just the same as an elastically designed structure, there will be no significant change in appearance. To the engineer it will mean a more rapid method of design. To the owner it will mean economy, because plastic design requires less steel. For the building authority it should mean a more rapid checking operation. To the steel industry it will mean more efficient use of its product, one consequence of which is its more frequent use. Thus, plastic design will mean that better use has been made of the material and intellectual resources with which Almighty God has so richly blessed us.

Acknowledgment is first due those organizations that have sponsored research at Lehigh University into the plastic behavior of structural members and frames. They include the American Institute of Steel Construction, American Iron and Steel Institute, Column Research Council, Welding Research Council, and the Navy Department. T. R. Higgins, Director of Engineering and Research of the American Institute of Steel Construction has been Chairman of the project committee representing these sponsors, and the author is sincerely appreciative for his continued support of the research and for his encouragement. Representatives of the other sponsors have been most helpful, too; these include William Spraragen (WRC), Samuel Epstein (AISI), John M. Crowley (ONR), John Vasta (BuShips), and LaMotte Grover (WRC). The results obtained in this research program provide the main background of material upon which this book is based. The investigation began at Lehigh University under the direction of Bruce G. Johnston. The writer further acknowledges the work of his colleagues—George C. Driscoll, Jr., Cyril D. Jensen, Robert L. Ketter, K. E. Knudsen, F. W. Schutz, Jr., and Bruno Thurlimann—who supervised phases of those studies and who have made substantial contributions to knowledge on the subject of plastic design. Also, the work of the numerous Research

Assistants who participated in that program and whose names appear in many of the references is gratefully acknowledged.

Appreciation is due T. R. Higgins, Edward R. Estes, Jr., Bruce G. Johnston, and Jonathan Jones for their reviews of portions of the manuscript, for their contributions to it, and for their encouragement throughout the project. The author is appreciative of the advice and support given him by Professor Wm. J. Eney, Head of the Civil Engineering Department and Director of Fritz Engineering Laboratory, Lehigh University.

The manuscript was typed by Mrs. Patricia Moscony, Mrs. Grace Check, Miss Judith Szakaly, and Miss Joan Szakaly. Sincere thanks are expressed to Paul R. Webster who was so generous with his time in assisting with the proofreading.

To all of these, the author is sincerely indebted.

LYNN S. BEEDLE

Bethlehem, Pa.
October, 1958

Contents

1 Introduction

1.1 PLASTIC DESIGN—THE CONCEPT

Plastic design is an advantageous replacement for conventional elastic design as applied to statically loaded structural steel frames of certain types. These are rigid-jointed frames, continuous or restrained beams and girders, and statically indeterminate structures in general which are stressed primarily (although not exclusively) in bending.

It is not suggested that plastic design be applied to steel frames with statically determinate beams and girders, nor to simple structures with effectively pin-connected members. The significance of this distinction between a determinate structure and an indeterminate one is as follows. A simply supported beam under a given set of loads has one point where the moment is a maximum. As the loading increases, this moment increases proportionately until the extreme fiber stress equals the yield point of the steel. If the loading is further increased, the material will deform more rapidly due to the ductility of steel, and the deflection will increase at a greater rate. Although there will be a modest increase in load above the elastic limit, this increase is accompanied by rapidly increasing deflections. It is therefore reasonable to consider the load that causes first yield as the critical load for a simple beam, the small plastic reserve of strength constituting an additional margin of safety.

A continuous beam or rigid frame behaves quite differently. Whereas the moment diagram for a simple beam contains but one maximum or "peak" point, an indeterminate structure has two or more points of peak moment. As load is applied to the structure, the cross section at the greatest of these peaks will reach the yield point. As more load is added, a zone of yielding develops there, but elsewhere the structure is still elastic and this fact serves to control the total deflection. Due to the ductility of steel the moment remains about constant at the first zone of yielding, and thus the structure must call upon its less heavily stressed portions to carry any further increase in load. Eventually, zones of yielding are formed at other points of peak moment; and when

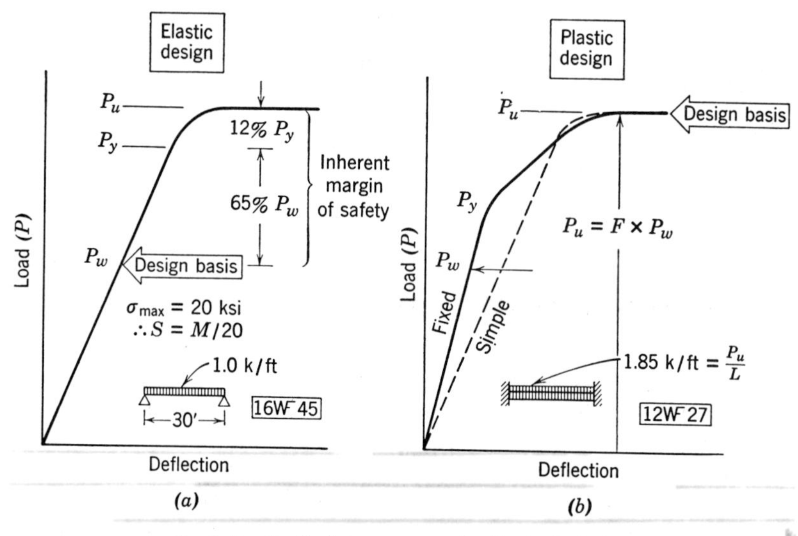

Fig. 1.1. Plastic design compared with elastic design.

a sufficient number of these yield zones have developed, the structure will fail.

Figure 1.1 illustrates the difference in behavior just described. The simply-supported beam of sketch (*a*) has but one point of maximum moment, and thus its true or ultimate load-carrying capacity is but little above the yield load P_y, as shown by the corresponding load-deflection curve. On the other hand, in the fixed-ended beam of sketch (*b*) there are three "peaks" in the moment diagram—the two at the ends which are equal and the one at the center. The corresponding load-deflection curve shown by the heavy solid line confirms that failure does not correspond to attainment of the elastic limit at the ends. Quite the contrary, there is a considerable reserve of load-carrying capacity beyond the yield load P_y. The failure (or "ultimate") load P_u is not reached until yield zones have developed not only at the ends, but also at the center. Through plastic design this reservoir of strength beyond the elastic limit may be utilized.

Therefore the essence of plastic design is to relate the safe or working load to the ultimate load thus found, arriving thereby at a somewhat smaller member than if first yield controlled the design as it does in elastic methods.

It is in order here to add two qualifying statements:

(1) The margin of safety proposed for plastic design is not less than that provided against failure of a simply supported beam, according to usual past practice.

(2) The reduction of member sizes made possible through plastic design will not be such as to deform the structure beyond limits that have proven suitable in the past.

In the remainder of this chapter, the foregoing simplified description of the plastic design concept will be examined and supported in some detail.

1.2 EARLY DEVELOPMENT OF PLASTIC ANALYSIS AND DESIGN

The concept of design based upon ultimate load as the design criterion is more than 40 years old. The application of plastic analysis to structural design appears to have been initiated by Dr. Gabor Kazinczy, a Hungarian, who published results of his tests of clamped girders as early as 1914.[1.1] He also suggested analytical procedures similar to those now current, and designs of apartment-type buildings were actually carried out. Early tests in Germany were made by Maier-Leibnitz [1.2] who showed that the ultimate capacity was not affected by settlement of supports of continuous beams. In so doing he corroborated the procedures previously developed by others for the calculation of maximum load capacity. The efforts of Van den Broek [1.3] in this country and J. F. Baker [1.4, 1.5] and his associates in Great Britain to actually utilize the plastic reserve strength as a design criterion are well known. Progress in theory of plastic structural analysis (particularly that at Brown University) has been summarized by Symonds and Neal.[1.6]

For many years the American Institute of Steel Construction, the Welding Research Council, the Navy Department, and the American Iron and Steel Institute have sponsored studies of the plastic behavior of structures at Lehigh University.[1.7, 1.8, 1.9] These studies have featured not only the verification of the plastic method of analysis through appropriate tests on large structures, but have also given particular attention to the conditions that must be met to satisfy important secondary design requirements.

Plastic design has now "come of age." It is already a part of certain specifications, and numerous structures both in Europe and North America have been constructed to designs based upon this method.[1.10, 1.11, 1.12, 1.13]

1.3 THE DUCTILITY OF STEEL

Steel possesses ductility—a unique property that no other structural material exhibits in quite the same way. Through ductility, structural steel is able to absorb large deformations beyond the elastic limit without the danger of fracture. It is this characteristic feature of structural

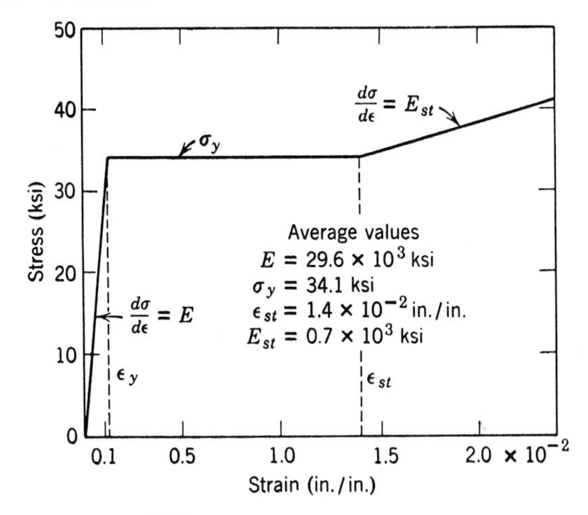

Fig. 1.2. Stress-strain curve of A7 structural steel, idealized. (*Proc. AISC Nat'l Engr. Conf.*, 1956.)

steel that makes possible the application of plastic analysis to structural design.

This ductility is evident from Fig. 1.2, which shows in somewhat idealized form the stress-strain properties of steel in the initial portion of the curve. This idealization is a very close approximation to the actual behavior of structural steel as revealed by measurements made with ordinary extensometers.[1.7, 1.14] The compressive and tensile stress-strain relationships are found to be practically identical, and are assumed so in the plastic theory.

The values shown in Fig. 1.2 are averages determined from many measurements made at Fritz Laboratory. For purposes of calculations to be made in the examples, the following values will be used:

$$E = 30.0 \times 10^3 \text{ ksi}$$

$$\sigma_y = 33.0 \text{ ksi}$$

$$\frac{\epsilon_{st}}{\epsilon_y} = 12.0$$

$$E_{st} = 700 \text{ ksi}$$

where E is the modulus of elasticity, σ_y is the yield stress level, ϵ_{st} is the strain at the onset of strain-hardening, ϵ_y is the strain at first attain-

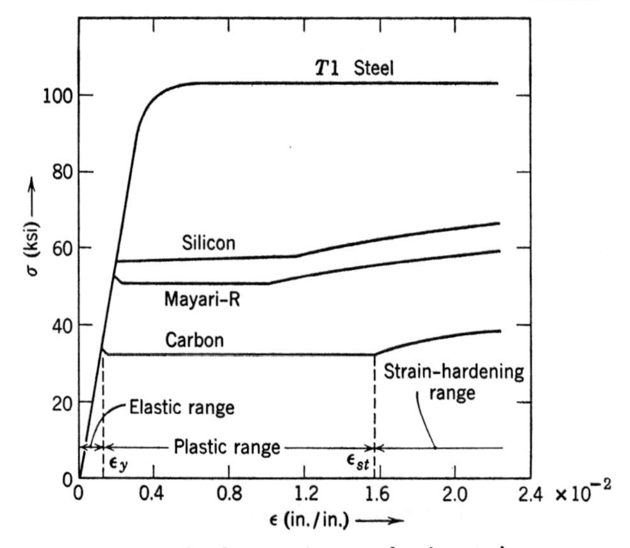

Fig. 1.3. Stress-strain curve of various steels.

ment of yield stress level, and E_{st} is the strain-hardening modulus. There will, of course, be variations from these values. Any variations, however, will not affect plastic analysis or design to any greater extent than similar variations have affected elastic procedures in the past.

Figure 1.3 shows partial tensile stress-strain curves for a number of different steels, the elastic range, the plastic range, and the strain-hardening range being indicated. Note that when the elastic limit is reached, elongations from 8 to 15 times the elastic limit take place without any decrease in load. Afterwards some increase in strength is exhibited as the material strain hardens. Although the first application of plastic design is to structures fabricated of structural grade steel, it is no less applicable to steels of higher strength as long as they possess the necessary ductility. Figure 1.3 attests to the ability of a wide range of steels to deform plastically with characteristics similar to ASTM A7 steel. It is, of course, from the strength of steel in the *plastic* range that the term *plastic* design evolved.

It is important to bear in mind that the strains shown in Fig. 1.3 actually are quite small. As shown in Fig. 1.4, for ordinary structural steel, final failure by rupture occurs only after a specimen has stretched

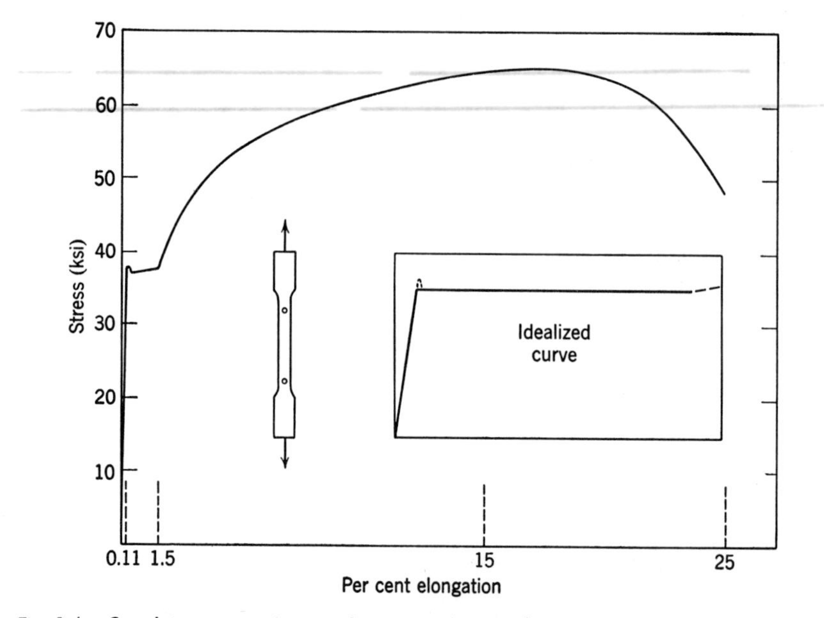

Fig. 1.4. Complete stress-strain curve for structural steel. (*Trans. ASCE*, 122, p. 1139, 1957.)

some 15 to 25 times the maximum strain that is encountered in plastic design. Even in plastic design, at ultimate load the critical strains will not have exceeded about 1.5% elongation. Thus the use of ultimate load as the design criterion still leaves available a major portion of the reserve ductility of steel which can be used as an added margin of safety. This maximum strain of 1.5% is a strain at ultimate load in the structure—not at working load. Further, this strain does not exist throughout the structure but only at a few critical sections and for a limited length in each section. In most cases, under working load the strains will still remain below the elastic limit.

Although structural steel does not supply an infinite amount of plastic strain, its ductility is adequate to meet the requirements of plastic design. The upper limit of deformability of other structural materials would require careful study prior to the application of plastic design to them.

1.4 TACIT ACCEPTANCE OF DUCTILE BEHAVIOR

All but the most recent texts and specifications in the field of structural steel design have required that the fiber stresses should nowhere exceed the yield point of the material when a specified overload is applied. As shown in Art. 1.1 this has been an appropriate criterion for simple beams because deformations start increasing rapidly at yield-point stress. But if it were argued that yielding could not be permitted in any part of a structure, then much of the past and present practice would be completely ruled out.

Both in buildings and in bridges, specifications allow the designer to use average stresses due to bending, shear, and bearing that result in actual local yielding. Such cases occur in pins and rivets and at local points. This local yielding results from stress concentrations that are neglected in the simple design formulas. Plastic action is thus depended upon to insure the safety of steel structures, and experience has shown that average or *nominal* maximum stresses form a satisfactory basis for design.

Perhaps the outstanding example of this variance between elastic design assumptions and the actual truth is to be found in the ordinary riveted or bolted joint. The assumption commonly is made that each fastener carries the same shear force. This is true for the case of two fasteners in a line. When more are added (Fig. 1.5), then as long as the joint remains elastic, the outer fasteners must carry the greater proportion of the load. For the example with four rivets, if each rivet transmitted the same load, then between rivets C and D one plate would carry perhaps three times the force in the other. Therefore it would stretch three times as much and would necessarily force the outer rivet (D) to carry more load than was assumed. The actual forces would look something like those shown under the heading "Elastic" in Fig. 1.5. What eventually happens is that the outer rivets yield, redistributing forces to the inner rivets until all forces are about equal as shown. Therefore the basis for design of a riveted joint is really its ultimate load and not the attainment of first yield.

To a greater extent than we may realize, the maximum *strength* of a structure has always been the dominant design criterion. When the usual permissible working stress has led to designs that were consistently too conservative, then that stress has been changed. Present design procedures disregard local over-stressing at points of stress-concentration, etc. and long experience with similar structures so-designed shows that this is a safe procedure. Thus, the stresses that are calculated for elastic design purposes often are not true maximum stresses at all; they simply provide an index for structural design.

A number of examples will now be given in which the ductility of steel has been counted upon in elastic design—knowingly or not—but certainly not through direct application of plastic design procedures. The following listing is in two categories: (1) factors that are neglected because of the compensating effect of ductility, and (2) instances in which the working stresses have been revised because the "normal" value was too conservative. Following the listing, several examples in each category will be discussed in further detail.

I. Factors that are neglected:
   (1) Residual stresses (in the case of flexure) due to cooling after rolling.
   (2) Residual stresses resulting from the cambering of beams.
   (3) Erection stresses.
   (4) Foundation settlements.
   (5) Over-stress at points of stress-concentration (holes, etc.).
   (6) Bending stresses in angles connected in tension by one leg only.
   (7) Over-stress at points of bearing.
   (8) Nonuniform stress-distribution in splices, leading to design of connections on the assumption of a uniform distribution of stresses among the rivets, bolts, or welds. (Discussed above.)
   (9) Difference in stress-distribution arising from the "cantilever" as compared with the "portal" method of wind stress analysis.

II. Revisions in working stress due to reserve plastic strength:
   (10) Bending stress of 30 ksi in round pins.
   (11) Bearing stress of 40 ksi in pins in double shear.
   (12) Bending stress of 24 ksi in framed structures at points of interior support.

Consider item (1) for example. All rolled members contain residual stresses that are formed due to cooling after rolling or due to cold-straightening. Figure 1.6 shows a typical WF shape (sketch *a*) with a characteristic residual stress pattern (sketch *b*). When load-carrying bending stresses are applied (sketch *c*), the resulting strains are additive to the residual strains already present. As a result, the "final stress" (sketch *d*) could easily involve yielding at working load. In the example of Fig. 1.6 such yielding has occurred both at the compression flange tips and at the center of the tension flange (sketch *e*). Thus it is seen

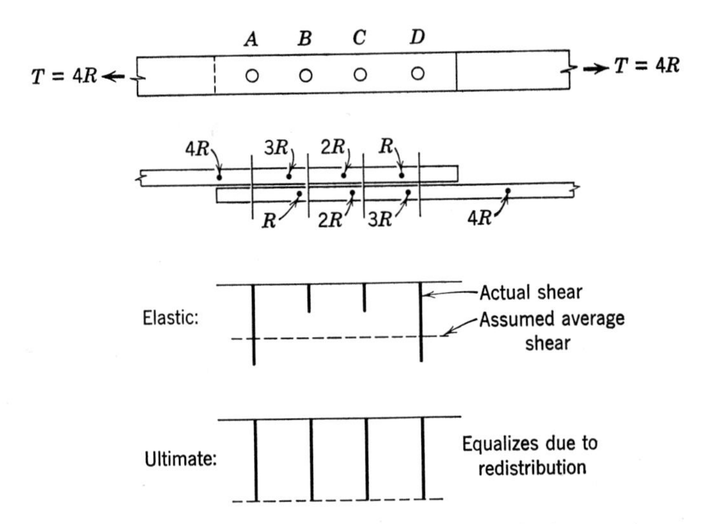

Fig. 1.5. Redistribution of shear in the fasteners of a lap joint.

that cooling residual stresses (whose influence is neglected and yet which are present in all rolled beams) may cause yielding in a beam even below the working load.

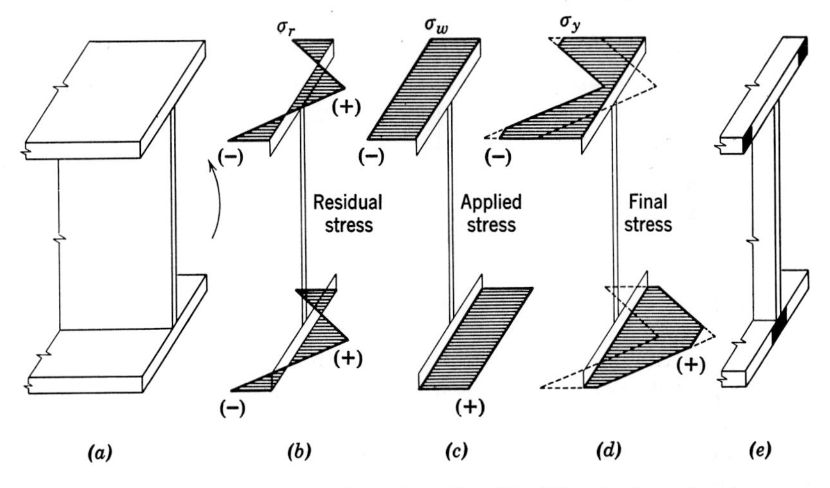

Fig. 1.6. Residual stresses due to cooling after rolling (*b*). When load-carrying stresses are applied (*c*), yielding may occur at elastic design working load (*d*) and (*e*).

As a justification for neglecting erection stresses (see item 3 on page 8), Fig. 1.7 shows how an erection force due to dimensional inaccuracy may introduce bending moments into a structure prior to the application of external load. (See the first line for $P = 0$. The load F is required to force the frame into position.) Although the yield-point load is reduced as a result of these "erection moments" (in the second line of the figure, the yield-point load has been reached for case b), *there is no effect whatever on the maximum strength.* The reason for this is that "redistribution of moment" (to be discussed in Chapter 2) follows the onset of yielding at the corners (case b) until the maximum moment

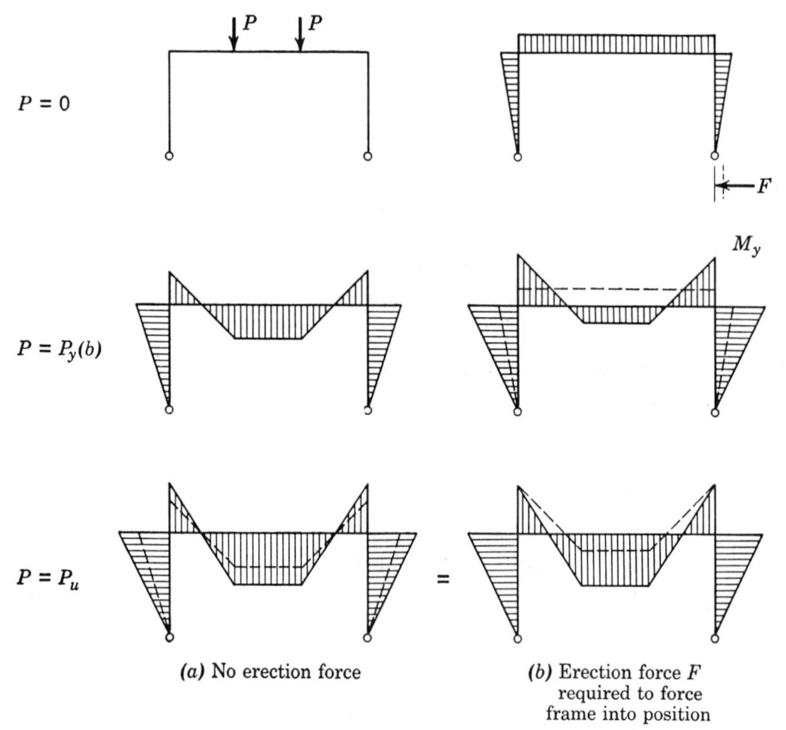

Fig. 1.7. Demonstration that erection stresses do not influence ultimate load.

is reached at the beam center (see the third line); therefore the ultimate load moment diagrams for cases a and b are identical.*

Figure 1.8 is a "revised working stress" example (see Item 10, page 8) and is concerned with the design of a round pin. In a simple beam of WF shape, when the maximum stress due to bending reaches the yield point, most of the usable strength has been exhausted. However, for some cross-sectional shapes, much additional load may be carried without excessive deflections. The relationship between bending moment

* Throughout this book the moment diagram is plotted on the side of the member that is in tension. See page 63 for discussion of sign convention.

and curvature for WF and for round beams is shown in Fig. 1.8. (Methods for determining these curves will be described in Chapter 2.) The upper curve is for the pin, the lower for a typical WF beam, the non-dimensional plot being such that the two curves coincide in the elastic range. The maximum bending strength of the wide-flange beam is $1.14M_y$, whereas that of the pin is $1.70M_y$. The permissible design stresses according to specifications of the American Institute of Steel

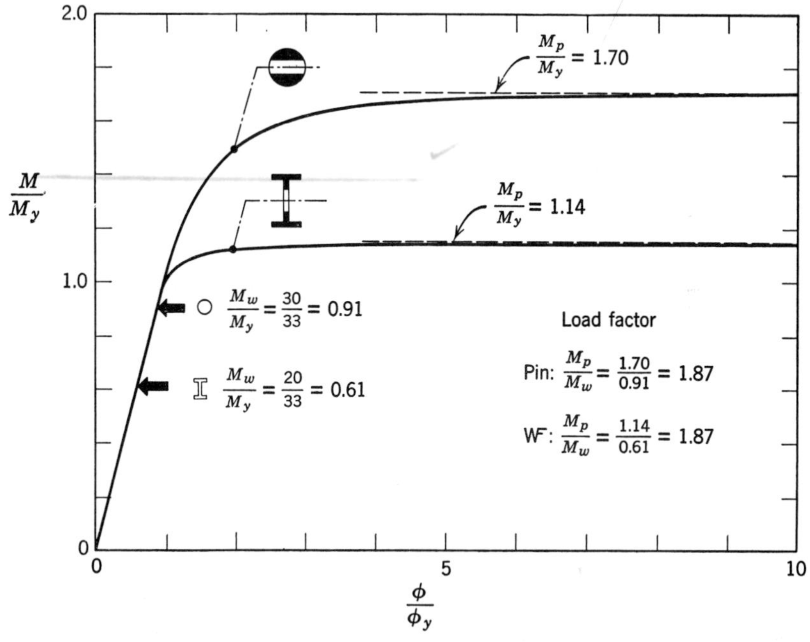

Fig. 1.8. The maximum strength of a round pin compared with that of a wide-flange beam.

Construction are 20 ksi for the WF beam and 30 ksi for the round pin. Expressing these stresses as ratios of yield point stress, for a WF beam

$$\frac{\sigma_w}{\sigma_y} = \frac{20}{33} = 0.61$$

for the round pin,

$$\frac{\sigma_w}{\sigma_y} = \frac{30}{33} = 0.91$$

For a simply supported beam the stresses, moments, and load all bear a linear relationship to one another in the elastic range and thus

$$\boxed{\frac{\sigma}{\sigma_y} = \frac{M}{M_y} = \frac{P}{P_y}}$$

Therefore, the moment at allowable working stress, M_w, in the WF beam is 61% of the yield moment, M_y; for the pin, on the other hand, the moment at allowable working stress is 91% of the yield moment. What is the true load factor of safety F for each case? As shown in the following tabulation, they are identical.

$$F_{\text{WF beam}} = \frac{P_{\max}}{P_w} = \frac{M_{\max}}{M_w} = \frac{1.14M_y}{0.61M_y} = 1.87$$

$$F_{\text{pin}} = \frac{P_{\max}}{P_w} = \frac{1.70M_y}{0.91M_y} = 1.87$$

The exact agreement between the true factors of safety with respect to ultimate load in the two cases, while somewhat of a coincidence, is

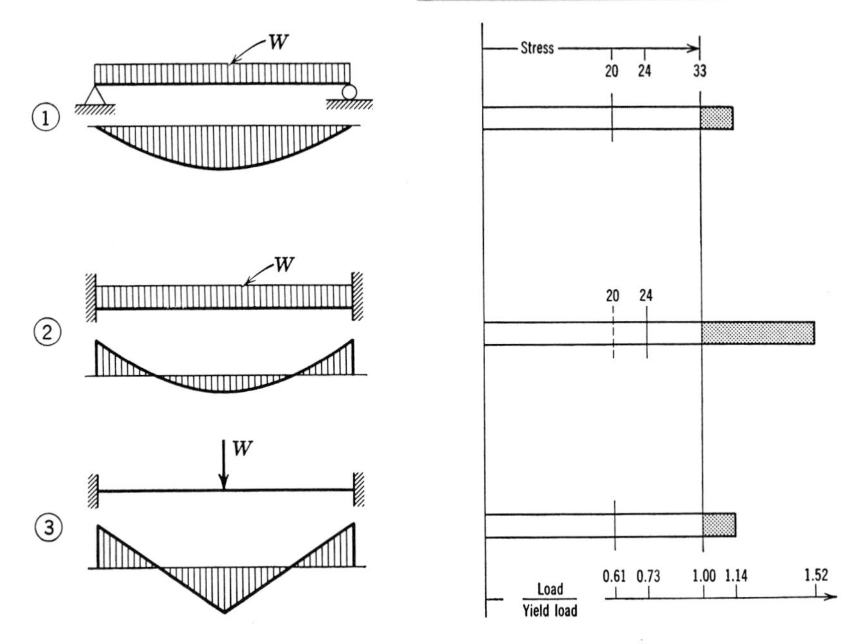

Fig. 1.9. Justification for the use of the "20% increase" in working stress for a fixed-ended beam, case 2.

indicative of the influence of long years of experience on the part of engineers which has resulted in different permitted working stresses for various conditions. Probably no such analysis as the foregoing influenced the choice of different unit stresses that give identical factors of safety with various sections; nevertheless, the choice of such stresses is fully justified on this basis. When years of experience and common sense have led to certain empirical practices, these practices can usually be justified on a scientific basis.

The permitting of a 20% increase in allowable working stress at points of interior support in continuous beams (item 12) represents another case in which both experience *and* a "plastic analysis" justify a revision in working stresses. Figure 1.9 demonstrates this by presenting the design of three structures: a simple beam (1), a uniformly loaded fixed-ended beam (2), and a fixed-ended beam with center concentrated load (3).* To the right, the bar graphs indicate on a stress scale (upper) and on a nondimensional load scale (lower) the theoretical strength of each structure. The common point of reference in each case is the load at computed yield, the stress scale being valid only up to that point. For case (1) the use of 20 ksi as the allowable stress provides a 65% margin of reserve strength up to the elastic limit and a further reserve of about 14% to ultimate. As is shown in case (2) the fixed-ended beam has a 52% reserve in strength *beyond first yield* (calculation follows in Chapter 2). Thus, the use of 24 ksi as the permissible design stress in this instance utilizes a portion of the reserve load-carrying capacity, and the load factor of safety is 1.52/0.73 = 2.04, a value greater than for the simple beam. Case (3) represents a situation in which the conventional design load is controlled by the stress in mid-span. This example therefore shows how a knowledge of plastic strength influenced the choice of permissible stresses in current U. S. design practice.

1.5 ULTIMATE LOAD AS THE DESIGN CRITERION

Once it is recognized that existence of yield-point stress at working load is no bar to further structural usefulness (provided that subsequent deformations be under control), then the way is open to study the usefulness of plastic design in which the design criterion is changed from unit stress to ultimate load.

It has been seen earlier that the method gives promise of reduction of member weight and hence of cost. As will be evident later, the method has a further advantage because it simplifies the design calculations. The reason for this is that the elastic condition of continuity need no longer be considered. The continuity condition is an essential and complicating feature of the elastic analysis of indeterminate frames, and yet in spite of its "exactness" it has often been demonstrated that elastic stress analysis cannot predict the real stress-distribution in a building frame with anything like the degree of accuracy that is assumed

* See footnote p. 10.

in the design. The work done in England by Prof. Baker and his associates as a forerunner to their studies of the maximum strength of structures clearly indicated this.[1.15]

Examples of "imperfections" that cause severe irregularity in measured stresses are: differences in beam-column connection fit-up and flexibility, spreading of supports, sinking of supports, residual stresses, flexibility assumed where actually there is rigidity (and vice-versa) and the presence

Fig. 1.10. Test of welded gabled frame of 40-ft span.

of stress concentrations. However such factors usually do not influence the maximum plastic strength and this simplifies the design.

As already implied the concept is more rational. Why should a simple beam have a margin of safety any different from that of a fixed-ended beam?

The final yardstick by which the suitability of a design method may be adjudged is the corroboration of theory by tests. Although the subject of design of steel frames on the basis of maximum strength is not new, it is only in recent years that sufficient tests of large-size structural members and frames have been performed to substantiate the theory. Typical is the frame shown in Fig. 1.10.[1.16] It represents a gabled frame of 40-ft span, loaded to simulate uniformly distributed vertical and horizontal loading (the frame is shown at the ultimate load). The results of the test are shown in Fig. 1.11. The loading is shown in the inset and the curve shows the relationship between vertical load and vertical deflection at the peak of the gable. The dotted line is the theoretical curve and the solid line is drawn connecting the test points. The ultimate load P_u and the allowable working load for plastic design P_w are also shown.

Figure 1.11 emphasizes two features of plastic design. At working loads, the structure is still within the elastic range and the deflections are still of the order of magnitude that are found in structures designed

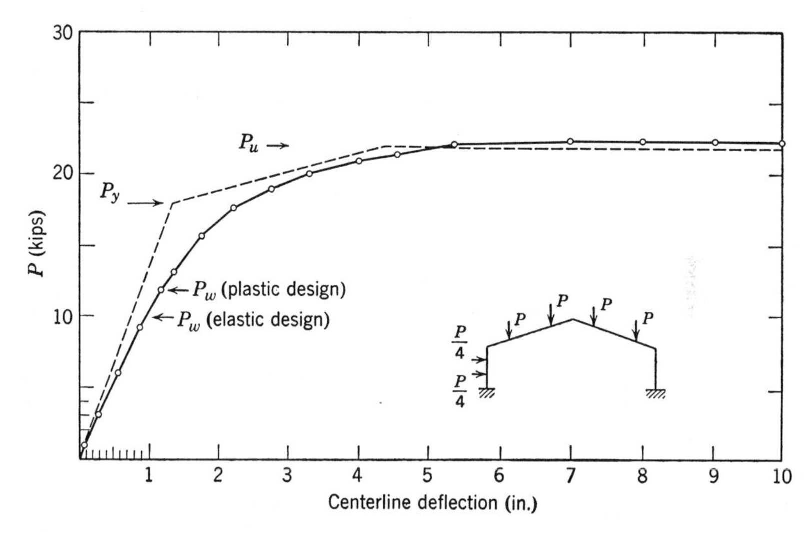

Fig. 1.11. Load-deflection curve of test frame shown in Fig. 1.10.[1.16]

by conventional methods. Secondly, the ultimate load (working load times load factor of safety) can be predicted by plastic analysis with an accuracy that far exceeds our ability to calculate the "elastic limit" by conventional elastic analysis. The agreement between the observed maximum load and the computed value is within three per cent; in contrast with this agreement, the observed load at first yield simply cannot be determined from the experimental curve.

Thus ultimate load is the appropriate design criterion for statically loaded frames of structural steel, continuous beams, single-story industrial frames, and for such other structures whose conditions of loading and geometry are consistent with the assumptions involved in the theory. Numerous applications will undoubtedly be made to other types of structures such as rings, arches, etc., but for the purposes of this book, the discussion will be limited to the types of structures indicated.

1.6 MAXIMUM STRENGTH OF ELEMENTARY STRUCTURES

On the basis of the ductility of steel (characterized by Fig. 1.2) it is now possible to quickly calculate the maximum carrying capacity of certain elementary structures. As a first example consider a tension member such as the eye bar shown in Fig. 1.12. The stress, σ, is

$$\sigma = \frac{P}{A}$$

where P is the applied load and A is the cross-sectional area. The

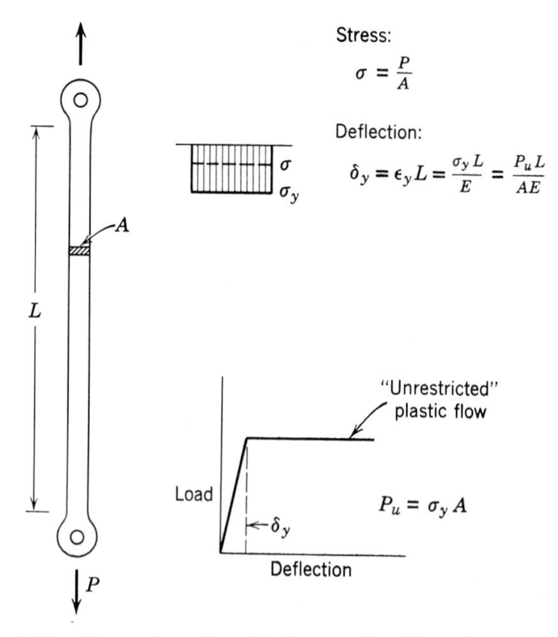

Fig. 1.12. The maximum strength of an eye bar (determinate structure).

load-versus-deflection relationship will be elastic until the yield point is reached. As shown in Fig. 1.12 the deflection at the elastic limit, δ_y, is given by

$$\boxed{\delta_y = \frac{P_u L}{AE}}$$

where P_u is the ultimate load and L is the length of the bar. Since the stress distribution is uniform across the section, unrestricted plastic flow will set in when the load reaches the value given by

$$\boxed{P_u = \sigma_y A}$$

Therefore this is the ultimate load for practical purposes and as postulated in design practice.

As a second example consider the three-bar structure shown in Fig. 1.13. The state of stress cannot be determined by statics alone and thus

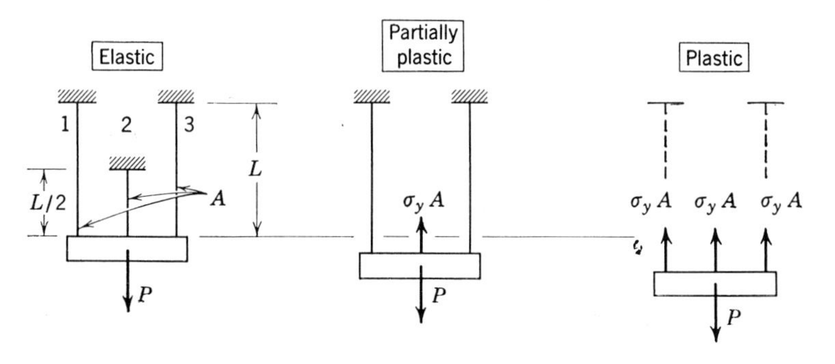

Equilibrium: $2T_1 + T_2 = P$

Continuity: $\Delta L_1 = \Delta L_2$

$$\frac{T_1 L_1}{AE} = \frac{T_2 L_2}{AE}$$

$$T_2 = 2T_1 = \frac{P}{2}$$

Yield limit: $P_y = 2T_2 = 2\sigma_y A$

Equilibrium: $P_u = 3\sigma_y A$

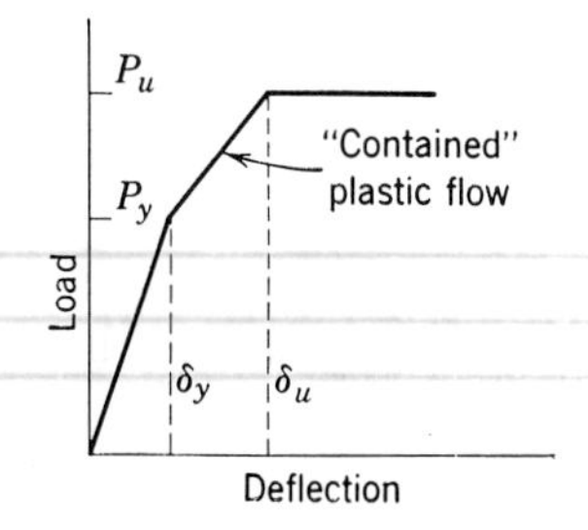

Fig. 1.13. Elastic and plastic analysis of an indeterminate system.

it is indeterminate. Consider first the elastic state. From the equilibrium condition there is obtained

$$2T_1 + T_2 = P \tag{1.1}$$

where T_1 is the force in bars 1 and 3, and T_2 is the force in bar 2. The next condition to consider is continuity. Assuming a rigid cross-bar, the total displacement of bar 1 will be equal to that of bar 2. Therefore,

$$\frac{T_1 L_1}{AE} = \frac{T_2 L_2}{AE} \tag{1.2}$$

$$T_1 = \frac{T_2}{2} \tag{1.3}$$

With this relationship between T_1 and T_2 obtained by the continuity condition, using Eq. 1.1 it is found that

$$T_2 = \frac{P}{2} \tag{1.4}$$

The load at which the structure will first yield may then be determined by substituting in Eq. 1.4 the maximum load which T_2 can reach, namely $\sigma_y A$. Thus

$$P_y = 2T_2 = 2\sigma_y A \tag{1.5}$$

The displacement at the yield load would be determined from

$$\delta_y = \epsilon_y L_2 = \frac{\sigma_y L}{2E} \tag{1.6}$$

Now, when the three-bar structure is partially plastic it deforms as if it were a two-bar structure except that an additional constant force equal to $\sigma_y A$ is supplied by bar 2 (the member which is in the plastic range). This situation continues until the load reaches the yield value in the two outer bars. Notice how easily one can compute the ultimate load:

$$P_u = 3\sigma_y A \tag{1.7}$$

The basic reason for this simplicity is that the continuity condition (or the indeterminacy due to continuity) need not be considered when the ultimate load in the plastic range is being computed.

The load–deflection relationship for the structure shown in Fig. 1.13 is indicated at the lower right. Even though yielding occurs in the center bar at the load P_y, the plastic flow is "contained" by virtue of the reserve capacity of the other parts of the structure, namely, the two long bars. Not until the load reaches that value computed by a plastic analysis (Eq. 1.7) do the deflections commence to increase rapidly. The deflection when the ultimate load is first reached, δ_u, can be computed from

$$\delta_u = \epsilon_y L_1 = \frac{\sigma_y L}{E} \tag{1.8}$$

The three important points involved in this simple problem in plastic analysis are as follows:

(1) Each portion of the structure (each bar) reaches the plastic yield condition.

f deflection rate which sets in when the yield stress is passed at the ιoment peaks at the beam ends. The deflection continues to be limited, owever, until at P_u a final zone of yielding develops at the center of ιe beam.

The important thing to note is that the factor of safety is chosen to e the same in the plastic design of the indeterminate structure as is ιown to be present in the elastic design of the simple beam. This ιherent margin of 1.85 having been accepted for so many years for that ιost common structural element, the simple beam, it seems logical to dopt the same margin as adequate for any indeterminate structure imilarly loaded.

1.8 THE CASE FOR PLASTIC DESIGN—SUMMARY

The goal in structural design is to provide a safe and enduring structure that incorporates maximum possible economy. If plastic analysis can be applied to design to realize these goals, it will be so applied, for the laws of evolution work as surely in the history of man-made structures as they do in a field such as biology.

The case for plastic design is supported by the following observations:

(1) Plastic design gives promise of economy in the use of steel, of saving in the design office by virtue of its simplicity, and of building frames more logically designed for greater over-all strength.
(2) The reserve in strength above the working loads computed by conventional elastic methods is considerable in indeterminate steel structures. Indeed, in some instances of elastic design, as much load-carrying capacity is disregarded as is used.
(3) Use of ultimate load as the design criterion provides at least the same margin of safety as is presently afforded in the elastic design of simple beams. (Fig. 1.1.)
(4) At working load the structure is normally in the so-called elastic range. (Fig. 1.1.)
(5) In most cases, a structure designed by the plastic method will deflect no more at working load than will a simply supported beam designed by elastic methods to support the same load. (Fig. 1.1.)

Plastic design is the realization of a goal that has been sought since the 1920's to see if some conscious design use could be made of the

(2) The equilibrium condition is satisfied at ultimat
(3) There is "unrestricted" plastic flow at the ultim

In principle, this simple example illustrates the essen
the plastic method; what is required to complete the
of an indeterminate beam or frame is to satisfy a plastic
an equilibrium condition, and an "unrestricted plas
dition.

Later in this book more complete study will be given
of analysis and design, and their simplicity as comp
elastic design methods for the same indeterminate stı
become evident.

1.7 MARGIN OF SAFETY

In the opening article it was pointed out that the m
of safety to be used in plastic design would be not less thar
in usual past practice against failure of a simply supporte
may be demonstrated as follows:

In conventional elastic design, a member is selected i
that the maximum allowable stress is equal to 20,000 pou
inch at the working load. As shown in Fig. 1.1, p. 2, a sim
beam has a reserve of elastic strength of 1.65 if the yield
33,000 pounds per square inch. Due to the ductility of
an additional reserve against failure which amounts to
the yield load for a wide flange shape. Thus the total inh
factor of safety is equal to $1.65 \times 1.12 = 1.85$. The design
by the open arrow.

Now in plastic design the selection of member sizes is b
ultimate load. This load P_u is computed by multiplying
load P_w by the same factor of safety that is *inherent* in
design (in this case, 1.85), and a member is selected t
support this factored load. In Fig. 1.1*b* is shown a fixec
designed plastically to support the same *working* load as tl
namely 1.0 k/ft. The corresponding ultimate load would
1.85 k/ft.

The load-versus-deflection curve for the restrained l
shown in Fig. 1.1*b*. The beam carries the same ultimate l
conventional design of the simple beam although its sect
is reduced, and it is elastic at working load. At P_y is noted

ductility of steel. This goal has been achieved because two important conditions have been satisfied. First, the theory concerning the plastic behavior of continuous steel frames has been systematized and reduced to simple design procedures. Secondly, every factor that might tend to limit the load-carrying capacity to something less than that predicted by the "simple plastic theory" has been investigated and design procedures have been formulated to safeguard against such limitations. In the following chapters the procedures of plastic analysis and design will be set forth and the necessary secondary design factors will be treated.

The reader who is interested in keeping abreast of theoretical and experimental developments in this field may do so through the American Society of Civil Engineers and the Welding Research Council. Reference 1.17 represents the first step in an effort to provide, on a continuing basis, a commentary on new provisions and procedures as they become available.

PROBLEMS

1.1 Define "ductility."

1.2 What size simple beam is required to support the load on the span of Fig. 1.1*a*. Explain. Does plastic design have application to simple beams? Why?

1.3 A rigid cross-beam is supported by three equally spaced tension rods. The two outside rods are of length L and area A and the center rod is of length $2L$ and area A. What maximum load P_u will the cross-beam support if the load is applied at the center? Draw the load-deflection curve. Compute P_y, δ_y, and δ_u.

1.4 Compute the ultimate load P_u for the structure shown in Fig. 1.14. For a rigid horizontal beam what is the load at first yield?

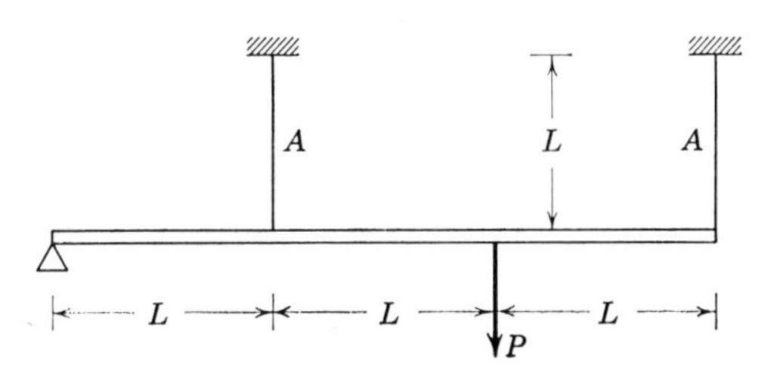

Fig. 1.14.

1.5 If $\frac{\epsilon_{st}}{\epsilon_y} = 12.0$, what will be the end deflection of the bar in Problem 1.4 when the load starts to increase above P_u?

1.6 If all the bars in Fig. 1.15 have equal areas, what is the ultimate load P_u if bars 2 and 3 are carbon steel and bar 1 is high-strength steel with a yield stress of three times that of plain carbon steel. Compute δ_u.

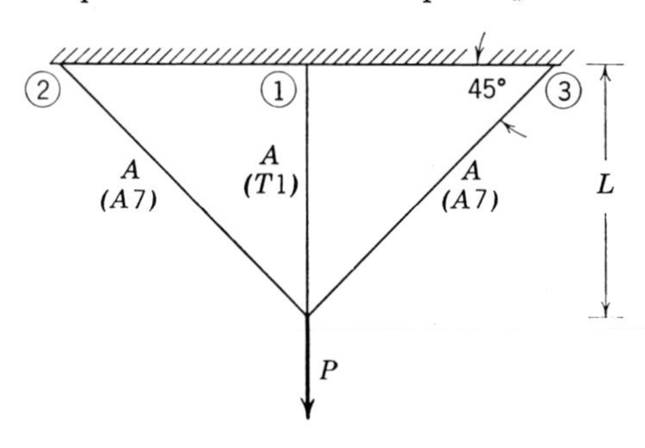

Fig. 1.15.

Flexure of Beams

2.1 INTRODUCTION

In Chapter 1 preliminary evidence was presented showing that full-size structures behave as predicted by the plastic theory. With the knowledge that this theory possesses certain advantages over conventional elastic theory and as a preliminary to studying methods of analyzing structures, consideration now will be given to determining the "maximum moment of resistance" of a beam. The development of "zones of yielding" in beams was described generally in Art. 1.1, and it was pointed out that the moment at such zones remained about constant while similar zones were developing elsewhere in the structure.

It is the first objective of this chapter to show how a beam deforms beyond the elastic limit under the action of bending moments, that is, to determine the moment–curvature relationship. Both rectangular and wide-flange beams are treated, and methods for computing the maximum moment (called the "plastic moment") are given. The zones of yielding that form as a result of plastic deformation are called "plastic hinges," and it is the next objective to describe the characteristics of a plastic hinge and to show how different factors can influence its formation and can affect the magnitude of the corresponding plastic moment. The moment transfer process which draws upon the reserve moment capacity of less highly stressed portions of the structure ("redistribution of moment") is next discussed. The chapter ends with a presentation of the results of experiments on beams.

2.2 ASSUMPTIONS AND CONDITIONS

Supplementing the elementary concepts of plastic analysis that were presented in Chapter 1 (Arts. 1.3 and 1.6), the following assumptions and conditions are used in the development of the moment–curvature (M–ϕ) relationship:

(1) Strains are proportional to the distance from the neutral axis (plane sections under bending remain plane after deformation).

(2) The stress-strain relationship is idealized to consist of two straight lines:

$$\begin{aligned} \sigma &= E\epsilon \qquad (0 < \epsilon < \epsilon_y) \\ \sigma &= \sigma_y \qquad (\epsilon_y < \epsilon < \infty) \end{aligned} \tag{2.1}$$

The "idealized" stress-strain diagram corresponding to Eq. 2.1 is shown in Fig. 2.1. It is emphasized again that the strains are quite small and,

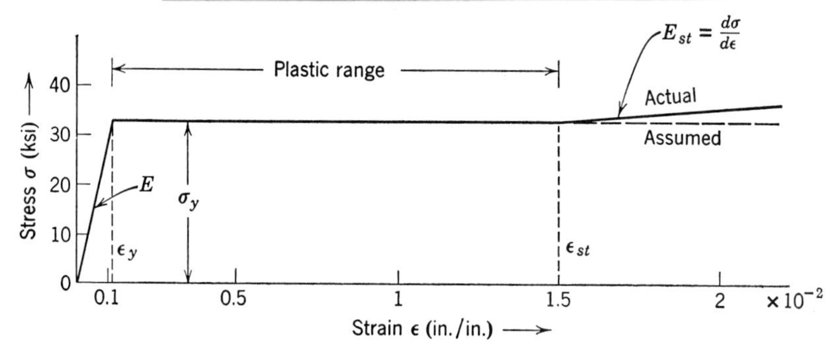

Fig. 2.1. Idealized stress-strain diagram for steel, showing pertinent nomenclature.

by comparison with Fig. 1.4, p. 6, it is evident that the major portion of the ductility is still available. The properties in compression are assumed to be the same as those in tension. Also, the behavior of fibers in bending is assumed to be the same as in tension or compression.

(3) Deformations are sufficiently small so that $\phi = \tan \phi$ (ϕ = curvature).

(4) The equilibrium conditions are as given by Eq. 2.2:

$$\begin{aligned} &\text{Normal force:} \quad P = \int_{\text{Area}} \sigma\, dA \\ &\text{Moment:} \quad M = \int_{\text{Area}} \sigma\, dA \cdot y \end{aligned} \tag{2.2}$$

where σ = stress at distance y from the neutral axis.

2.3 BENDING OF A RECTANGULAR BEAM

As a background and for later comparison with the inelastic case, the equations for elastic bending are:

$$\phi = \frac{1}{\rho} = \frac{\epsilon}{y} = \frac{\sigma}{Ey} \tag{2.3}$$

and

$$M = EI\phi \tag{2.4}$$

and

$$M_y = \sigma_y S \tag{2.5}$$

where ϕ = curvature
ρ = radius of curvature
ϵ = strain
y = distance from neutral axis to fiber
σ = stress at distance y from neutral axis
σ_y = yield stress level
E = modulus of elasticity
I = moment of inertia
S = section modulus (I/c)

These equations and terms all should be familiar to the reader. Any strength of materials text may be consulted for review.

In the plastic range the moment–curvature relationship and the magnitude of the maximum plastic moment are derived by following the same procedures as in elastic analysis, that is, by considering the deformed structure and obtaining the corresponding curvature and moment. Figure 2.2 shows the development of strain, stress, and yield distribution as a rectangular beam is bent in successive stages beyond the elastic limit (stage 1) and up to the plastic limit (stage 4). The strain distribution is first selected or assumed and this establishes the stress-distribution.

At the top of Fig. 2.2 is replotted for reference purposes the stress-strain relationship. At stage 1, as shown in the next line of Fig. 2.2 the strains have reached the yield strain. When more moment is applied (say to stage 2), the extreme fiber strains are twice the elastic limit value. The situation is similar for stage 3 ($\epsilon_{max} = 4\epsilon_y$). Finally at stage 4 the extreme fiber has strained to ϵ_{st}.

What are the stress distributions that correspond to these strain diagrams? These are shown in the next line of Fig. 2.2. As long as the strain is greater than the yield value ϵ_y, then from the stress-strain curve it is evident that the stress remains constant at σ_y. The stress

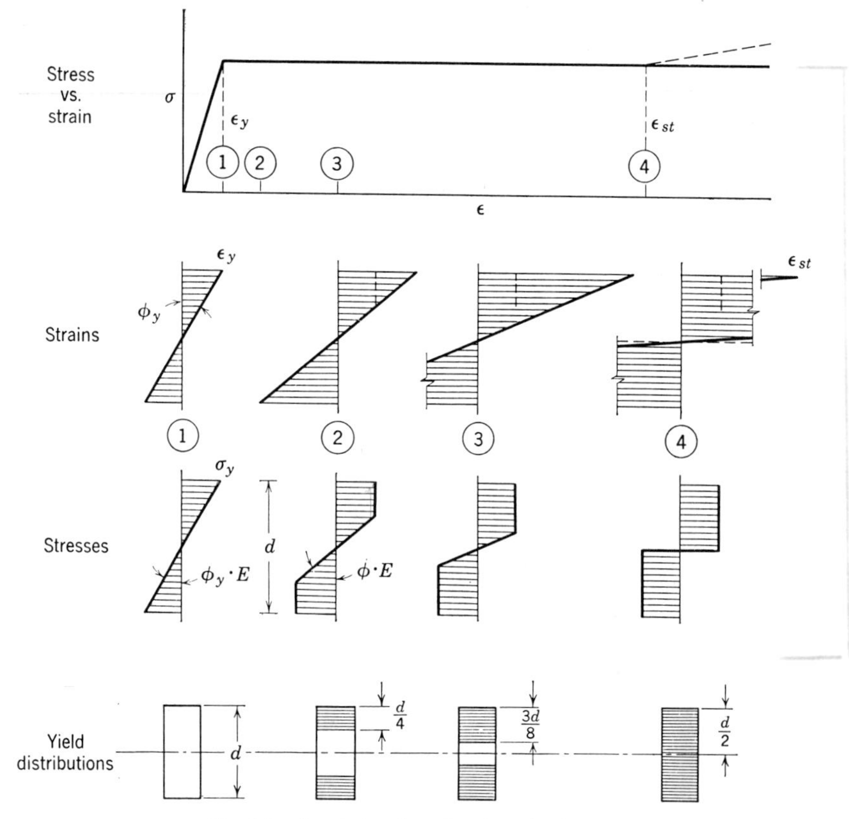

Fig. 2.2. Plastic bending of rectangular beam.

distributions, therefore, follow directly from the assumed strain distributions. As a limit the "stress block" is obtained—a rectangular pattern that is very close to the stress distribution at stage 4.

Figure 2.2 also shows the curvature, which is the relative rotation of two sections a unit distance apart. According to the first assumption (as in elastic bending)

$$\phi = \frac{1}{\rho} = \frac{\epsilon}{y} = \frac{\sigma}{Ey} \tag{2.3}$$

Just as it is of basic importance to elastic analysis, the relationship of bending moment to the curvature, ϕ, is a basic concept in plastic analysis.

The expressions for curvature and moment (and, thus, the resulting M–ϕ curve) are derived by reference to Fig. 2.3 as follows:

(a) Curvature. Curvature at a given stage is obtained from the particular stress distribution. Whereas curvature reflects strain-distribution, rather than stress-distribution, the curvature of a partially plastic section is controlled by the deformations of the still-elastic interior fibers; the yielded outer fibers offer no additional resistance to deformation.

Thus the curvature at stage 2 of Fig. 2.2 is derived from the pattern of Fig. 2.3, in which y_o is the ordinate to the furthest still-elastic fiber

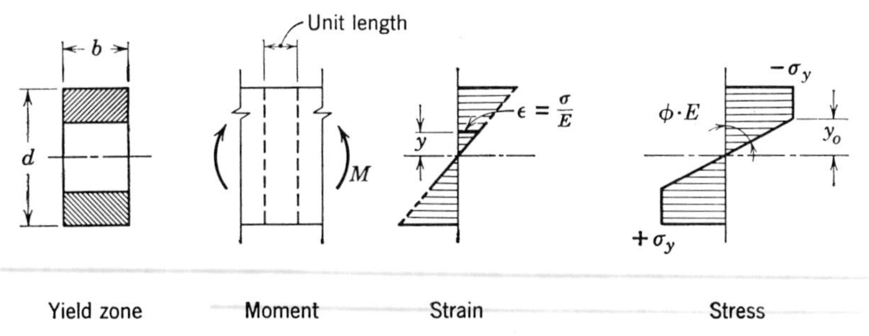

Fig. 2.3. Stress-distribution in a partially plastic rectangular cross section.

and the angular rotation is as indicated by arrows. The maximum elastic strain is given by

$$\epsilon = \frac{\sigma_y}{E}$$

Then

$$\tan \phi = \phi = \frac{\epsilon}{y_o}$$

or

$$\phi = \frac{\sigma_y}{E y_o} \tag{2.6}$$

(b) *Moment.* The moment at a given stage is obtained by integrating the stress areas. Thus the moment at stage 2 of Fig. 2.2 is derived from Fig. 2.4, in which the stress distribution of Fig. 2.3 is restated in three parts that are chosen only for convenience of calculation. Then

$$M = \sigma_y S_e + \sigma_y Z - \sigma_y Z_e \tag{2.7}$$

where S_e is the section modulus of the elastic interior portion I/c
Z is the "plastic modulus" (defined below) that would apply if the entire cross section were plastic and
Z_e is the plastic modulus that would apply to the interior portion that is in fact not plastic but elastic.

Then

$$M = \sigma_y (S_e + Z_p) \tag{2.7a}$$

Zp = Z - Ze

where $Z_p = Z - Z_e$. The subscripts e and p refer to the elastic and plastic portions, respectively, of the cross section.*

The function Z, called the plastic modulus, corresponds in application to the elastic section modulus, S. It is easily found from Fig. 2.4 to be

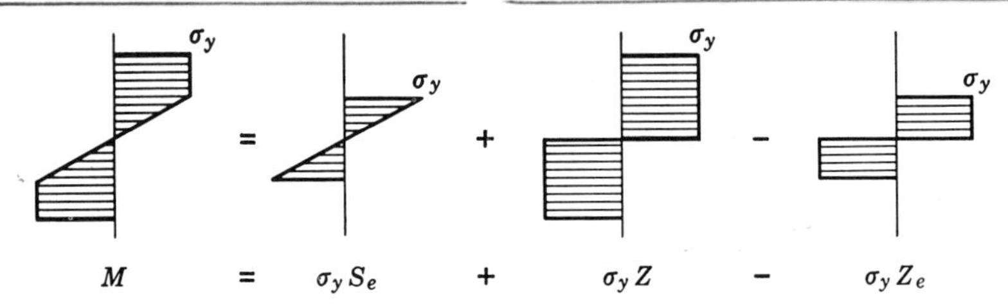

Fig. 2.4. Stress elements of a partially plastic section.

equal to the sum of the statical moments (taken about the neutral axis) of the plastic sectional areas above and below that axis. General methods for computing Z will be discussed later. Z_e is similarly calculated from the statical moments of the areas remaining elastic.

For the rectangular section, necessary values for section modulus S_e and partial plastic modulus Z_p for use in Eq. 2.7 may be derived from Figs. 2.3 and 2.4 and are:

$$Z_e = 2by_o\frac{y_o}{2} = by_o^2$$

$$S_e = \frac{b(2y_o)^2}{6} = \frac{2}{3}by_o^2 = \frac{2}{3}Z_e$$

$$Z = \frac{bd^2}{4} \tag{2.9}$$

* Eq. 2.7*a* may also be derived directly from Eq. 2.2. Referring to Fig. 2.3, for symmetrical cross sections,

$$M = \int_A \sigma\, dA \cdot y$$

$$= 2\int_o^{y_o} \sigma \cdot b\, dy \cdot y + 2\int_{y_o}^{d/2} \sigma_y \cdot b\, dy \cdot y \tag{2.8}$$

$$= 2\int_o^{y_o} \sigma_y \cdot \frac{y}{y_o} \cdot b\, dy \cdot y + 2\int_{y_o}^{d/2} \sigma_y \cdot b\, dy \cdot y$$

$$= \sigma_y \frac{2\int_o^{y_o} y^2 \cdot b\, dy}{y_o} + \sigma_y 2\int_{y_o}^{d/2} y \cdot b\, dy$$

$$= \sigma_y S_e + \sigma_y Z_p$$

Thus the bending moment in terms of Z is given by

$$M = \sigma_y(S_e + Z_p) = \sigma_y(\tfrac{2}{3}Z_e + Z - Z_e)$$

$$= \sigma_y\left(Z - \frac{Z_e}{3}\right) \tag{2.10}$$

The maximum moment is obtained when the elastic part is reduced to zero or

$$\boxed{M_p = \sigma_y Z} \tag{2.11}$$

M_p is called the "plastic moment."

From the equations just derived for curvature and moment, it is now possible to write the desired moment–curvature relationship. Substituting $Z_e = by_o^2$ in Eq. 2.10,

$$M = \sigma_y\left(Z - \frac{by_o^2}{3}\right) \tag{2.12}$$

The expression may be written in terms of ϕ by substituting for y_o the value obtained from Eq. 2.6, giving

$$M = \sigma_y\left(Z - \frac{b\sigma_y^2}{3E^2\phi^2}\right) \qquad (\phi_y < \phi < \infty) \tag{2.13}$$

A nondimensional relationship is obtained by dividing both sides of Eq. 2.13 by $M_y = \sigma_y S$. This yields

$$\frac{M}{M_y} = \frac{Z}{S} - \frac{b\sigma_y^2}{3SE^2\phi^2}$$

Using Eq. 2.6 with $y_o = d/2$ and $\phi = \phi_y$, and substituting $Z = bd^2/4$ and $S = bd^2/6$, the following is obtained,

$$\frac{M}{M_y} = \frac{3}{2}\left\{1 - \frac{1}{3}\left(\frac{\phi_y}{\phi}\right)^2\right\} \qquad (\phi > \phi_y) \tag{2.14}$$

The resulting nondimensional M–ϕ curve for a rectangle is shown in Fig. 2.5. The numbers in circles correspond to the "stages" of Fig. 2.2. Stage 4, approached as a limit, represents complete plastic yield of the cross section, where $M_p = \sigma_y Z$. Note that for the rectangular cross-section there is a 50% increase in strength above the computed elastic limit (stage 1) due to the "plastification" of the cross section. This represents one of the sources of reserve strength beyond the elastic limit of a rigid frame.

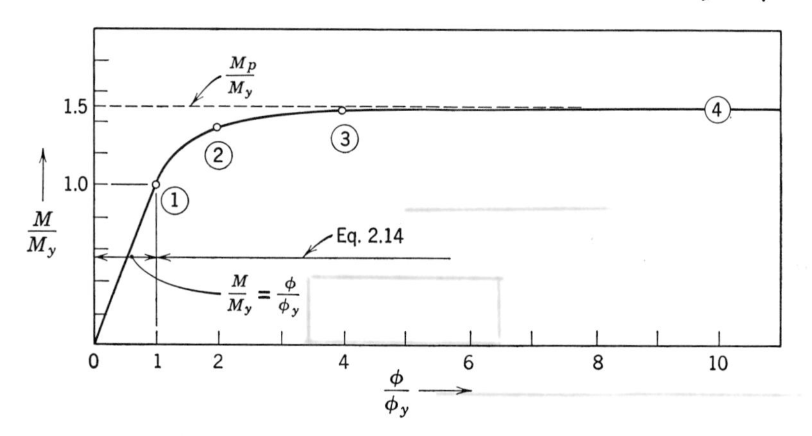

Fig. 2.5. Nondimensional moment–curvature relationship for rectangular beam.

The ratio of the plastic moment M_p to the yield moment M_y, representing the increase of strength due to plastic action, will be a function of the cross-sectional form or shape. Thus the "shape factor" is given by

$$f = \frac{M_p}{M_y} = \frac{\sigma_y Z}{\sigma_y S} = \frac{Z}{S} \tag{2.15}$$

For the rectangle being considered, $f = bd^2/4 \div bd^2/6 = 1.50$, as indicated in Fig. 2.5.

2.4 BENDING OF WF BEAM

Figure 2.6 shows a simple approximation to the action of a WF beam under bending moment. It assumes that all of the material in the beam is concentrated in two lines representing the flanges. After the elastic limit has been reached the compression flange of such a beam shortens at constant load and the tension flange lengthens at constant load. The resisting moment therefore remains constant under increasing deformation; the member rotates as a hinge except that there is always present this constant resisting moment equal to the "plastic hinge moment."

Figure 2.7 shows a more realistic picture of the moment–curvature relationship of a WF shape. At point 1 the yield stress, σ_y, is reached; at point 2 (a slight additional moment) the flanges and part of the web are plastic; and at point 3 (very slight additional moment) the cross section approaches a condition of full plastic yield.

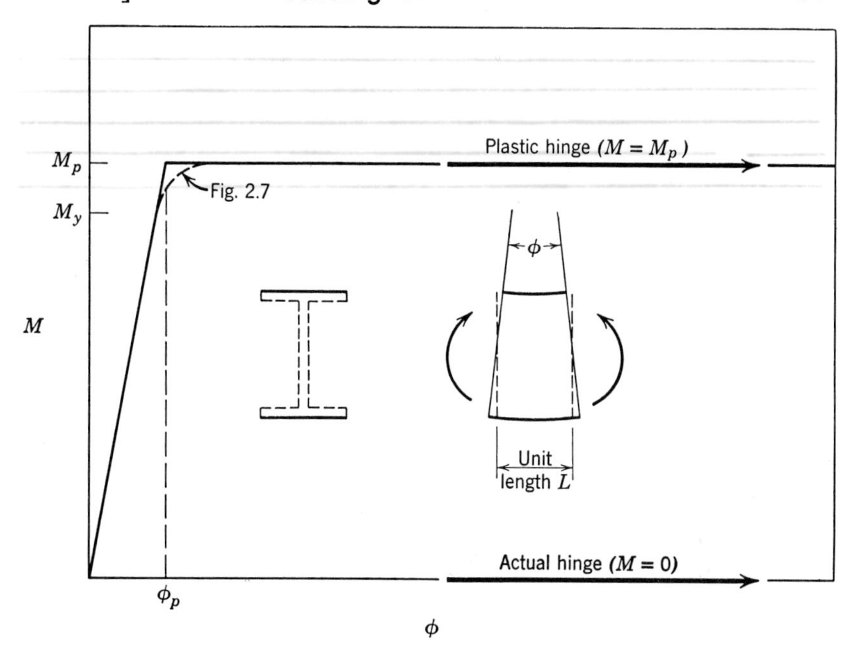

Fig. 2.6. Idealized moment–curvature relationship for WF beam. (*Trans. ASCE*, 122, p. 1139, 1957.)

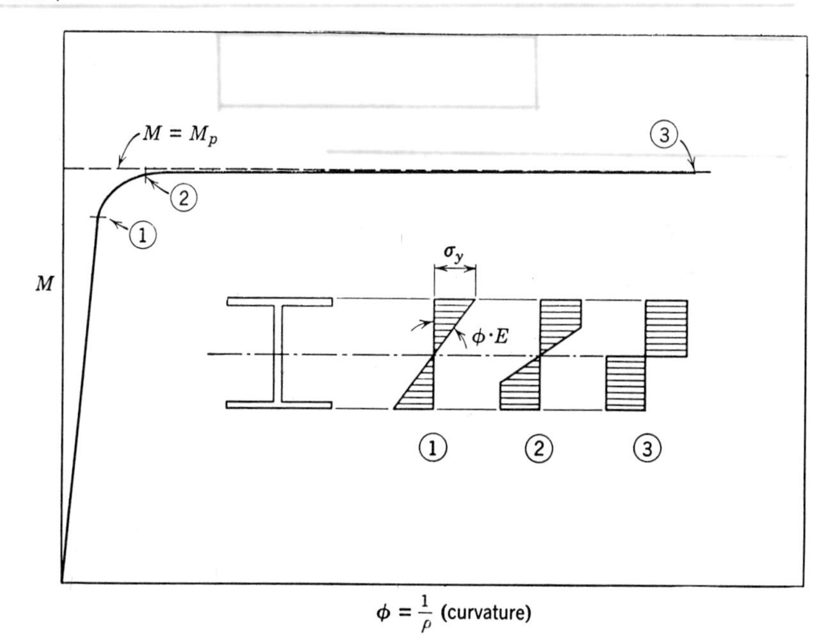

Fig. 2.7. Typical theoretical moment–curvature relationship for WF beam.[2.2]

The magnitude of the moment, M_p, may be computed directly from the stress distribution shown for point 3 of Fig. 2.7. As shown in Fig. 2.8, it is equal to the couple created by the tensile and compressive forces, $\sigma_y A/2$. The moment due to each of these forces is equal to the product of the yield stress, σ_y, and the area above the neutral axis, $A/2$, multiplied by the distance $\bar{y}$ measured to the center of gravity of that area. Thus,

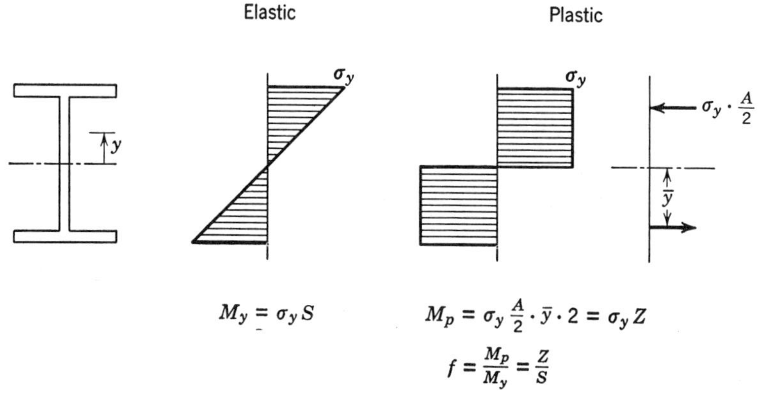

Fig. 2.8. Elastic and plastic limit moments.

$$M_p = 2\sigma_y \frac{A}{2} \bar{y} = \sigma_y A\bar{y} \tag{2.16}$$

The quantity $A\bar{y}$ is the "plastic modulus," defined and given the symbol Z in Art. 2.3 above; therefore,

$$M_p = \sigma_y Z \tag{2.11}$$

as was obtained previously.

The moment–curvature relationship may be developed for WF shapes by the same procedure as outlined for a rectangular cross section.

When the yield zone penetrates part way through the flanges, as

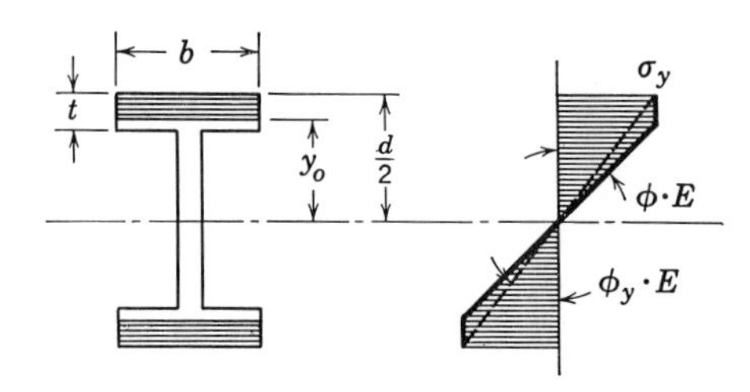

Fig. 2.9. Plastic stress distribution in WF beam—case 1: partial yielding of flange.

shown in Fig. 2.9, the nondimensional M–ϕ curve becomes

$$\frac{M}{M_y} = \frac{\phi}{\phi_y}\left(1 - \frac{bd^2}{6S}\right) + \frac{bd^2}{4S}\left\{1 - \frac{1}{3}\left(\frac{\phi_y}{\phi}\right)^2\right\} \qquad \left(1 < \frac{\phi}{\phi_y} < \frac{d/2}{\left(\frac{d}{2} - t\right)}\right) \tag{2.17}$$

When the curvature increases to the extent that yielding penetrates

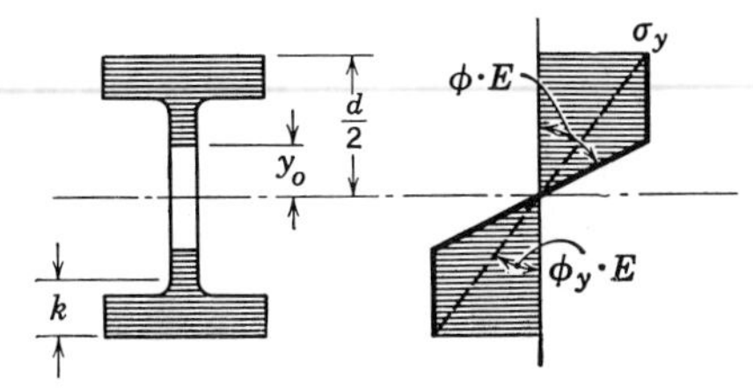

Fig. 2.10. Plastic stress distribution in WF beam—case 2: partial yielding of web.

through the flanges and into the web (Fig. 2.10), the nondimensional M–ϕ relationship becomes

$$\frac{M}{M_y} = f - \frac{wd^2}{12S}\left(\frac{\phi_y}{\phi}\right)^2 \qquad \left(\frac{d/2}{\left(\frac{d}{2} - k\right)} < \frac{\phi}{\phi_y} < \infty\right) \tag{2.18}$$

The curve resulting from Eqs. 2.17 and 2.18 is shown in Fig. 2.11 for

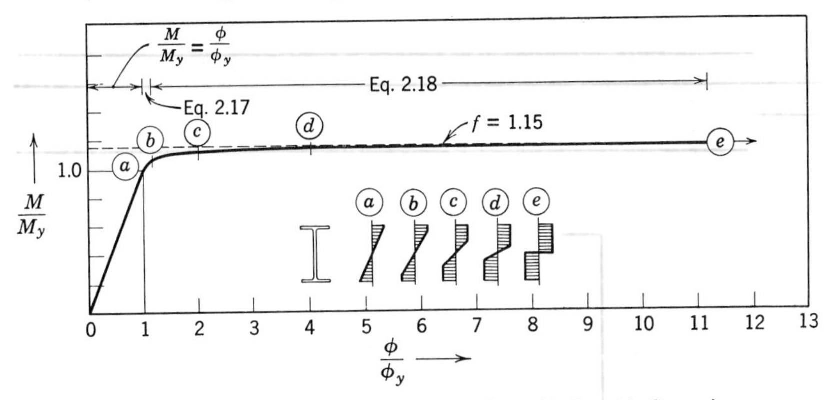

Fig. 2.11. Nondimensional moment–curvature relationship for wide-flange beam.

a typical WF shape. The stress-distributions a to e correspond to the lettered points on the M–ϕ curve. The yield distributions for the WF shape at the same stages of deformation are shown in Fig. 2.12.

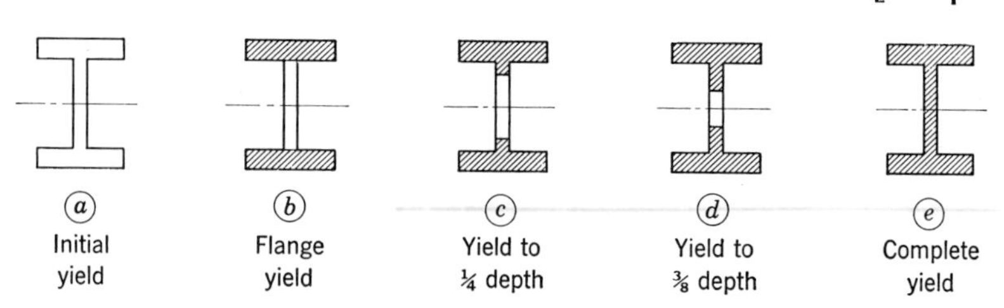

Fig. 2.12. Distribution of yield zones in WF beam at various stages of deformation.

It will be noted that the shape factor is smaller than for the rectangle. (Compare with Fig. 2.5), the average value of f for all WF beams being 1.14. Correspondingly there is a more rapid approach to M_p when compared with the rectangle. As a matter of fact when the curvature is twice the elastic limit value (stage c of Fig. 2.11) the moment is within 2% of the full M_p value.

2.5 THE PLASTIC HINGE

The reason a structure will support the computed ultimate load is that plastic hinges are formed at certain critical sections. What is the plastic hinge? What factors influence its formation? What is its importance?

The M–ϕ curve (Fig. 2.11) is characteristic of the plastic hinge. The following two features are particularly important:

(1) After the elastic limit is reached, the curve approaches very rapidly to the horizontal line corresponding to the plastic moment value.

(2) There is an indefinite increase in curvature at constant moment.

These two features are expressed in the following idealization of Fig. 2.11 through the use of two straight lines,

$$M = EI\phi \qquad (0 < \phi < \phi_p)$$

$$M = M_p \qquad (\phi > \phi_p) \tag{2.19}$$

where

$$\phi_p = \frac{M_p}{EI}$$

Fig. 2.6 is a representation of Eq. 2.19. According to it, the member remains elastic until the moment reaches M_p. Thereafter, rotation occurs at constant moment; i.e., the member acts as if it were hinged except with a constant restraining moment, M_p. By comparing Figs. 2.6 and 2.7 it is seen that the only effect of the "idealization" is to wipe out a short piece of curve in the actual M–ϕ relationship.

The Plastic Modulus, Z

Closely related to the plastic hinge is the plastic modulus, Z. It has already been defined for symmetrical sections as twice the statical moment taken about the neutral axis of the half-sectional area.* This result may also be obtained from the second of Eqs. 2.2. Referring to Fig. 2.8,

$$M_p = 2\int_{A/2} \sigma_y \, dA \cdot y = \sigma_y \cdot \underbrace{2\int_{A/2} y \, dA}_{Z} \tag{2.20}$$

As noted earlier,

$$Z = A\bar{y} \tag{2.21}$$

For American WF beams, the quantity $\bar{y}$ may be determined directly as $d/2 - y_1$ from the tabulated property y_1 of split tees[2.1] and thus

$$Z_{WF} = A\left(\frac{d}{2} - y_1\right) \tag{2.22}$$

where y_1 is the distance from the flange to the center of gravity (Fig.

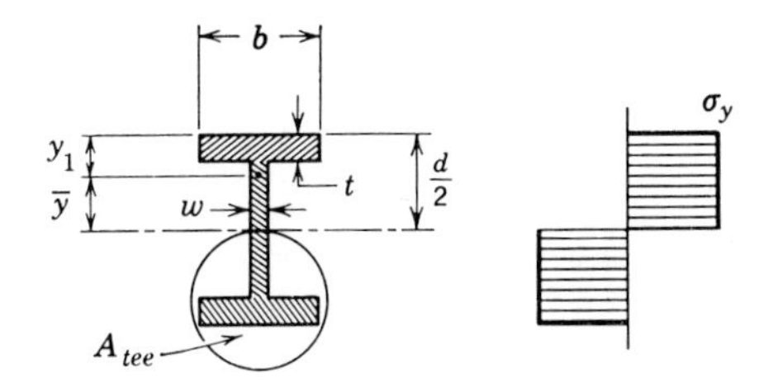

Fig. 2.13. Dimensions of WF beam in terms of split tee properties.

2.13). When data from split tees is not available the value of Z may

* It is necessary to compute the statical moment of the half-sectional area because $\int y \, dA = 0$.

be computed with sufficient accuracy from the equation

$$Z = (A - wd)\left(\frac{d - t}{2}\right) + \frac{wd^2}{4} \tag{2.23}$$

The plastic modulus has been computed for WF and I shapes according to Eqs. 2.21 and 2.23, and a table of these values is presented in Appendix 2.

A useful approximation for welded *H* cross sections (neglecting fillets) is

$$Z \cong bt(d - t) + \frac{w}{4}(d - 2t)^2 \tag{2.24}$$

Another approximation in terms of total flange area ($A_f = 2bt$), web area ($A_w = wd$), and total area A, is

$$Z = A_f(d/2) + A_w(d/4) = \frac{d}{2}(A - wd/2) \tag{2.25}$$

The shape factor, already defined as $f = Z/S$, varies for WF shapes used as beams from 1.10 to 1.18. The most frequent value (mode) is

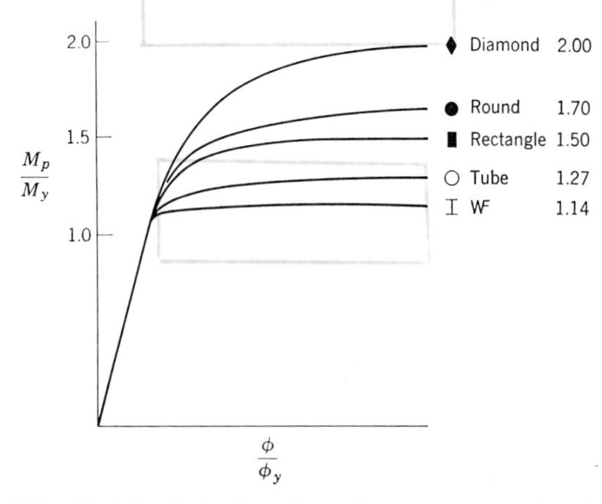

Fig. 2.14. Variation in the shape factor for various cross-sectional forms.

1.12, the average being 1.14. A few heavy columns have shape factors as high as 1.23. The average value of f for American Standard I-shapes is 1.18. Examples of the ratio of $Z/S = f$ for symmetrical shapes other than the wide flange are shown in Fig. 2.14. From Eqs. 2.15 and 2.21 the following useful equation for computing the value of f for symmetrical shapes is obtained:

$$\boxed{f = \frac{Z}{S} = \frac{A\bar{y}}{I/c}} \tag{2.26}$$

where $\bar{y}$ is the distance from the neutral axis to the centroid of the half area.

For sections which only have symmetry about an axis in the plane of bending, the position of the neutral axis at the plastic moment condition must first be computed before Z can be determined. This position follows directly from Eq. 2.2. Since $P = 0$, and $\sigma = \sigma_y$, then the area above the neutral axis must equal that below if equilibrium of horizontal forces is to be maintained. In other words, the *neutral axis divides the section into two equal areas.* Therefore Z may be defined more generally as the *combined statical moment of the cross-sectional areas above and below the neutral axis*, and the equation may be written,

$$\boxed{Z = \frac{A}{2}a} \tag{2.27}$$

As an illustration, for a triangular cross section of depth d, with symmetry about a vertical axis through the toe, the elastic neutral axis is at a distance of $\frac{2}{3}d$ from the toe, whereas the "plastic" neutral axis is at a distance of $d/\sqrt{2}$ from that point.

Factors Affecting Bending Strength and Stiffness

In addition to the shape factor whose influence on the strength of a cross-section has already been described, several other factors influence the ability of members to form plastic hinges. Some of these are important from the design point of view and are treated in Chapters 4 and 7. The remainder are only briefly discussed here and reference is made to other sources for further information on the subject.

Material properties. Any variation in the yield stress level, σ_y, will have a direct effect upon the magnitude of the plastic moment, M_p (see Eq. 2.11). The yield stress level is subject to considerable variation and is affected by such things as the thickness of the material, the rate of strain used in the material test, the position of the test specimen in the cross section, and the prior strain history of the material.[1.7, 1.8, 2.2] It is because of just these factors that the results of ASTM acceptance type tests (mill tests) are consistently higher than the results of "coupon tests" conducted as a part of laboratory research. This difference usually amounts to from 15 to 20%. More recently, it has been shown that "stub column" tests best reveal the basic yield stress level of the material.[1.14, 2.3] This average value for ASTM A7 steel is so close to the minimum value called for in the specifications that 33.0 ksi is appropriate for use in design. The variation from this average value for A7 steel may range from 25 ksi to 48 ksi.[2.3]

The proportional limit is of no practical significance insofar as plastic bending is concerned.[2.2] It simply reduces the proportional limit in bending but has no effect on the plastic moment. The principal factor

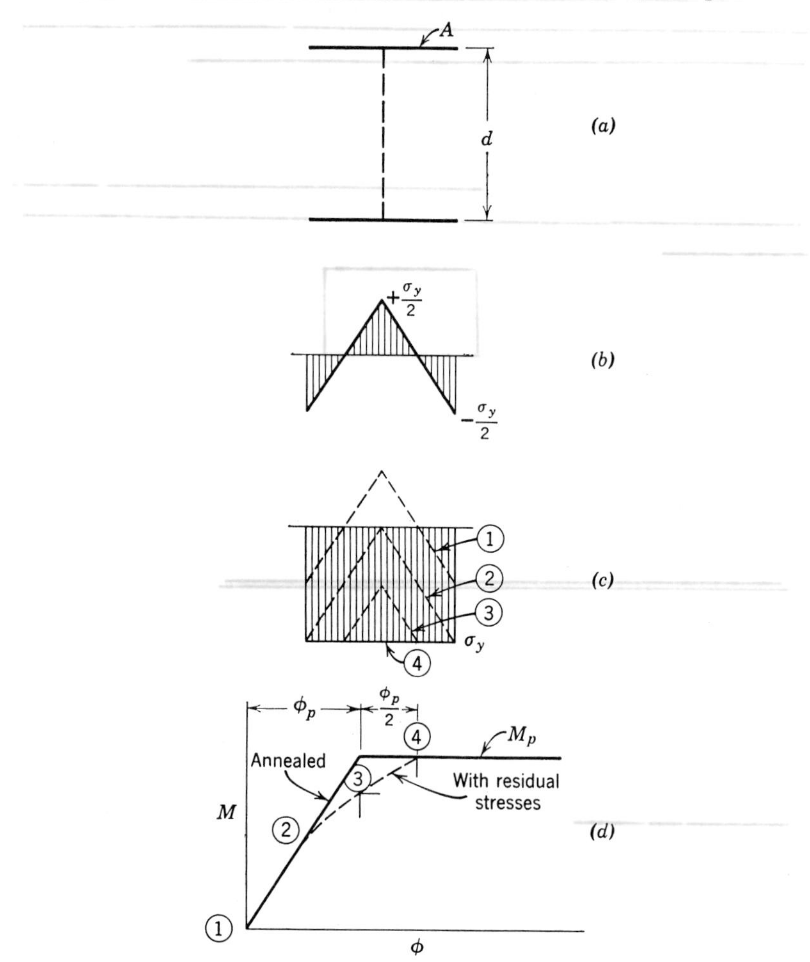

Fig. 2.15. Representation of the influence of residual stresses upon the moment–curvature relationship of a beam in bending (idealized).

affecting the proportional limit is the presence of residual stress, a topic next to be discussed.

Residual stress. In general, the effect of residual stress in a steel member is twofold: First, residual stress causes an initiation of yield at loads lower than predicted by usual stress analysis. Secondly, residual stresses may lower the ultimate capacity by inducing either local or general buckling of a compression element or column. In Art. 1.4 and in Fig. 1.6, p. 9, it was shown that residual stresses constitute one of the factors that are ignored in past conventional design procedures. Although residual stresses due to cooling, cold-bending, or welding reduce the proportional limit in bending they have a negligible effect on the maximum bending strength of a member.[1.8, 2.2]

In order to see the influence upon the M–ϕ curve of residual stresses in rolled beams, consider an idealized WF shape as shown in Fig. 2.15*a* with all of the material concentrated at the flanges. Assume that the residual stress pattern in the flanges is as shown in Fig. 2.15*b* with compression in the flange tips of $-\sigma_y/2$, and tension at the flange centers of $+\sigma_y/2$. The total force in one flange at any stage is given by the first of Eqs. 2.2,

$$P = \int_{\text{Area}} \sigma \, dA$$

Prior to the application of any external moment, the flange must be in equilibrium ($P = 0$) and from Fig. 2.15*b* and Eq. 2.2 it is evident that this is true.

Now, when uniform strain is applied across the flanges as the member is bent by pure moment, the stress distribution will change from stage 1 ($M = 0$) to stage 2 at which point the extreme fiber stress reaches the yield stress level, Fig. 2.15*c*. Beyond stage 2 the behavior is inelastic as shown in Fig. 2.15*d*; stage 3, for example, represents a partially plastic case, the flange being yielded halfway to the flange centerline. Finally at stage 4 the entire flange has yielded, and from the second of Eqs. 2.2

$$M = \int_{\text{Area}} \sigma \, y \, dA = Pd$$

from which

$$M = \sigma_y A d$$

This quantity is equal to the full plastic moment of the idealized WF shape. Thus the presence of residual stress has no effect on the moment capacity. Although the nonlinear portion of the M–ϕ curve in Fig. 2.15 may be computed without difficulty, such a departure from the straight-line relationship is entirely inconsequential when compared with the larger rotations that occur when the plastic hinge forms. The action of the plastic hinge is just the same as in a member that had not contained any such stresses.

The conclusion just drawn cannot be applied to a member whose load capacity is limited by instability. For example in the case of columns supporting primarily axial loads, the effect of residual stress is of predominant importance and enters in a significant way into the column formulas that will be suggested in Art. 4.6.

Stress concentrations. The effect of stress concentrations is similar to that of residual stresses.[2.2] Due to a nonuniform elastic stress-distribution, yielding will commence at a lower stress than otherwise would be expected and at the most highly stressed fiber. If the applied stress on the element is uniform, subsequently the entire element will be stressed into the plastic region and will then support the same load as a similar element without the concentration.

Stress concentrations that involve a reduction in cross-sectional area may be ignored if such reductions are small, because the effect of strain-hardening will overcome the corresponding loss of strength. Otherwise the computed plastic modulus may be reduced in proportion to the statical moment of the area removed (see p. 249).

Strain-hardening. After structural steel has been strained through the plastic region to about 15 times the elastic limit strain, further deformation is accompanied by an increase in the stress capacity of the material. This is strain-hardening and was illustrated in Figs. 1.2, 1.4, and 2.1. Insofar as its influence on the bending strength is concerned, strain-hardening is a beneficial effect. However, the complications of taking it into account either in ultimate load or in deflection calculations are such that it is neglected in the simple plastic theory. Computation of loads and deflections including this effect may be found in Ref. 2.4.

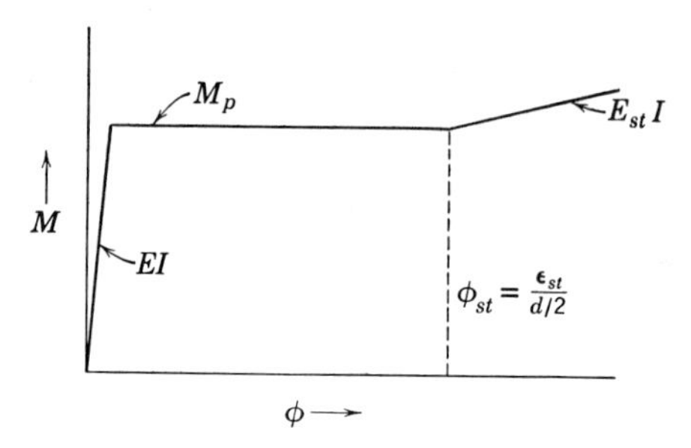

Fig. 2.16. Influence of strain-hardening upon the moment–curvature relationship.

An idealized moment–curvature relationship may be obtained using the same assumptions as those leading to Fig. 2.6 and is shown in Fig. 2.16. For ϕ greater than the point of strain-hardening,

$$M = M_p + E_{st}I(\phi - \phi_{st}) \qquad (\phi > \phi_{st}) \tag{2.28}$$

where E_{st} is the strain-hardening modulus (about 700,000 psi for structural steel), and $\phi_{st} = \dfrac{2\epsilon_{st}}{d}$.

Axial force and shear force. The presence of axial force and/or shear force tend to reduce the moment-capacity of a member. In some cases these forces are important design considerations and therefore they are treated more extensively in Chapter 4, Arts. 4.2 and 4.3.

Local buckling and lateral buckling. The simple plastic theory presumes that failure by local buckling of flange or web elements or by lateral buckling of a member does not occur prior to the attainment of the computed ultimate load. Eventually, one or both of these phenomena will occur, resulting in a subsequent loss of moment capacity. Therefore, rules are needed to prevent their premature onset prior to the development of all necessary hinges in different parts of the structure. The subject is discussed in Chapter 4, Arts. 4.4 and 4.5.

Unsymmetrical cross section. Sections with symmetry about an axis lying in the plane of bending can be expected to develop a plastic hinge with a moment capacity given by Eq. 2.11. Although a section that is unsymmetrical about this axis might develop the appropriate plastic moment value, combined bending and torsion would be introduced. Therefore, only sections that are symmetrical about an axis in the plane of bending will be considered, the plastic modulus Z being computed according to Eq. 2.21 or 2.27 as the case may be.

Encasement. The moment of resistance of a beam as it is found in a structure will usually be greater than M_p because of the additional restraint provided by the cladding.[2.5] Any beneficial effect from this source is ordinarily neglected in the simple plastic theory.

Brittle fracture. If improper attention is paid to workmanship, design details, and material, then the ability of a member to rotate in the plastic region and form a plastic hinge might be limited due to brittle fracture. Suggestions for use in design are noted in Art. 4.7.

Distribution of the Plastic Hinge

For the idealized M–ϕ curve of Fig. 2.6, the plastic hinges form at discrete points at which all plastic rotation occurs. Thus the length of the hinge approaches zero. In actuality the hinge extends over a length of member that is dependent on the loading and the geometry. For example in the rectangular beam of Fig. 2.17*a* ($M_y = 0.67\ M_p$) the hinge length is equal to one-third of the span. For a wide-flange beam with a shape factor of 1.14 and loaded as shown in Fig. 2.17*b*, the hinge length is $L/8$. In other words, the hinge length ΔL is the length of the beam over which the moment is greater than the yield moment M_y.*

* See page 63 for sign conventions.

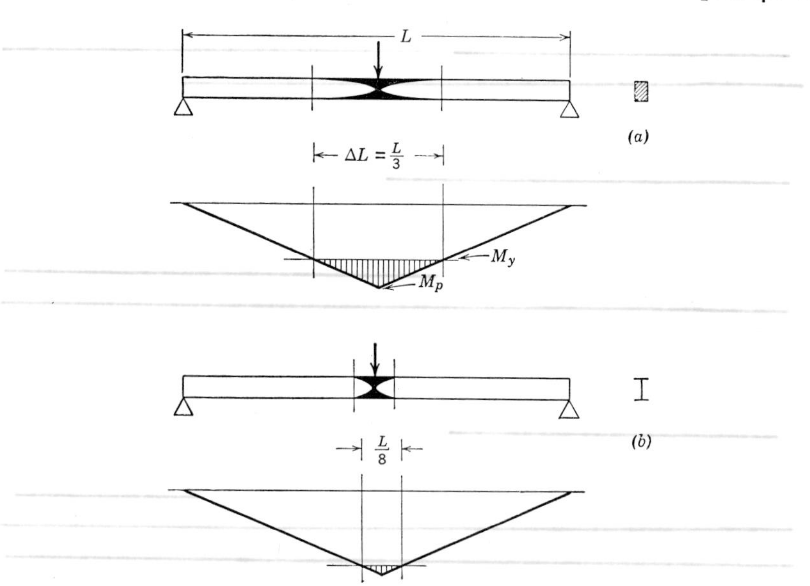

Fig. 2.17. Theoretical length of yielded portion of (a) rectangular and (b) WF beam with central concentrated load.

The definitions or principles given below are in summary of this section and are important to a later understanding of plastic analysis:

(1) *A plastic hinge is a zone of yielding due to flexure in a structural member.* Although its length depends on the geometry and loading, in most of the analytical work it is assumed that all plastic rotation occurs at a point. At those sections where plastic hinges are located, *the member acts as if it were hinged, except with a constant restraining moment, M_p (Fig. 2.6).*

(2) *Plastic hinges form at points of maximum moment.* Thus in a framed structure with prismatic members, it would be possible for plastic hinges to form at points of concentrated load, at the end of each member meeting at a connection involving a change in geometry, and at the point of zero shear in a span under distributed load.

(3) The plastic moment M_p equals $\sigma_y Z$.

(4) The shape factor ($f = Z/S$) is one source of reserve strength beyond the elastic limit.

Application of the plastic hinge concept to analysis is illustrated in the next article.

2.6 REDISTRIBUTION OF MOMENT

In addition to the modest increase of load that results from the formation of a plastic hinge, a second factor contributing to the reserve of strength of an indeterminate structure loaded beyond the elastic limit is called "redistribution of moment." It is a consequence of the action of the plastic hinges. As load is added to a structure eventually the

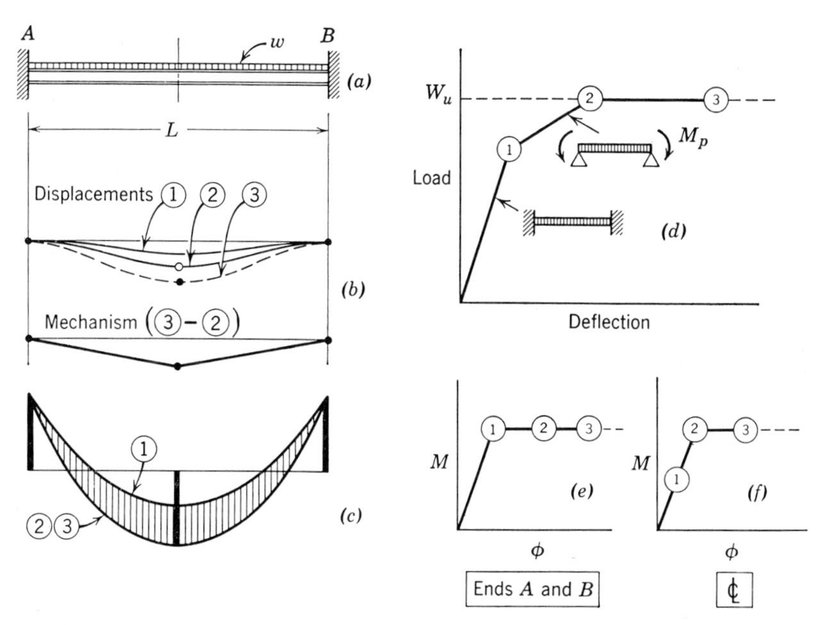

Fig. 2.18. Redistribution of moment in a fixed-ended beam with uniformly distributed load. (*Proc. AISC Nat'l Engr. Conf.*, 1956.)

plastic moment is reached at a critical section—the section that is most highly stressed in the elastic range. As further load is added, this plastic moment value is maintained while the section rotates. Other less highly stressed sections maintain equilibrium with the increased load by a proportionate increase in moment. This process of moment transfer due to the successive formation of plastic hinges continues until the ultimate load is reached.

The fixed ended, uniformly loaded beam of Fig. 2.18*a* will be used to illustrate how plastic hinges allow a structure to deform under load beyond the elastic limit, permit a redistribution of moment and, thereby, an increase in load capacity. The numbers ①, ②, and ③ in Fig. 2.18

represent three phases of loading:

① Attainment of first yield.
② First attainment of computed ultimate load.
③ An arbitrary deflection obtained by continued straining at the ultimate load.

The idealized moment–curvature relationship of Fig. 2.6 is assumed. In Fig. 2.18*b* are curves of the deflected shape; in Fig. 2.18*c* are the moment diagrams at the three loading phases; in Fig. 2.18*d* is shown the load-vs.-deflection curve; and in Figs. 2.18*e* and 2.18*f*, the M–ϕ action at the ends and at the center, respectively.

By an elastic analysis, the moment diagram of sketch (*c*) of Fig. 2.18 can be determined when yielding commences (phase ①). The center moment would be $wL^2/24$ and the end moment would be $wL^2/12$. (w is the distributed load per unit length. L is the span length). On the load-vs.-deflection curve of sketch (*d*) the load has reached point ①. The moment-capacity has been used up at the ends (sketch *e*); however, from sketch (*f*), since at the center of the beam $M = \frac{1}{2}M_p$ at phase ①, additional moment capacity is still available there. Therefore, as load increases beyond phase ①, "hinge action" will start at the ends and the beam now behaves as if it were simply supported, except that the end moment remains constant at M_p. The deflection increases at a somewhat faster rate (the rate of increase being that of a simply supported beam of length L).

At phase ② the beam reaches its maximum load since the moment capacity at the beam center is exhausted. Beyond phase ②, the beam will continue to deform under constant load (phase ③). The shaded portion of sketch (*c*) represents the simple beam moment diagram that, due to redistribution of moment, is superimposed upon the existing moment diagram (phase ①) and corresponds to the increase of load between phases ① and ② (sketch *d*).

It is evident from the load–deflection curve shown in Fig. 2.18*d* that the formation of each plastic hinge acts to remove one of the indeterminates in the problem, and the subsequent load-deflection relationship will be that of a new (and simpler) structure. For example, in the elastic range, the deflection under load can be determined for the completely elastic beam. Starting from point ① the segment ①–② represents the load-deflection curve of the beam in the sketch but loaded within the elastic range.

In principle, the process just described is precisely that which took place in the case of the three-bar truss of Fig. 1.13, p. 17, except that tensile forces were involved instead of moments. When the force in bar 2 reached the yield condition it remained constant there while the forces continued to increase in bars 1 and 3. The ultimate load was reached when all critical bars became plastic.

2.7 MECHANISMS

After a sufficient number of plastic hinges have formed to accomplish all of the transfer of moment that is possible, any further displacement occurs at constant load. The segments of the beam between plastic hinges are able to move without an increase of load—and this system of members is called a *mechanism*. Actually it is a pseudo- or quasi-mechanism because the real hinges that would be present in a linkage

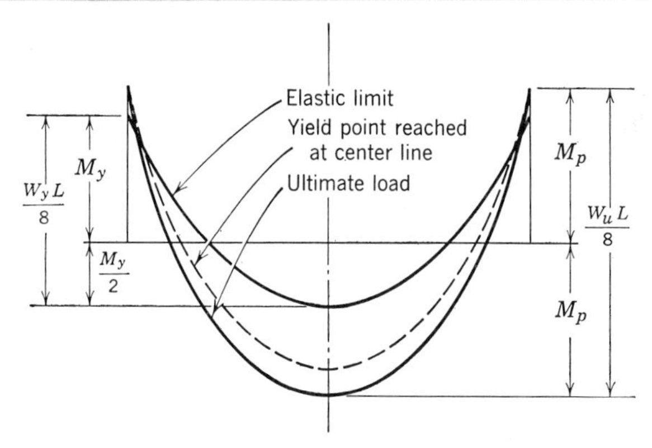

Fig. 2.19. Moment diagram at various stages of loading for fixed-ended beam with uniformly distributed load.

are replaced by plastic hinges; further, the members are bent between plastic hinges due to the applied moments, M_p, instead of being straight bars; but after all hinges are formed the *further* motion is just like that of a linkage. Indeed, if one gives an arbitrarily small displacement to the structure beyond the ultimate load and then subtracts the deflections that occurred up to that point, the resulting deformed shape of all members would be straight lines. This is shown to a greatly expanded scale in Fig. 2.18*b*. Stage ② shows the deformed shape when the ultimate load is first reached. *Mechanism motion* occurs between stages ② and ③; and subtracting the displacements of ② from those of ③ gives the mechanism shown in the lower part of sketch (*b*). Thus there is no further deformation within the member itself; it all occurs at the plastic hinges. This fact will be very useful later when methods of plastic analysis are described.

It will be of interest to compare the ratio of the ultimate load W_u to the load at first yield W_y for the fixed-ended beam just considered; and for this purpose the moment diagrams are redrawn in Fig. 2.19. The yield load and ultimate load may be computed from this moment diagram. Thus,

$$\frac{W_y L}{8} = \frac{3}{2} M_y \quad \text{or} \quad W_y = \frac{12 M_y}{L} \tag{2.29}$$

$$\frac{W_u L}{8} = 2\,M_p \quad \text{or} \quad W_u = \frac{16 M_p}{L} \tag{2.30}$$

from which

$$\frac{W_u}{W_y} = \frac{16 M_p / L}{12 M_y / L} = \frac{4}{3} \frac{M_p}{M_y} \tag{2.31}$$

Thus the reserve strength due to redistribution of moment is one-third. Considering the average shape factor of WF shapes, the total reserve strength due to redistribution and the shape factor (plastification) is

$$\frac{W_u}{W_y} = \left(\frac{4}{3}\right)(1.14) = 1.52$$

For this particular problem, then, the ultimate load is 52% greater than the load at first yield, representing a considerable margin that is disregarded in conventional elastic design.

In further summary:

(1) Plastic hinges are reached first at sections subjected to greatest deformation (curvature).
(2) Formation of plastic hinges allows a subsequent redistribution of moment until M_p is reached at each critical (maximum) section.
(3) The maximum load is attained when a mechanism forms.

2.8 EXPERIMENTS ON BEAMS

As a demonstration that the theoretical maximum moment is attained through plastification of the cross section, Fig. 2.20*a* shows an M–ϕ curve obtained from a beam in pure bending.[1.7] The dotted line is the

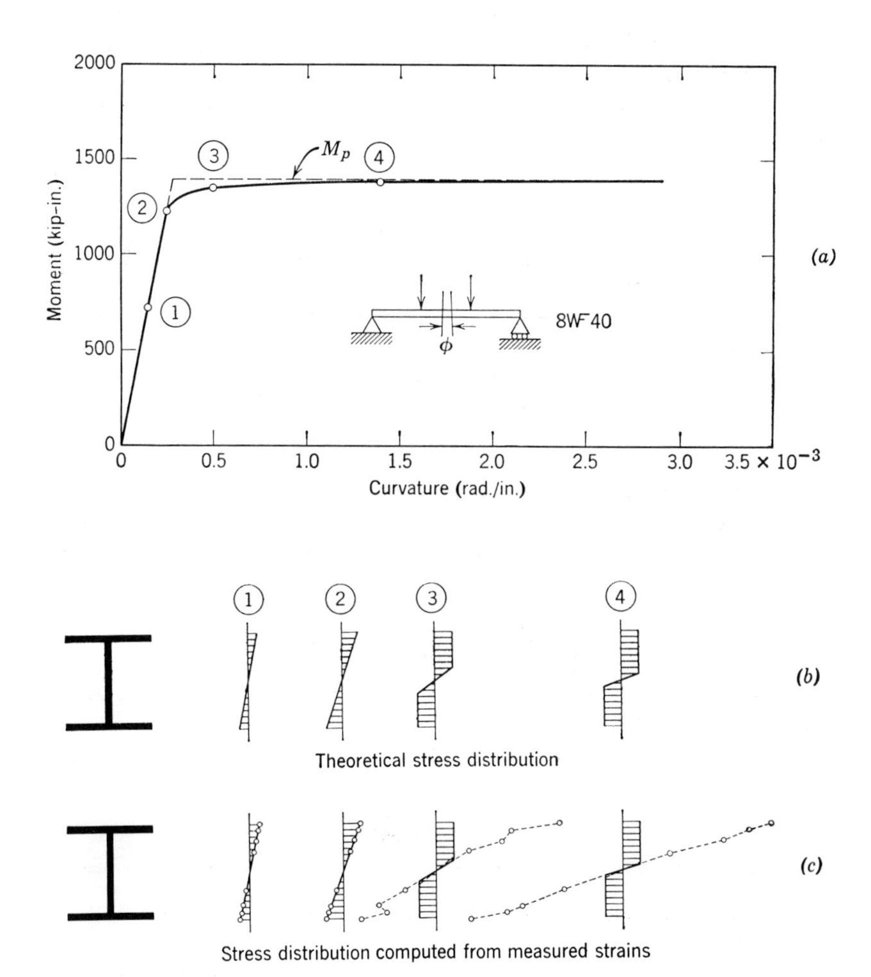

Fig. 2.20. Experimental verification of plastification of the cross-section and the formation of a plastic hinge.[1.7]

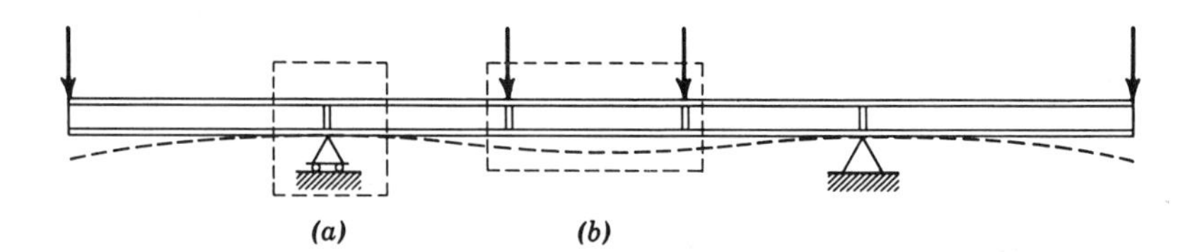

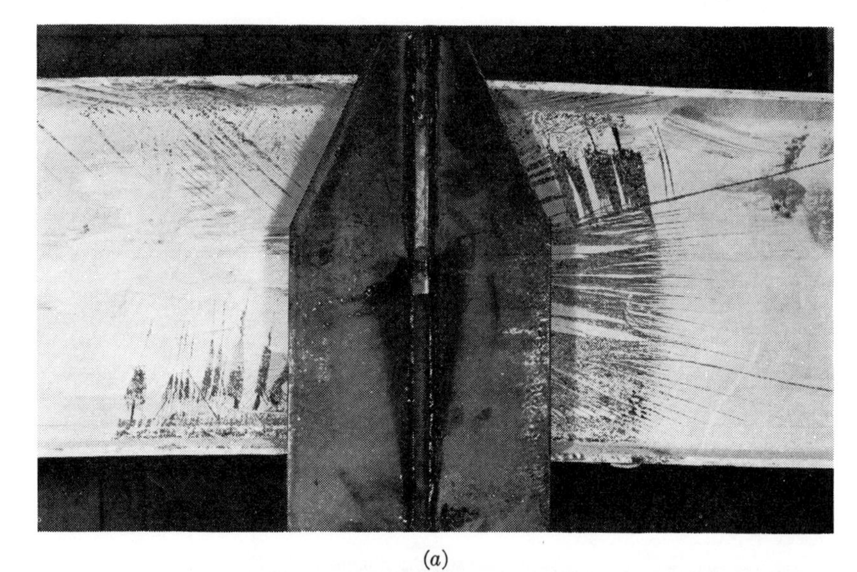

(a)

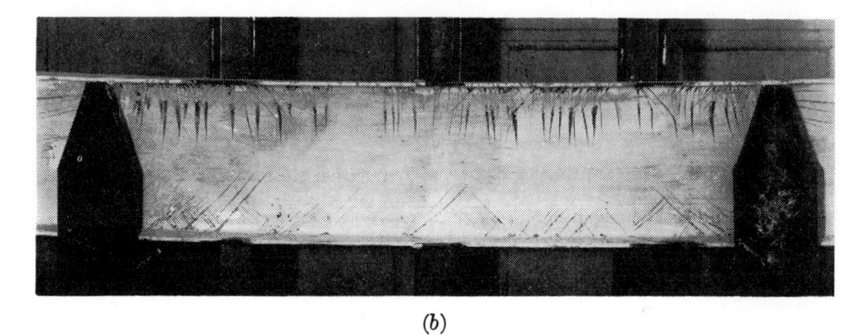

(b)

Fig. 2.21. "Plastic hinges" in a 14WF30 continuous beam (*a*) at a point of support and (*b*) in a region of pure moment between two load points. (*Proc. AISC Nat'l Engr. Conf.*, 1956.)

assumed simplified curve and the solid line through the circles represents the test result. The theoretical stress distributions (according to the simple plastic theory) at different stages of bending are shown in Fig. 2.20*b*. In Fig. 2.20*c* are shown the corresponding stress distributions computed from SR-4 strain measurements. It will be seen that plastification of the cross section *does* occur, and that the magnitude of the bending moment corresponding to this condition is the full plastic moment as computed from the equation $M_p = \sigma_y Z$.

The many tests conducted on rolled shapes indicate that the curve shown in Fig. 2.20 is not an exceptional case but that wide-flange beams will develop the strength predicted by the plastic theory and that a plastic hinge (characterized by rotation at near-constant moment) actually does form.[1.7, 1.16, 2.2, 2.6]

Plastic hinges may form at load points or at supports in continuous beams. Examples are shown in Fig. 2.21. Photograph *a* is a detail taken at the support as indicated in the upper sketch, the details being somewhat obliterated by the auxiliary loading stiffener. Photograph *b* shows the region between two load points. (The loading is such that the moment is uniform in the center portion.) The yield patterns differ depending on the magnitude of shear force that is present.

Further experimental evidence is given in Fig. 2.22. Shown in this figure is a series of photographs taken at four different stages in a bending test (the four stages are shown in the sketch). At stage ① the extreme fibers in the flanges are yielded. In stage ② the yield lines have penetrated into the web. At stage ③ the computed plastic moment condition has been reached; and, finally, in stage ④ the beam has been deformed through the plastic region and strain-hardening has commenced. Eventually, the moment capacity would drop off because the beam does not retain its cross-sectional shape and buckles laterally.

In Fig. 2.23 is shown the process of redistribution of moment—both theoretically and as confirmed experimentally. Use is made of an example similar to that used in Fig. 2.18. A continuous beam was tested to simulate the condition of third point loading on a fixed-ended beam.[2.2] The beam and its component behavior are shown at four stages:

① Near the computed elastic limit.

② After the plastic hinge has theoretically formed at the ends and the load is increasing towards the ultimate value.

③ When the theoretical ultimate load is first reached, and

④ After deformation has been continued through an arbitrary displacement.

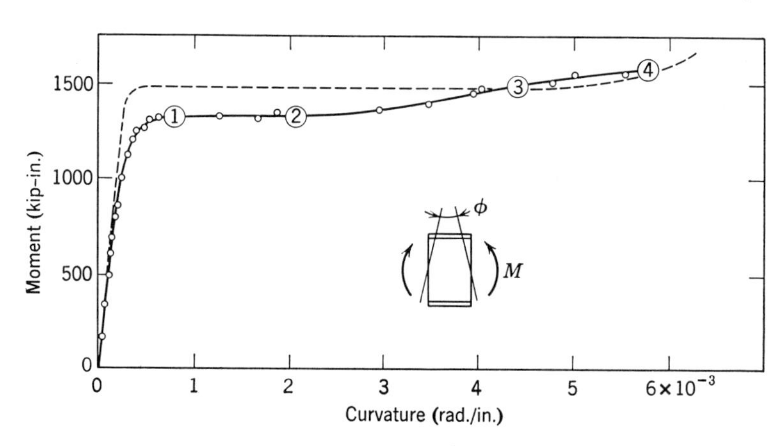

Fig. 2.22. The behavior of a WF beam in bending showing four stages during the development of a plastic hinge. (*Proc. AISC Nat'l Engr. Conf.*, 1956.)

The figure shows the loading, deflected shape at the two phases, the moment diagram, the load-deflection curve, and the moment–curvature relationship near the ends and at the center. The solid lines are the experimental results and the dotted lines are the theoretical relationships.

In the elastic range it will be seen that the beam behaves just as assumed by the theory, the moment at the center being one-half the moment at the fixed ends. (Fig. 2.23*c* and *d*.) As the moment at the ends approaches the yield moment, the curvature ϕ commences to increase more rapidly and the plastic hinge begins to form (sketch *e*). Because of this "hinge action," the additional moments due to increase in load are distributed in a different ratio, most of the increase going to the center and a small amount going to the ends as plastification occurs (sketch *e*). So the beam actually behaves somewhat more flexibly than before (sketch *d*) and at stage ② the elastic moment capacity near the center is practically exhausted. It is quite evident from Fig. 2.23 that all of the moment capacity has been substantially absorbed by the time stage ③ is reached (ultimate load). Beyond this point the beam simply deforms as a mechanism with the moment diagram remaining largely unchanged, the plastic hinges at the ends and center developing further (stage ④).

Clear evidence is therefore available that redistribution of moment occurs through the formation of plastic hinges, allowing the structure to reach (and usually exceed) its theoretical ultimate load.

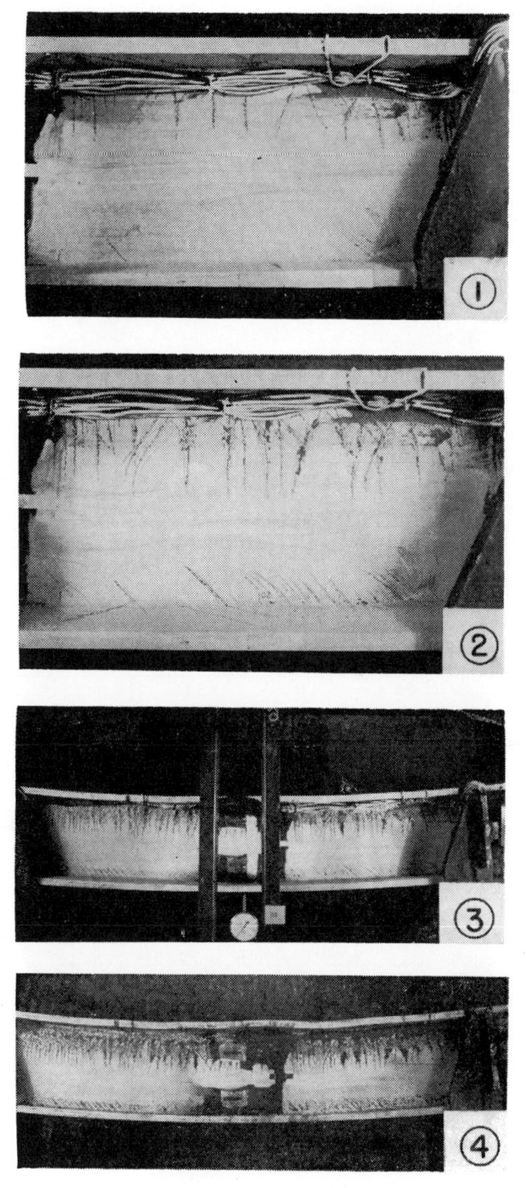

Fig. 2.22. (Continued.)

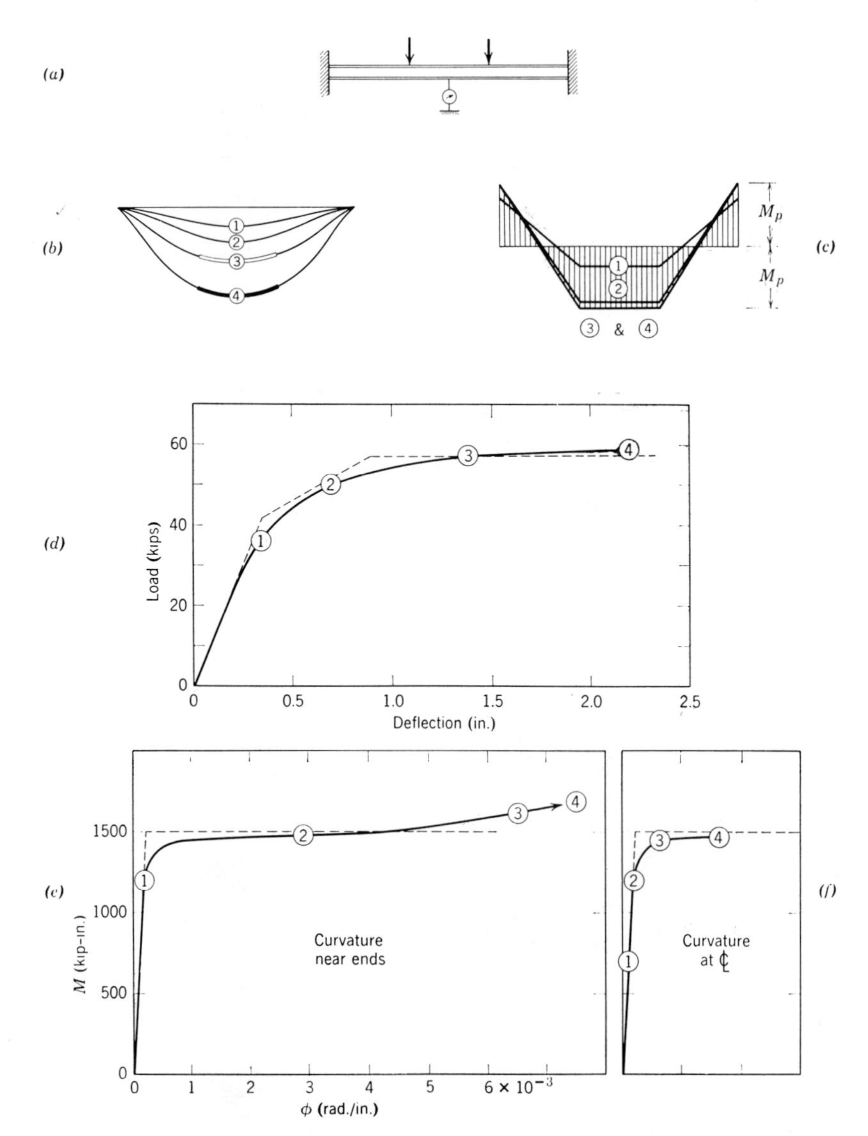

Fig. 2.23. Redistribution of moment as revealed by test on a restrained beam. (*Proc. AISC Nat'l. Engr. Conf.*, **1956.)**

PROBLEMS

2.1 Derive the equations of elastic bending (Eqs. 2.3, 2.4, and 2.5).

2.2 (*a*) Derive the nondimensional M–ϕ relationship for a diamond shape, bent about an axis of symmetry.
(*b*) Plot the curve, indicating coordinates at $\phi/\phi_y - 1.0, 2.0, 4.0, 10.0$.
(*c*) On the same curve plot the moment–curvature relationship for a rectangle.

2.3 Find the shape factor of an isosceles triangle bent about an axis parallel to the base.

2.4 Discuss the effect of the idealization of the M–ϕ curve to consist of two straight lines (Fig. 2.6).

2.5 Derive Eq. 2.17.

2.6 Derive Eq. 2.18.

2.7 Compute the nondimensional M–ϕ curves to the point of strain-hardening for 14WF426 and 21WF112 shapes. Plot points at the "limits" for flange and web and at ϕ/ϕ_y of 2.0, 3.0, 4.0, 5.0, 6.0. Sketch the idealized curve for each.

2.8 Compute Z for a 12I 50.0 shape taking into account the taper of the flanges. Compare with the approximate expression for Z.

2.9 Compute Z for a 14WF150 shape. Compare the result with the approximate expression.

2.10 Compute Z, S, and f for a box section with outside depth of 16 in., wall thickness of $\frac{1}{2}$ in., and a width of 8 in. Flexure is about the strong axis.

2.11 What is the distribution of the plastic hinge in a rectangular simply-supported beam under uniform load? Of a WF beam with $f = 1.15$?

2.12 For problem 1.4, what is the required M_p in order that the computed ultimate load will be reached?

2.13 For a fixed-ended cantilever beam of rectangular cross section loaded at the end with a concentrated load, derive the equation of the elastic-plastic interface (line separating the elastic from the plastic part) when $M = M_p$. Sketch.

2.14 For a uniformly loaded, simply supported beam of rectangular cross section, derive the equation of the elastic-plastic interface when $M = M_p$. Sketch result.

2.15 Do problem 2.14 except for 24WF100 beam of 20-ft length, fixed at the ends. Sketch all plastic zones.

2.16 Compute the residual stress at the extreme fiber of a rectangular section after it has been bent to the plastic moment condition and the moment subsequently released.

2.17 What are the two sources of reserve strength beyond the elastic limit in an indeterminate steel frame? Describe each.

2.18 Define "mechanism." How does a "plastic mechanism" differ from a real mechanism?

2.19 Explain why erection forces influence the elastic stress-distribution in a structure, but have no effect on the ultimate load. (Refer to the illustration in Chapter 1.) What is the difference between two structures at ultimate load, both identical in every respect except that erection forces are present in one and not in the other.

2.20 A fixed-ended beam has a concentrated load at the left third-point. Assuming the idealized M–ϕ relationship (Fig. 2.6), draw the moment diagram as each of the plastic hinges forms. Sketch the load-deflection curve. Compute P_u.

2.21 Given a fixed-ended beam of length L with uniformly-distributed load w. If the allowable working stress is 20.0 ksi, the yield point is 33.0 ksi, and the shape factor is 1.15, calculate:

(*a*) allowable working load (elastic design);
(*b*) yield load;
(*c*) ultimate load;
(*d*) factor of safety against yielding;
(*e*) factor of safety against ultimate load;
(*f*) give two reasons why (*c*) is greater than (*b*).

2.22 For a fixed-ended beam loaded at the third points, plot the moment vs. load relationship at the ends and at the center. A precise relationship is desired. Assume idealized M–ϕ curve.

3 Analysis of Structures for Ultimate Load

3.1 INTRODUCTION

With the evidence presented in Chapter 2 that full-size continuous beams behave as predicted by plastic theory, it is appropriate to consider next the methods of plastic analysis. The objective of this chapter is to describe briefly the fundamental principles upon which plastic analysis rests and then to show how these principles are used in analyzing continuous beams and frames.

The basis for computing the ultimate load (or maximum plastic strength) is the strength of steel in the plastic range. As shown in Art. 1.3, structural steel has the ability to deform plastically after the yield-point is reached. The resulting flat stress-strain characteristic assures dependable plastic strength, on the one hand, and provides an effective "limit" to the strength of a given cross section theoretically making it independent of further deformation. Thus, when certain parts of a structure reach the yield stress, they maintain that same stress under increasing deformation while other less highly stressed parts deform elastically until they, too, reach the yield condition. Since all critical sections eventually reach the yield condition, the analysis is considerably simplified because only this fact need be considered. It is of no importance *how* the moments are redistributed; one need only recognize that they *are*. The "continuity" condition is no longer applicable.

Since the analysis depends on plastic characteristics, it is probable that certain details must be re-evaluated. Elastic analysis rests upon the elastic behavior of all the parts. Such behavior now becomes relatively unimportant and the emphasis on details must be that the plastic action of that detail does not otherwise prevent the structure from reaching the ultimate load (say by local crippling). Therefore "modifications" to the simple plastic theory must be considered and these will be studied in later chapters.

3.2 PLASTIC ANALYSIS COMPARED WITH ELASTIC ANALYSIS

Although plastic and elastic analysis were compared at the outset from the design point of view, it is of interest now to compare them with regard to the fundamental conditions satisfied by each method.

An analysis according to the plastic method must satisfy three conditions that may be deduced from what has already been said:

(1) Mechanism condition (the ultimate load is reached when a mechanism forms)

(2) Equilibrium condition (summation of forces, and moments, is equal to zero)

(3) Plastic moment condition (the moment may nowhere be greater than M_p)

Actually these conditions are similar to those in elastic analysis which requires a consideration of the *continuity*, the *equilibrium*, and the *limiting stress* conditions.

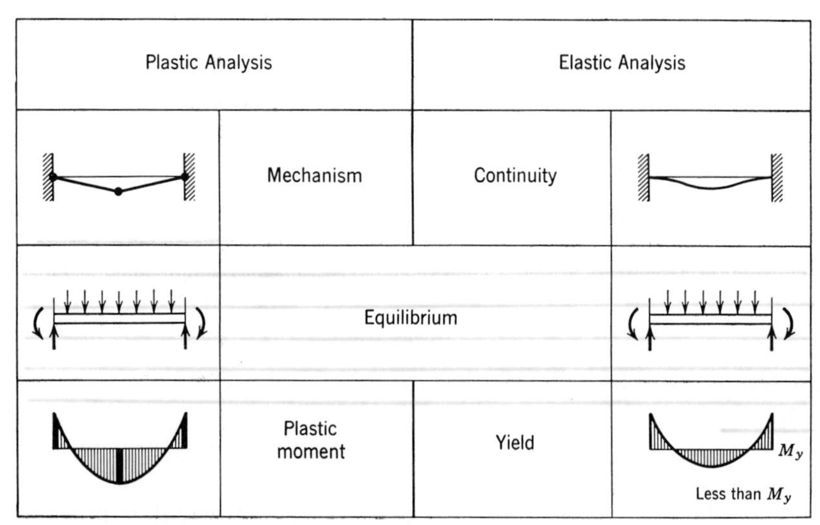

Fig. 3.1. The conditions necessary for a correct plastic analysis as compared with those necessary for a correct elastic analysis.

This similarity is demonstrated in Fig. 3.1. With regard to continuity, the situation in plastic analysis is just the reverse of that which exists in elastic analysis: Theoretically plastic hinges interrupt continuity, so the requirement is that sufficient plastic hinges form to allow the structure (or part of it) to deform as a mechanism. This could be termed a *mechanism* condition. The *equilibrium* condition is the same, namely, the load must be supported. Instead of initial yield, the limit of usefulness is the attainment of plastic hinge moments, not only at one cross section but at each of the critical sections; this will be termed a *plastic moment* condition.*

As will be discussed in articles that follow, two useful methods of plastic analysis in steel take their names from the particular conditions being satisfied:

(a) Mechanism Method	→ satisfies	Mechanism condition Equilibrium condition
(b) Statical Method (Equilibrium)	→ satisfies	Equilibrium condition Plastic moment condition

In the first method, a mechanism is assumed and the resulting equilibrium equations are solved for the ultimate load. This value is only correct if the plastic moment condition is also satisfied. On the other hand, in the statical or "equilibrium" method, an equilibrium moment diagram is drawn in such a manner that $M \leq M_p$. The resulting ultimate load is only the correct value if sufficient plastic hinges were assumed to create a mechanism.

* It is important to make a clear distinction between plastic design in steel and ultimate strength design in concrete, and from the preceding discussions of necessary and sufficient conditions it is now possible to make such a distinction. For plastic design in steel one makes use not only of the plastic theory of bending (the "plastic moment condition") but also of redistribution of moment (which is the basis of the "mechanism condition"). In ultimate strength design for concrete structures, on the other hand, an elastic analysis is necessary to determine the elastic distribution of moments throughout the frame. Then the cross sections of the individual members are proportioned according to the ultimate bending strength required of each cross section. Therefore, ultimate strength design in concrete only makes use of an "inelastic moment condition" but does not take into account redistribution of moment.

3.3 FUNDAMENTAL PRINCIPLES

Virtual Displacements

The principle of virtual displacements is useful in expressing the equilibrium condition. It may be stated as follows: *

> If a system of forces in equilibrium is subjected to a virtual displacement, the work done by the external forces equals the work done by the internal forces.

If the internal work is called W_I and the external work is called W_E, this principle may be expressed in the form

$$W_E = W_I \tag{3.1}$$

Application of this equation will be given in Art. 3.5.

Upper and Lower Bound Theorems

Usually it is not possible to satisfy all three of the necessary conditions (mechanism, equilibrium, and plastic moment) in one operation. Although the *equilibrium* condition will always be satisfied, a solution arrived at on the basis of an assumed *mechanism* will give a load-carrying capacity that is either correct or *too high.* On the other hand, one that is arrived at by drawing a statical moment diagram that does not violate the *plastic moment* condition will either be correct or *too low.* Thus, depending on how the solution to the problem is started, one will obtain an upper "limit" or "bound" below which the correct answer must lie, or one will determine a lower limit or bound which is less than the true load capacity.

These important upper and lower bound theorems or principles were proved by Greenberg and Prager.[3.2] When both theorems are satisfied in any given problem, then the solution is in fact the correct one. The two principles will be stated and illustrated.

> UPPER BOUND THEOREM
>
> A load computed on the basis of an assumed mechanism will always be greater than or at best equal to the true ultimate load.

* Ref. 3.1 contains an excellent discussion of the principle of virtual displacements (p. 1 ff.).

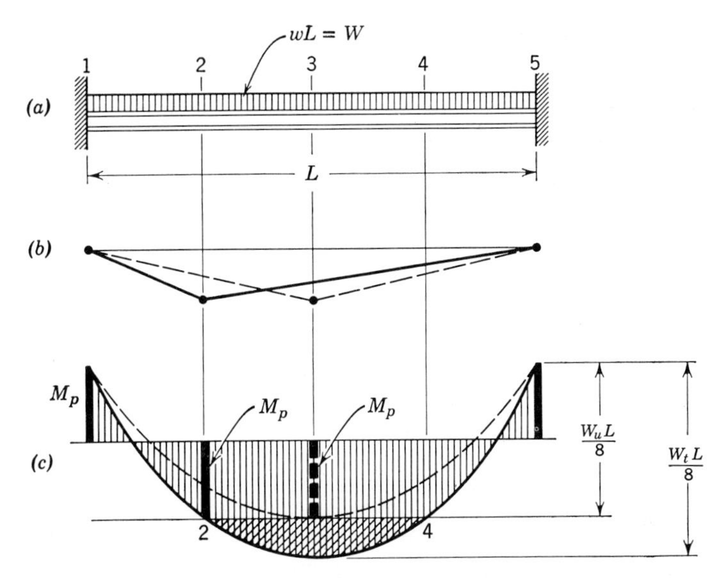

Fig. 3.2. An illustration of the upper bound theorem.

Consider the fixed-ended beam of Fig. 3.2. If a mechanism is assumed as shown by the solid lines in Fig. 3.2*b* on the basis of a guess that the plastic hinge in the beam forms at 2, then the equilibrium moment diagram would be as shown by the solid line in (*c*).* Since M_p is exceeded from 2 to 4 the beam would have to be reinforced over this length in order to carry the "trial" load W_t; the load is too great. Only when the mechanism is selected in such a way that the plastic moment value is nowhere exceeded (see the dotted lines) is the correct (lowest) value obtained.

> LOWER BOUND THEOREM
>
> A load computed on the basis of an assumed equilibrium moment diagram in which the moments are not greater than M_p is less than or at best equal to the true ultimate load.

Illustrating with the fixed-ended beam of Fig. 3.3, if the redundants are selected so that the moment is never greater than M_p, then the corresponding trial load W_t may be less than W_u (Fig. 3.3*b*). The full load capacity of the beam has not been used because the centerline moment

* As before, the moment diagram is plotted on the tension side of the member.

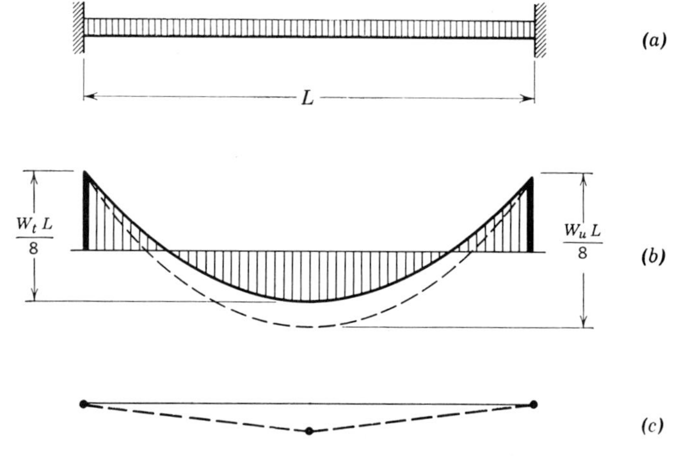

Fig. 3.3. An illustration of the lower bound theorem.

is less than M_p. Only when the load is increased to the point that a mechanism is formed (dotted) will the correct value be obtained.

Thus, if the problem is approached from the point of view of assuming a mechanism, an upper bound to the correct load will be obtained. But this could violate the plastic moment condition. On the other hand, if the problem is attacked by making an arbitrary assumption of the moment diagram, then the load might not be sufficiently great to create a mechanism. It is seen, then, that the *statical* method of analysis is based on the lower bound theorem. The *mechanism* method, on the other hand, represents an upper limit to the true ultimate load.

Further Assumptions

In addition to the assumptions of Art. 2.2, the following further assumptions are necessary:

(1) The theory considers only first order deformations. The deformations are assumed to be sufficiently small so that equilibrium conditions can be formulated for the undeformed structure (just as in the case of elastic analysis).

(2) Instability of the structure will not occur prior to the attainment of the ultimate load. (This is assured through attention to secondary design considerations, Chapter 4.)

(3) The connections provide full continuity so that the plastic moment M_p can be transmitted. (See Chapter 5.)

(4) The influences of normal and shearing forces on the plastic moment M_p are neglected (see Chapter 4 for necessary modifications).

(5) The loading is proportional, that is, all loads are such that they increase in fixed proportions to one another. However, independent increase can be allowed, provided no local failure occurs. (See Chapter 4 for repeated loading.)

These assumptions, coupled with a moment–curvature relationship that asymptotically approaches a limiting or plastic moment, are the essence of what has been termed the "simple plastic theory."

3.4 STATICAL METHOD OF ANALYSIS

As noted in Art. 3.3, the *statical* method of analysis is based on the lower bound theorem. The procedure is first described and then several examples are solved.

STATICAL METHOD

The objective is to find an equilibrium moment diagram in which $M \leq M_p$ such that a mechanism is formed. The procedure is as follows:

(1) Select redundant(s).
(2) Draw moment diagram for determinate structure.
(3) Draw moment diagram for structure loaded by redundant(s).
(4) Sketch composite moment diagram in such a way that a mechanism is formed (sketch mechanism).
(5) Compute value of ultimate load by solving equilibrium equation.
(6) Check to see that $M \leq M_p$.

EXAMPLE Fixed-Ended, Uniformly Loaded Beam (Fig. 3.4)

The problem, already treated in Art. 2.6, is to find the ultimate load W_u that the beam of moment capacity M_p will support. The beam shown in Fig. 3.4*a* is indeterminate to the second degree; for the redundants one could select the end moments. The resulting determinate structure and the beam loaded by the redundants are shown in sketches (*b*) and (*c*). The moment diagram for the determinate structure is shown in sketch (*d*), the moment at the center being given by

$$M_s = \frac{WL}{8} \tag{3.2}$$

The moment diagram for the structure loaded by the redundants is shown in sketch (*e*) the moment M_1 being an unknown.*

The next step is to combine the two moment diagrams (sketches *d* and *e*) in such a way that a mechanism is formed. This will be accomplished if the "fixing line" (designated as *A*) is drawn in such a way that the moment at sections 1 and 3 is equal to that at section 2. The resulting composite moment diagram is drawn in sketch (*f*). Notice that if this "fixing line" had been drawn elsewhere, as at *B*, then no mechanism would have been formed at ultimate load,

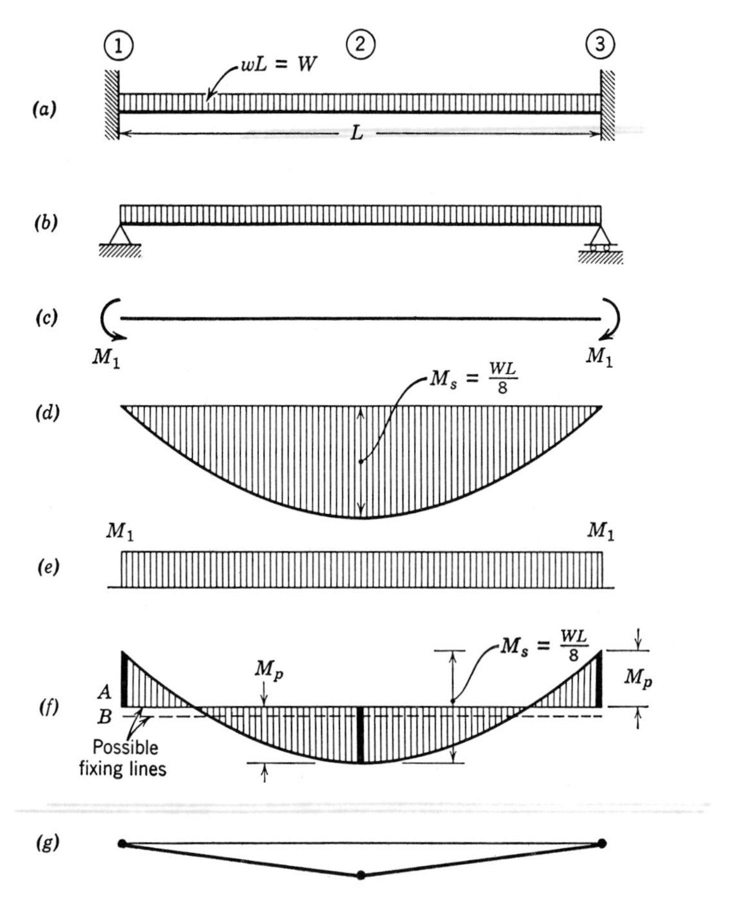

Fig. 3.4. Plastic analysis of a fixed-ended uniformly loaded beam by the statical method.

W_u, since the moment at 2 would not equal the moment at 1 and 3 and hinges could have formed only at the latter points. Only when the fixing line is drawn at *A* does a mechanism form as shown in sketch (*g*). $M = M_p$ at the three sections of maximum moment.

* For sign convention, see page 63.

The equilibrium equation, from Fig. 3.4*f* (section 2), is

$$\frac{W_u L}{8} = M_p + M_p$$

and the ultimate load is given by

$$W_u = \frac{16M_p}{L} \tag{3.3}$$

EXAMPLE Two-Span Continuous Beam with Concentrated Load (Fig. 3.5)

This structure is indeterminate to the first degree; the redundant is selected as the moment at 3, M_3. The resultant loadings are shown in Figs. 3.5*a* and 3.5*b*, the end reactions being omitted.

Moment diagrams due to loads and redundants are shown in sketches (*c*) and (*d*).

The composite moment diagram is drawn in sketch (*e*) in such a way that the necessary mechanism is formed, sketch (*f*), with maximum moments M_p at sections 2, 3, and 4.

The equilibrium equation is obtained by summing the moments at section 2,

$$\frac{P_u L}{4} = M_p + \frac{M_p}{2}$$

$$P_u = \frac{6M_p}{L} \tag{3.4}$$

Since all three of the necessary conditions are satisfied (mechanism, equilibrium, and plastic moment), this is the correct answer.

* Further examples of the use of this method are given in Chapters 8 and 9, Plates I to V and VII to IX.

Sign Conventions

The moment diagram convention adopted in this book is that the moment is plotted on the side of the member that is in tension. This means that the moment diagram for continuous or restrained beams will appear to be "reversed" to many readers, but the diagrams for frames will appear in the customary form. As a further point of interest, it will be noted that it is simpler to use a discontinuous base line for the moment diagram than to redraw the moment diagram about a single straight line as is customary in conventional analysis. Thus in sketch (*e*) of Fig. 3.5, moment diagrams (1) and (2) are identical, the base line being *a-b-c* in each case. The first form will be used in this book, since it is obtained directly in the statical method.

As a matter of fact, in routine design, one would eliminate sketches (*a*), (*b*), (*c*), and (*d*) of Fig. 3.5. The total construction required for the solution is that shown in sketch (*e*–1).

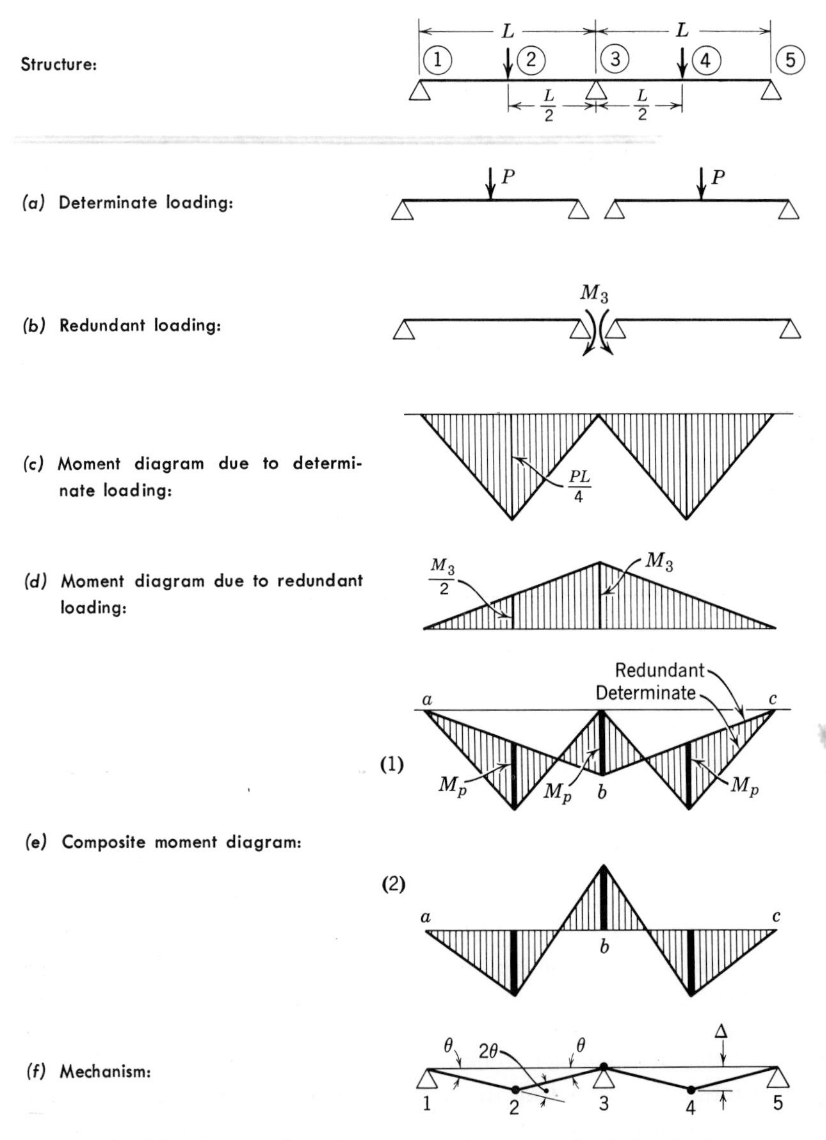

Fig. 3.5. Plastic analysis of two-span continuous beam (statical method).

In addition to the above sign convention for constructing the moment diagram, one other sign convention will be necessary. When considering the equilibrium equations (Table 7.3, Procedure 8) *the moment convention says that positive moments tend to rotate the end of a member in a clockwise direction and moments within a beam are positive when they produce tension in the lower fiber.* In the moment-balancing method for making the plastic moment check (Art. 3.8) the same convention is used. Also in Chapter 6, the sign convention for both moments and rotations is that "clockwise is positive."

The student should make it a practice to study all sign conventions carefully. A helpful exercise would be to work several of the later problems using an alternate convention.

3.5 MECHANISM METHOD OF ANALYSIS

General Procedure

As the number of redundants increases, the number of possible failure mechanisms also increases. Thus it may become more difficult to construct the correct equilibrium moment diagram. For such cases the *mechanism method* of plastic analysis may be used, and various "upper bounds" to the correct load will be obtained for the different possible mechanisms. The correct mechanism will be the one which results in the lowest possible load (upper bound theorem) and for which the moment does not exceed the plastic moment at any section of the structure (lower bound theorem). Thus the objective is to find a mechanism such that the plastic moment condition is not violated.

The procedure is next outlined and is followed by several examples.

MECHANISM METHOD

Find a mechanism (independent or composite) such that $M \leq M_p$:

(1) Determine location of possible plastic hinges (load points, connections, point of zero shear in a beam span under distributed load).
(2) Select possible "independent" and "composite" mechanisms.
(3) Solve equilibrium equation (virtual displacement method) for the lowest load.
(4) Check to see that $M \leq M_p$ at all sections.

EXAMPLE Two-Span Continuous Beam (Fig. 3.5)

As a first illustration consider the same problem that was solved in the example on p. 63. The first step is to determine the location of possible plastic hinges, and for this problem hinges could form at sections 2, 3, and 4. Since the structure and loading are symmetrical, there is only one possible mechanism and this is shown in sketch (*f*).

Suppose that after the ultimate load is reached the structure is allowed to move through a small additional displacement. The mechanism of sketch (*f*) is the result, the deflection at any point representing the *increment* of virtual motion at ultimate load (see Art. 2.7). If the rotation at 1 is called θ, then by geometry all of the remaining angles can be computed, and the rotation at each of the plastic hinges turns out to be 2θ.

Using the principle of virtual displacements it is now possible to write an equilibrium equation. The external work, W_E, is the work done as the two loads P move vertically with the hinges at sections 2 and 4. This distance Δ may be computed in terms of the same angle θ, and thus

$$\Delta_2 = \Delta_4 = \frac{L}{2}\theta$$

The total external work is thus given by

$$W_E = (P)\left(\frac{L}{2}\theta\right) + P\left(\frac{L}{2}\theta\right) \tag{3.5}$$

The internal work in the structure will be the sum of the virtual work done at each of the plastic hinges. The work done at each hinge is equal to the plastic moment at that hinge times the angle through which it rotates. Thus the total internal work, from sketch (*f*), is given by

$$W_I = \underset{\substack{\text{Plastic}\\\text{moment at}\\\text{Section 2}}}{M_p} \times \underset{\substack{\text{Mechanism}\\\text{angle at}\\\text{Section 2}}}{2\theta} + \underset{\substack{\text{Plastic}\\\text{moment at}\\\text{Section 3}}}{M_p} \times \underset{\substack{\text{Mechanism}\\\text{angle at}\\\text{Section 3}}}{2\theta} + \underset{\substack{\text{Plastic}\\\text{moment at}\\\text{Section 4}}}{M_p} \times \underset{\substack{\text{Mechanism}\\\text{angle at}\\\text{Section 4}}}{2\theta} \tag{3.6}$$

Using the principle of virtual displacements, one can equate expressions 3.5 and 3.6, and obtain

$$2\frac{PL}{2}\theta = 3M_p(2\theta)$$

The term θ cancels and thus

$$P_u = \frac{6M_p}{L} \tag{3.4}$$

which is, of course, the same answer as obtained previously. A moment check is not necessary because there was only one possible mechanism.

EXAMPLE Unsymmetrical Two-Span Beam (Fig. 3.6)

In the previous example, there was only one possible solution to the problem because of the symmetrical arrangement chosen. In this illustration there is more than one possible mode of failure. The first step in the mechanism method is to determine the position of possible plastic hinges. There are four, and they may form at sections 2, 3, 4, and 5 of Fig. 3.6*a*.

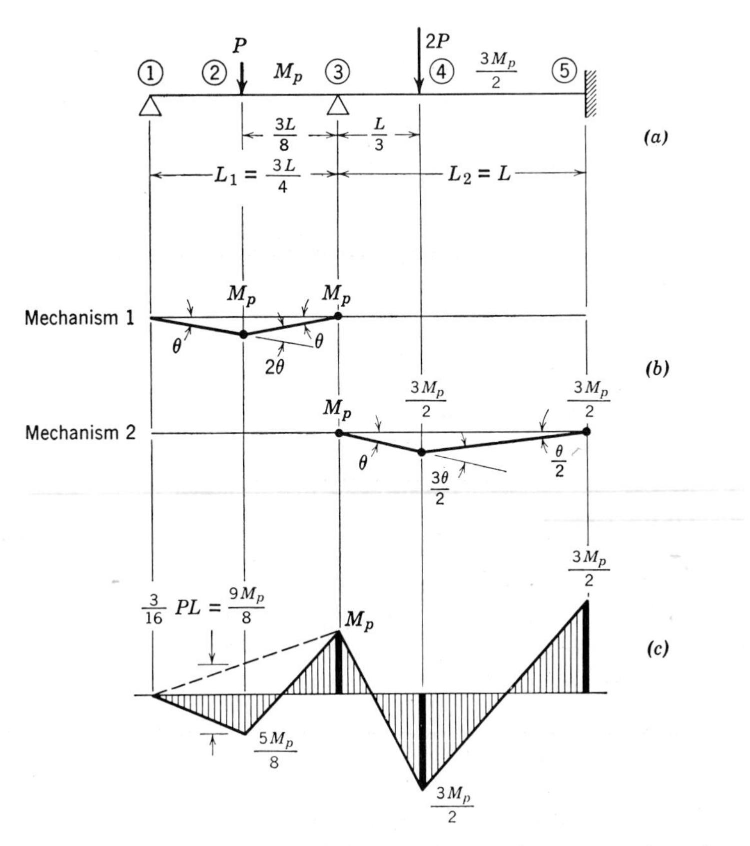

Fig. 3.6. Mechanism method of analysis applied to an unsymmetrical two-span continuous beam.

There are two possible mechanisms and these are shown in sketch (*b*). One involves local failure of beam 1–3, the other involves failure of beam 3–5. The next step is to work out the geometry of the mechanism motion for each case. For mechanism 1, if θ is the virtual angle of rotation at section 1, then the rotation at 3 is also equal to θ; the angular discontinuity at section 2 is therefore 2θ. The vertical displacement of hinge 2 is equal to the rotation at 1 times the distance from 1 to 2 or $\Delta = \theta(3L/8)$. For mechanism 2 if θ is the angle of rotation at 3, then the rotation at 5 will equal one-half of this value, or $\theta/2$. The angle at section 4 is equal to the sum of the angles at 3 and 5, or $3\theta/2$. The vertical displacement of hinge 4 is equal to $\theta L/3$. Note that the plastic moment value at section 3 is equal to M_p due to the change in section there.

From the principle of virtual displacements (Eq. 3.1) the external and the internal work may be equated, and for mechanism 1 the following equation is obtained

$$\underset{\substack{\text{Load}\\\text{at 2}}}{P} \times \underset{\substack{\text{Vertical}\\\text{displace-}\\\text{ment at 2}}}{\theta\frac{3L}{8}} = \underset{\substack{\text{Plastic}\\\text{moment}\\\text{at 2}}}{M_p} \times \underset{\substack{\text{Rota-}\\\text{tion}\\\text{at 2}}}{2\theta} + \underset{\substack{\text{Plastic}\\\text{moment}\\\text{at 3}}}{M_p} \times \underset{\substack{\text{Rota-}\\\text{tion}\\\text{at 3}}}{\theta} \tag{3.7}$$

from which

$$P_u = 8M_p/L$$

For mechanism 2, the virtual work equation is

$$\underbrace{\underset{\substack{\text{Load}\\\text{at 4}}}{2P} \times \underset{\substack{\text{Deflec-}\\\text{tion}\\\text{at 4}}}{\frac{\theta L}{3}}}_{W_E} = \underbrace{\underset{\substack{\text{Plastic}\\\text{moment}\\\text{at 3}}}{M_p} \times \underset{\substack{\text{Rota-}\\\text{tion}\\\text{at 3}}}{\theta} + \underset{\substack{\text{Plastic}\\\text{moment}\\\text{at 4}}}{\frac{3}{2}M_p} \times \underset{\substack{\text{Rota-}\\\text{tion}\\\text{at 4}}}{\frac{3\theta}{2}} + \underset{\substack{\text{Plastic}\\\text{moment}\\\text{at 5}}}{\frac{3}{2}M_p} \times \underset{\substack{\text{Rota-}\\\text{tion}\\\text{at 5}}}{\frac{\theta}{2}}}_{W_I} \tag{3.8}$$

from which

$$P_u = 6.0\frac{M_p}{L} \tag{3.9}$$

Since the lowest load is that corresponding to mechanism 2 (Eq. 3.9), then this is the correct answer.

A check of the plastic moment condition would not be required to determine if we had selected the correct mechanism, because all possibilities were tried. As a check, however, the moment diagram is drawn and is shown in Fig. 3.6*c*. The moment at 2 is found by static equilibrium and is equal to $\frac{5}{8}M_p$. Since a mechanism is formed and since the plastic moment value of the beams is not exceeded at any point, then this is the correct answer.

Suppose the critical mechanism in the foregoing example had turned out to be mechanism 1? No hinges would have formed at sections 4 and 5 and the beam would have been partially redundant at failure. It would seem at first that the equations of elastic continuity would have to be solved in order to obtain the moment diagram and thus make certain that the plastic moment condition was not violated at sections 4 or 5. However, the *exact* magnitude of the moment in the redundant portion of the beam is not of interest, and it develops that simple techniques are available for obtaining a possible moment diagram when the structure is partially redundant at failure. These methods are described in Art. 3.8.

For statical Equilibrium:

The Ms =

EXAMPLE Rectangular Portal Frame (Fig. 3.7)

Given a rectangular frame of uniform section whose plastic moment capacity is M_p, what is the ultimate load it will carry?

In the frame shown in Fig. 3.7*a* locations of possible plastic hinges are at sections 2, 3, and 4. Now, in the previous examples the possible failure mechanisms were obvious. However, in this problem there are several possibilities. "Elementary" or "independent" mechanisms 1 and 2 shown in Figs. 3.7*b* and *c* correspond to the separate action of the two loads, whereas mechanism 3, Fig. 3.7*d*, is a "composite" mechanism formed by combination of mechanisms 1 and 2 to eliminate a plastic hinge at section 2. Which is the correct one? It is the one which results in the lowest critical load P_u, since any greater load would mean that the plastic moment condition would be violated. ?

The method of virtual displacements may be used to compute the ultimate load. After the ultimate load is reached, the frame is allowed to move through a small *additional* displacement such as shown by Δ in Fig. 3.7*b*. For equilibrium, the *external work* done by the loads as they move through small displacements must equal the *internal work* absorbed at each hinge as it rotates through a corresponding small angle (Eq. 3.1).

The following equations are then obtained for the various mechanisms:

	External work		Hinge at 2	Hinge at 3	Hinge at 4

Mechanism 1:
$$P\Delta = M_p\theta + M_p(2\theta) + M_p\theta \tag{3.10}$$
$$P\left(\frac{L\theta}{2}\right) = M_p(4\theta)$$
$$P_1 = \frac{8M_p}{L} \tag{3.11}$$

Mechanism 2:
$$\frac{P}{2}\Delta = M_p(\theta + \theta) \tag{3.12}$$
$$\frac{P}{2}\left(\frac{L\theta}{2}\right) = 2M_p\theta$$
$$P_2 = \frac{8M_p}{L} \tag{3.13}$$

Mechanism 3:
$$P\Delta_1 + \frac{P}{2}\Delta_2 = M_p(2\theta) + M_p(2\theta) \tag{3.14}$$
$$P\left(\frac{L\theta}{2}\right) + \frac{P}{2}\left(\frac{L\theta}{2}\right) = 4M_p\theta$$
$$P_3 = \frac{16}{3}\frac{M_p}{L} = P_u \tag{3.15}$$

The lowest value is P_3, which is therefore the true ultimate load P_u.

To make sure that some other possible mechanism was not overlooked it is necessary to check the plastic moment condition to see that $M \leq M_p$ at all sections. The complete moment diagram is drawn as shown in Fig. 3.7*e*, the moments being plotted on the tension side of each member. The moment at section 2 is determined as follows:

$$H_5 = \frac{M_p}{L/2} = \frac{2M_p}{L}$$

$$H_1 = \frac{P}{2} - H_5 = \left(\frac{16M_p}{3L}\right)\frac{1}{2} - \frac{2M_p}{L} = \frac{2}{3}\frac{M_p}{L}$$

$$M_2 = H_1\frac{L}{2} = \left(\frac{2}{3}\frac{M_p}{L}\right)\left(\frac{L}{2}\right) = \frac{M_p}{3}$$

Since the moment is not greater than M_p at any section, the correct answer has been obtained, and the problem is solved.

In the example on p. 69 the virtual work equation was solved anew for the composite mechanism. An alternate procedure for computing the ultimate load for a composite mechanism is to add together the virtual work equations for each mechanism in the combination, being careful to subtract the internal work done in an elementary mechanism at any hinges being eliminated by the combination. Using this procedure for the example on p. 69 there is obtained from the previous set of equations:

Mechanism	Virtual Work Equation	Hinges Cancelled
Beam mechanism (1):	$\frac{PL}{2}\theta = 4M_p\theta$	$-M_p\theta$
Panel mechanism (2):	$\frac{PL}{4}\theta = 2M_p\theta$	$-M_p\theta$
Combination (Mechanism 3)	$\frac{3PL}{4}\theta = 6M_p\theta$	$-2M_p\theta$

$$P_3 = \frac{16M_p}{3L} \quad (3.16)$$

This is the same answer as obtained in Eq. 3.15.

In the previous examples there were a sufficiently small number of possible mechanisms so that the combinations were almost obvious. Further, the geometry of the deformed position could be developed with no difficulty. A number of guides and techniques will now be discussed that are useful in solving more involved problems.

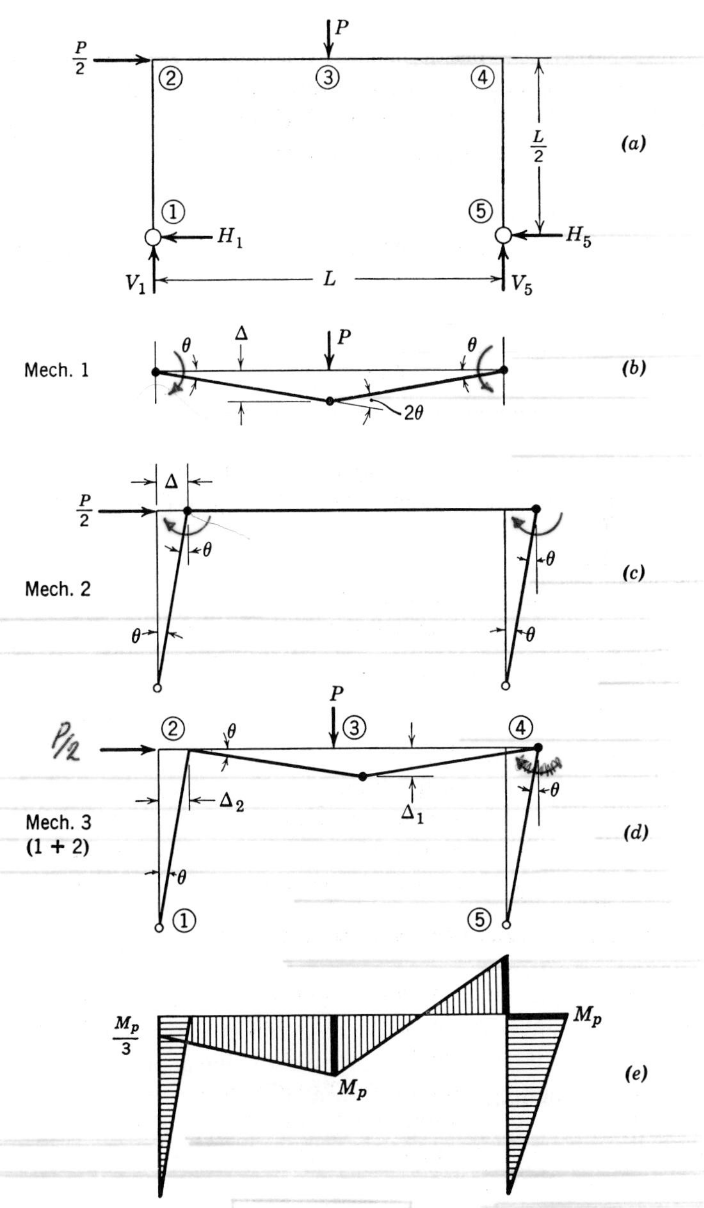

Fig. 3.7. Mechanism method of analysis applied to a rectangular portal frame with pinned bases.

Types of Mechanisms

For convenience in referring to different modes of failure, there are the following types of mechanisms which are illustrated in Fig. 3.8 using the structure shown in sketch (*a*) thereof:

Beam mechanism — Sketch (*b*)

(Four examples are given here of the displacement of single spans under load)

Panel mechanism — Sketch (*c*)

(This motion is due to side-sway)

Gable mechanism — Sketch (*d*)

(This is a characteristic mechanism of gabled frames, involving spreading of the column tops with respect to the bases)

Joint mechanism — Sketch (*e*)

(This independent mechanism forms at the junction of three or more members and represents motion under the action of an applied moment)

Composite mechanism — Sketches (*f*), (*g*)

(Various combinations of the independent mechanisms may be made. A composite mechanism may be "partial" as indicated by sketch (*f*) and for which the frame is still indeterminate at failure; or it may be a "complete" composite mechanism, in which case the frame is determinate at failure.)

Number of Independent Mechanisms

If it were known in advance how many independent mechanisms existed, then combinations could be made in a systematic manner and there would be less likelihood of overlooking a possible combination. Fortunately a simple procedure is available for determining this.

If the number of possible plastic hinges is N and if the number of redundancies is X, then the number of possible independent mechanisms n may be found from

$$\boxed{n = N - X} \tag{3.17}$$

Thus, in Fig. 3.7 there are 3 possible plastic hinges (at sections 2, 3, 4). The frame is indeterminate to the first degree, and therefore, there are

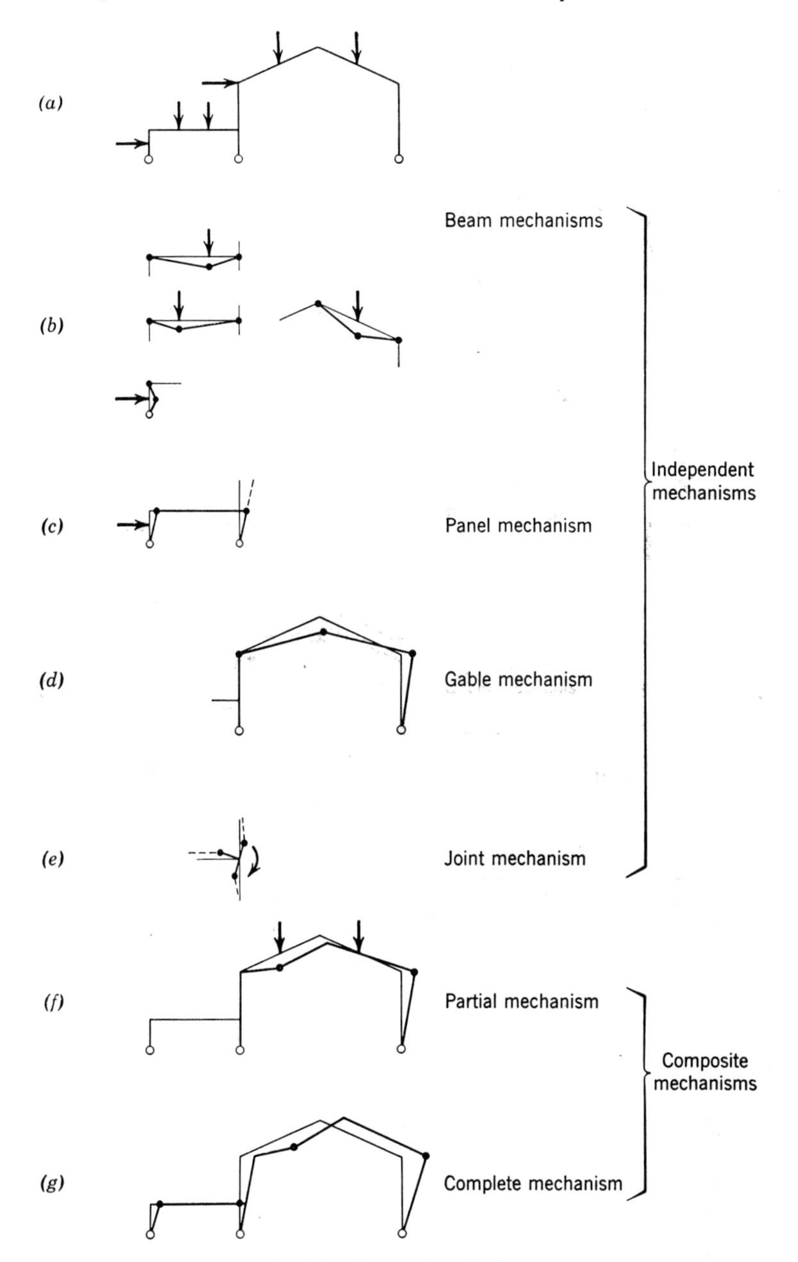

Fig. 3.8. Types of mechanisms.

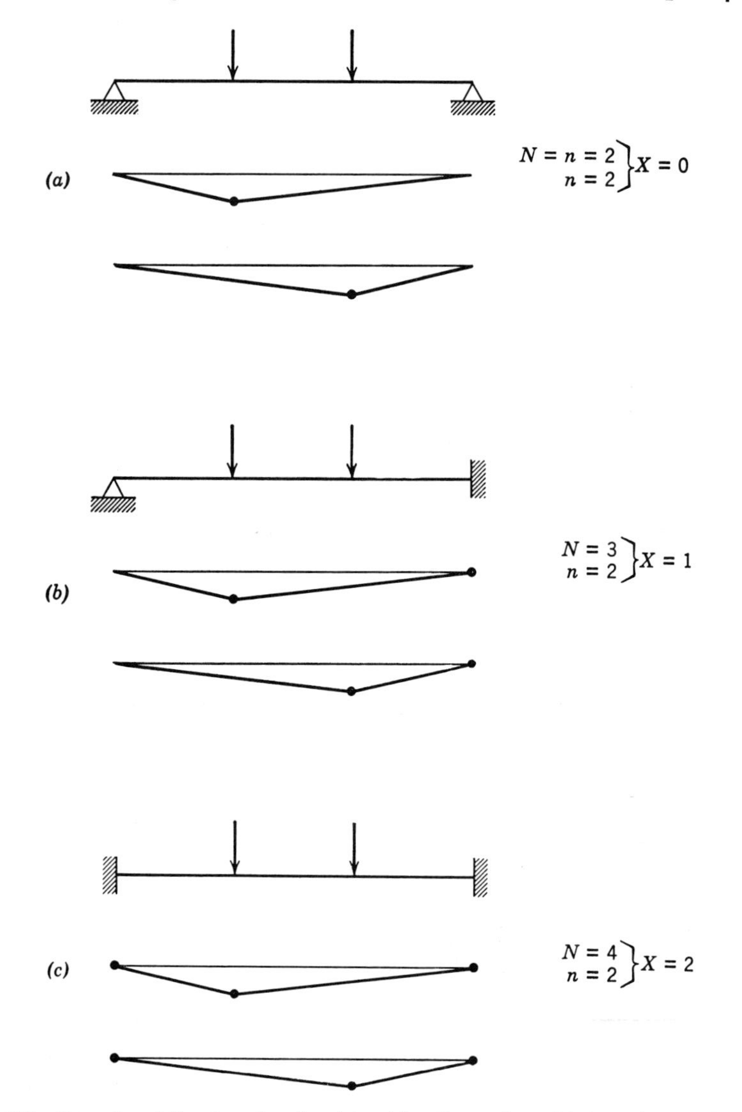

Fig. 3.9. Examples of the procedure for determining the number of independent mechanisms.

two independent mechanisms (beam mechanism 1 and panel mechanism 2).*

This correlation is no coincidence because each independent mechanism corresponds to the action of a different loading system. Said in

* An exception to this rule exists when haunched connections or tapered members are used. See discussion on p. 174 of Art. 5.4.

another way, each mechanism corresponds to an independent equation of equilibrium. In Fig. 3.7 mechanism 1 corresponds to equilibrium between applied vertical load and vertical shear. Mechanism 2 corresponds to equilibrium between applied horizontal load $P/2$ and horizontal shear in the two columns. These force systems are "elementary" or "independent" and hence the term.

This relationship (Eq. 3.17) is illustrated in another way. For a determinate system, if a plastic hinge develops, the structure becomes a mechanism. Thus, for each possible plastic hinge in the determinate structure there corresponds a mechanism; if there are N possible plastic hinges, there will be N mechanisms. (See Fig. 3.9*a*). As the structure becomes more redundant a plastic hinge is added for each redundant but the number of mechanisms does not change. Where the member was free to deform beforehand (at a real hinge), it is now restrained; however, the number of basic mechanisms remains unchanged (see for example Fig. 3.9*b* and *c*). Thus the number of possible plastic hinges N equals the number of mechanisms n, plus the number of redundants X, or $n = N - X$.

Composite Mechanisms

Equation 3.17 is useful because it enables one to set out all the possible "elements" from which combinations later may be made. These combinations are to be selected in such a way as to make the external work a maximum or the internal work a minimum, since by this means the lowest possible load P_u is obtained. Therefore, the procedure generally is to make combinations that involve mechanism motion by as many loads as possible and the elimination of plastic hinges—as was done in composite mechanism 3 of Fig. 3.7. As noted previously, these composite mechanisms may be either the "partial" or the "complete" type.

Indeterminacy

In order to determine the number of redundants X for use in Eq. 3.17 it is merely necessary to cut sufficient supports and structural members such that all loads are carried by simple beam or cantilever action. The number of redundants is then equal to the number of forces and moments required to restore continuity. (In the example shown in Fig. 3.7 cutting the horizontal reaction at section 5 removes the last restraint to simple beam action and thus $X = 1$.) In Fig. 3.10 are shown two additional examples; cuts made in the indicated members will reduce each structure to either cantilever or simply supported elements. The fixed-ended beam requires a shear force and moment to restore continuity at the cut section and thus $X = 2$. In the two-story structure a horizontal force, a vertical force, and a moment are required at each section for continuity, and thus $X = 12$.

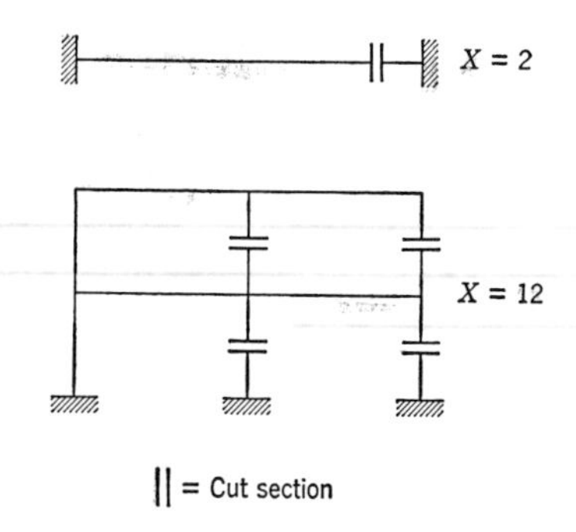

Fig. 3.10. Illustration of method for determining the number of redundants.

Geometry of Mechanism Motion (Instantaneous Centers)

As will be evident later in cases involving sloping roofs (Fig. 3.8*f*) computation of the geometrical relationship of the displacement in the direction of the load as the structure moves through the mechanism motion may become somewhat tedious. In such cases the motion of the structure and of its elements may be found by using one of the methods of basic mechanics, namely that of "instantaneous centers."

Although the use of instantaneous centers was not needed in the solution of the example on p. 69, consider its application to mechanism 3 of that problem (Figs. 3.7*d* and 3.11). As the structure moves, segment 1–2–3 pivots around the base at 1. Member 5–4 pivots about point 5. About what center does segment 3–4 move? The answer is obtained by considering how the *ends* of that segment move.

Point 4 is constrained to move perpendicular to line 4–5 and thus its center of rotation (as part of segment 3–4) must be somewhere along line 5–4 extended. Point 3, on the other hand, moves about point 1, since it is a part of segment 1–2–3. Therefore it moves normal to line 1–3 and its center of rotation as part of segment 3–4 must be along 1–3 extended. Point I satisfies both conditions and therefore segment 3–4 rotates about point I, that point being its "instantaneous center" of rotation.

What are the mechanism angles ("kink angles") at the plastic hinges? The rotation at both column bases is θ. The horizontal motion of point 4 is thus $\theta(L/2)$. Since the length I–4 is also equal to $L/2$, then the rotation of 3–4 about I is $\theta(L/2)/(L/2) = \theta$. The total rotation at 4, therefore, equals 2θ and that at 3 is also 2θ, since the lengths 1–3 and 3–I are equal.

What is the vertical motion of the load at point 3? Since no hinge forms in joint 2, it remains a right angle and the rotation of 2–3 with respect to the horizontal is also equal to θ. The vertical motion is

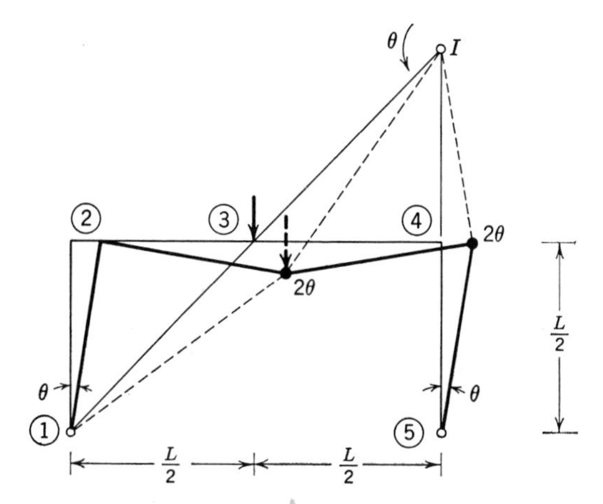

Fig. 3.11. Location of instantaneous center for the rectangular frame mechanism of Fig. 3.7.

therefore $\theta L/2$. This answer for vertical displacement and that in the previous paragraph for mechanism angles are identical, of course, with those obtained in the example on p. 69.

EXAMPLE Gabled Frame Mechanism (Fig. 3.12)

The use of "instantaneous centers" is more appropriate for gabled frames. For the structure shown in Fig. 3.12, the value θ will be assigned to the arbitrarily small rotation of member 6–7 about point 7. Segment 1–2–3 will rotate about point 1 an amount yet to be determined. To find the instantaneous center of segment 3–4–6, find the common point about which both *ends* rotate. Point 6, being constrained to move normal to line 7–6 will have its center along that line. Similarly, the center of 3 will be along 1–3 extended. Thus point I is located.

By geometry the length I–7 is equal to $4(5L/4) = 5L$. 6–I therefore equals $4L$. Since the horizontal displacement of point 6 is θL, the rotation at $I = \theta/4$.*

* Note that the rotation at I is in general equal to the rotation at the column base times the ratio of the distances 7–6 to 6–I.

By similar triangles, the ratio of 3–I to 1–3 is 3:1. Thus the rotation at 1 is given by

$$\frac{\theta}{4}\left(\frac{3}{1}\right) = \frac{3\theta}{4}$$

Mechanism angles and displacements in the direction of load may now be computed. The rotation at 6 $= \theta + \theta/4 = (\frac{5}{4})\theta$. The rotation at section 3 $= \theta/4 + \frac{3}{4}\theta = \theta$. The displacements of the loads in the direction of application are as follows:

$$\begin{aligned} &\text{Horizontal load:} && \Delta_1 = (\tfrac{3}{4})(\theta)(L) \\ &\text{Left vertical load:} && \Delta_2 = (\theta/4)(3L) \\ &\text{Right vertical load:} && \Delta_3 = (\theta/4)(L) \end{aligned} \tag{3.18}$$

The correctness of the last two equations may be seen in two ways. If the loads are imagined as being hung from the dashed positions shown, then it is evident that the vertical displacements are as shown above and in Fig. 3.12*a*. Alternatively, working out the geometry on the basis of similar triangles as shown in Fig. 3.12*b*, the vertical component of the mechanism motion of point 3, for ex-

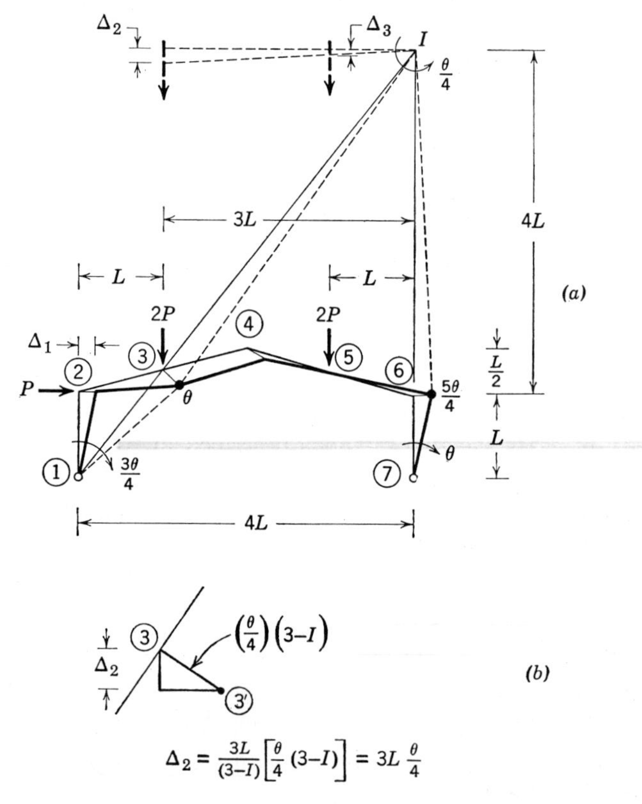

Fig. 3.12. Location of instantaneous center for a gabled frame mechanism.

ample, is equal to the rotation about the appropriate instantaneous center times the distance to that center measured normal to the line of action.

To complete the problem, the ultimate load *for this mechanism* is given by

$$\underset{\text{Horizontal load}}{P\left(\frac{3\theta L}{4}\right)} + \underset{\text{Vertical load, } L}{2P\left(\frac{3\theta L}{4}\right)} + \underset{\text{Vertical load, } R}{2P\left(\frac{\theta L}{4}\right)} = M_p\left(\underset{\text{Section 3}}{\theta} + \underset{\text{Section 6}}{\frac{5\theta}{4}}\right) \tag{3.19}$$

and

$$P_u = \frac{9}{11}\frac{M_p}{L} \tag{3.20}$$

Construction of the moment diagram shows that the moment is nowhere greater than M_p, so this is the correct answer.

Precisely the same answer would have been obtained in the foregoing example if the deformation at the various joints had been worked out through a consideration of the frame geometry in the deformed position. The convenience of the use of "instantaneous centers" should be evident, however.

Further illustrations of the use of instantaneous centers are to be found in the examples of Chapter 9; numerous additional examples of the mechanism method will also be found there.

3.6 FURTHER METHODS OF ANALYSIS

In addition to the statical and mechanism methods of analysis, there are additional techniques for determining the ultimate load a structure will support. Two methods in particular are the "Method of Inequalities" [3.3] and a pseudo "moment distribution" technique.[3.4,3.5] Although the methods are of interest for certain structural problems, in a great majority of cases the two methods described in Arts. 3.4 and 3.5 will be sufficient. The interested reader is referred to the indicated references for information on the "method of inequalities" and on the "plastic moment-distribution" procedure.

3.7 DISTRIBUTED LOAD

A slight modification of procedure is necessary in case the load is distributed. In the event that a mechanism involves formation of a hinge within the beam (i.e., between supports) the precise location of that hinge may not be known in advance.

Take the case shown in Fig. 3.13: a portion of a continuous beam in which the M_p values are as shown in the circles. If the load is actually distributed along the member, then the correct value of the ultimate load is obtained by determining the distance to the point of maximum moment (and zero shear). The distance x can be computed by writing the virtual work equation in terms of x and either minimizing the loads by differentiation or by solving for x by making a few trials. Alternatively x may be found by plotting the uniform load parabola, A-B-C-D, from the base line A-D.

To illustrate the computations, from the mechanism of Fig. 3.13b the virtual work equation gives

$$W\frac{x}{2}\theta^* = 3M_p\theta\left(1 + \frac{x}{L - x}\right) + 2M_p\theta\frac{x}{L - x}$$

$$W = \frac{2M_p\left(3 + \dfrac{3x}{L - x} + \dfrac{2x}{L - x}\right)}{x}$$

$$W = \frac{2M_p}{x}\left(3 + \frac{5x}{L - x}\right) \tag{3.21}$$

Selecting values of x and solving for W, the magnitude of x to give the minimum value is

$$x = 0.44L$$

and

$$W_u = \frac{31.3M_p}{L} \tag{3.22}$$

The graphical method was used in Fig. 3.13c and a value $x = 0.45L$ was obtained. Convenient charts have been developed by Horne[3.4] as an aid in solving problems involving distributed load. These charts are also available in Ref. 1.5.

With errors that are usually slight, the analysis could be made on the basis that the distributed load is replaced by a set of equivalent concentrated loads. Thus in Fig. 3.14, if the distributed load $wL = P$ is concentrated in the various ways shown, the uniform load parabola is always circumscribed (giving the same maximum shear). The result is always conservative because the *actual* moment in the beam is always

* The external work for a mechanism under distributed load may be written conveniently as the load/unit length times area swept during mechanism motion. In this example, "area" $= (L)(\theta x)(\frac{1}{2})$.

less than or at most equal to the assumed moment. Of course, the more concentrated loads assumed, the closer is the approximation to the real problem.

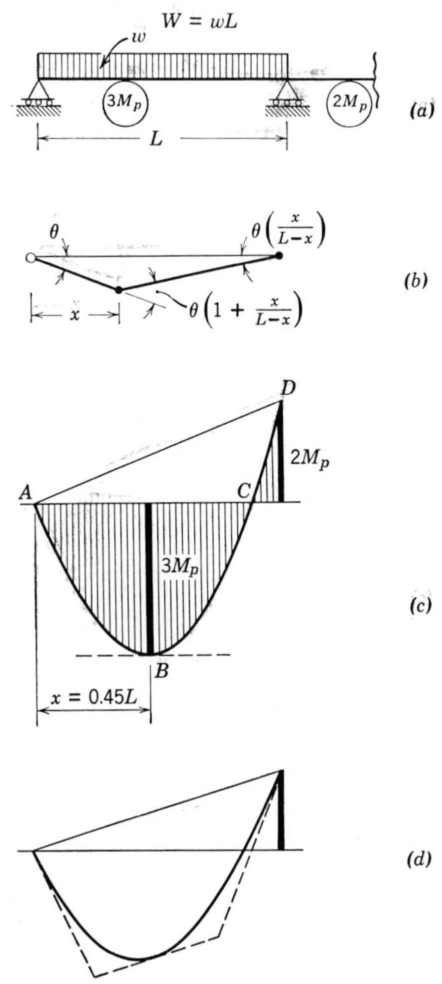

Fig. 3.13. Position of hinges in beam with distributed load.

If the distributed load is actually brought to the main frame through purlins and girts, the uniform load may be converted, at the outset, to actual purlin reactions (on the basis of assumed purlin spacing). The analysis is then made on the basis of the actual concentrated loads. The only difficulty with this procedure is that numerous additional possible plastic hinges are created—one at each purlin. And for every

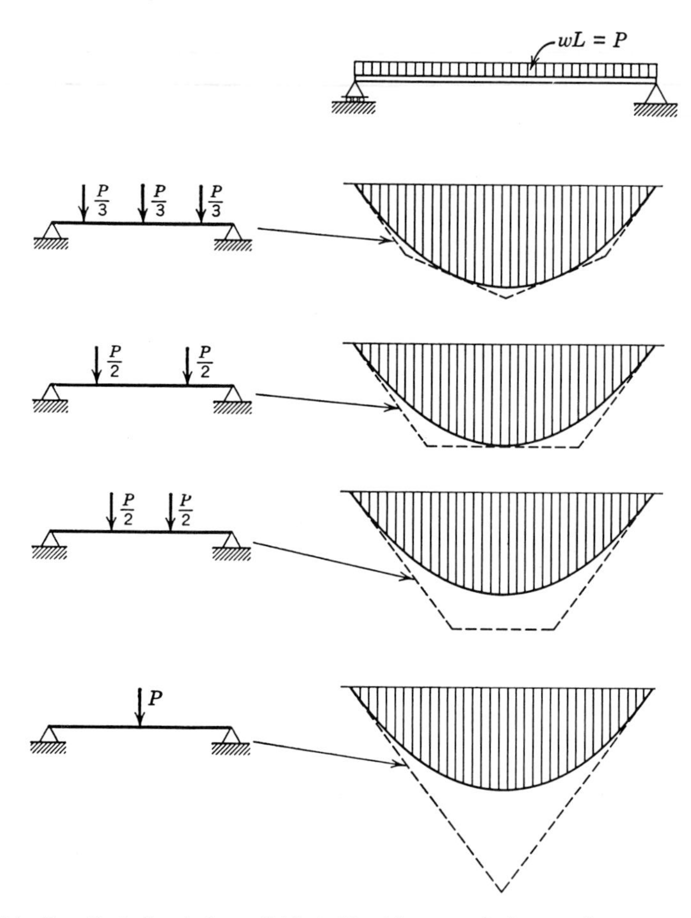

Fig. 3.14. The effect of replacing a distributed load by an equivalent set of concentrated loads.

possible hinge position there is another possible mechanism. With experience the designer will be able to tell how many of these mechanisms he should investigate.

Still another method is to assume that the hinge forms at mid-span and then to revise the design after the "correct" mechanism has been found.

3.8 MOMENT CHECK

One of the conditions that a "plastic" solution must satisfy is that the moment is not greater than the plastic moment at any section

(Art. 3.2). In the case of the statical method (Art. 3.4), there is no particular problem, because the moments used in the equations of equilibrium presumed $M \leq M_p$. However, in the mechanism method the solution leads to an upper bound and it is necessary therefore to see if the solution also satisfies equilibrium with $M \leq M_p$ throughout the frame. Otherwise it is possible to overlook a more favorable combination of mechanisms which would have resulted in a lower load.

Two cases must be considered:

(1) Structure is determinate at ultimate load;
(2) Structure is indeterminate at ultimate load.

For the first case, the equations of simple statics are all that are necessary to determine the moments in all parts of the frame. In the second case, however, an elastic analysis would be required to determine precisely the moments in those segments that do not contain plastic hinges at their ends. However, in solutions by plastic analysis, the *precise* magnitude of moment at a section that remains elastic is not of interest. If a mechanism has already been created, it is only necessary to show that moments elsewhere are not greater than M_p. As a result, approximations may be used to find a possible equilibrium moment diagram. If the plastic moment condition is met, then the solution satisfies the lower bound theorem, and the computed load must be the correct value.

Prior to considering the partially indeterminate cases further, it must be pointed out that a design that leads to such a condition (that is, part of the structure indeterminate) is probably not the best design. The design objective is to make *all* of the structure perform as efficiently as possible. If the frame is still indeterminate at ultimate load, it should be obvious that material can be saved somewhere in the structure, bringing moments up to their plastic values. What this means is that simple statics will usually be adequate for making the "moment check." As a routine procedure it will not be required to carry out what would otherwise be a more complicated checking operation, because a structure that turns out to be partially redundant would be redesigned for lighter weight.

Further examples of the moment check do not appear necessary here for the determinate cases. The examples of Chapter 9 are illustrative.

The first step in the case of an indeterminate structure is to check on the redundancy. The following rule indicates the number of redundancies that remain in a structure when the ultimate load is reached: [1.9]

$$I = X - (M - 1) \tag{3.23}$$

where X = number of redundancies in the original structure
M = number of plastic hinges necessary to develop a mechanism *

In Fig. 3.15 are shown three continuous beams and a two-span, fixed base frame and Eq. 3.23 correctly indicates the number of remaining

(a)

$X = 2$
$M = 3$
$I = X - (M - 1) = 2 - 2 = 0$

(b)

P_1 P_2

$X = 1$
$M = 2$
$I = X - (M - 1) = 1 - 1 = 0$

(c)

$X = 2$
$M = 2$
$I = X - (M - 1) = 2 - 1 = 1$

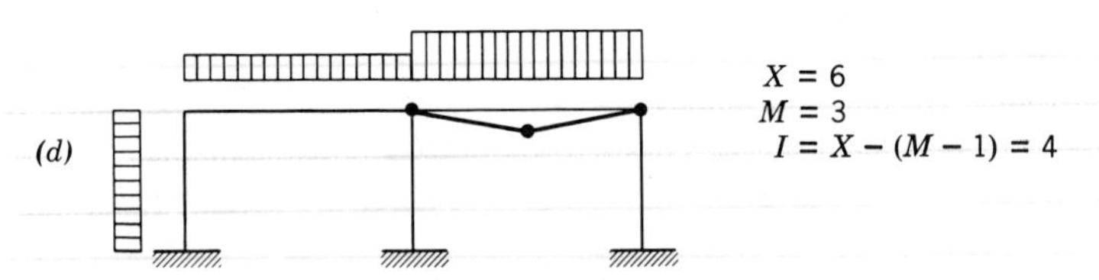

(d)

$X = 6$
$M = 3$
$I = X - (M - 1) = 4$

Fig. 3.15. Examples of the procedure for determining the number of remaining redundancies in a structure.

redundancies. In sketches (*c*) and (*d*) the structures are indeterminate at failure.†

* The quantity N in Eq. 3.17 is the number of *possible* plastic hinges whereas M in Eq. 3.23 is the *actual* number of hinges necessary for failure.

† It is emphasized that the quantity M is the *necessary* number of hinges for failure. Should more than one beam mechanism form in a frame that is otherwise redundant, then I would be found correctly on the assumption that only the minimum number of hinges actually were formed. (The other hinge positions would be assumed as slightly strengthened.) For example, in sketch (*d*) if a beam mechanism formed in the left span simultaneously with the one shown, then the value of M does not equal 6, but 3. The quantity I would still be 4, the subsequent calculations being made as if the three hinges had not formed.

If now, the frame is redundant, several methods are convenient for determining a *possible* equilibrium configuration:

(*a*) Trial and error method
(*b*) Moment-balancing method
(*c*) Semi-graphical method

None of these methods give the "exact" moment diagram in the redundant portion of the structure. However, although the continuity condition is violated, they give a bending moment diagram that is in equilibrium with the loads and does not violate the plastic moment condition. It thus represents a lower bound and is therefore adequate for these purposes.

Trial and Error Method

When there are only one or two remaining redundancies (as determined from Eq. 3.23), the "trial and error" method is most suitable for making the plastic moment check. Values for the remaining I moments are guessed and the equilibrium equations solved for the other unknowns.

EXAMPLE Three-Span Continuous Beam (Fig. 3.16)

Given the three-span continuous beam of uniform section M_p and with concentrated loads in each span, it is desired to obtain a possible moment diagram. Suppose the answer has been obtained on the basis of the assumed mechanism

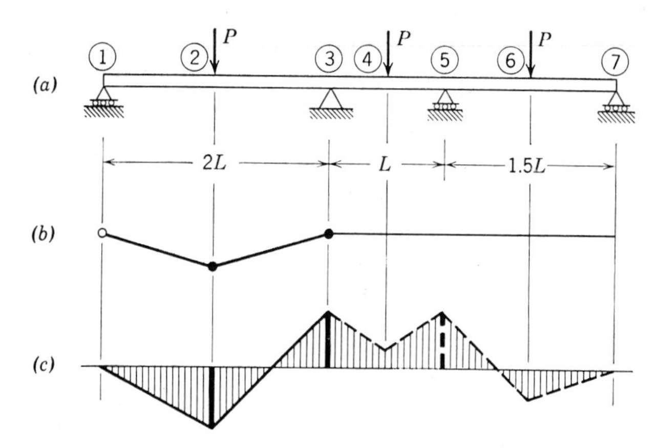

Fig. 3.16. Moment check using the trial and error method.

shown in Fig. 3.16*b*. For this case,

$$P_u = \frac{3M_p}{L} \tag{3.24}$$

The remaining redundancies from Eq. 3.23 are, $I = X - (M - 1) = 2 - (2 - 1) = 1$ (namely the moment at section 5).

The next step is to assume a value for this moment (say, $M_5 = M_p$). Solving the equilibrium equation for spans 3–5 and 5–7,*

$$M_4 = \frac{M_3}{2} - \frac{M_5}{2} + \frac{PL}{4} = -\frac{M_p}{2} - \frac{M_p}{2} + \frac{3M_p}{4}$$

$$M_4 = -\frac{M_p}{4}$$

$$M_6 = \frac{M_5}{2} - \frac{M_7}{2} + \frac{1.5PL}{4} = -\frac{M_p}{2} + \frac{9}{8}M_p$$

$$M_6 = +\tfrac{5}{8}M_p$$

The resulting moment diagram is shown by the dashed lines. Since $M \leq M_p$ throughout, the trial solution is correct and $P_u = 3M_p/L$. Quite evidently, more efficient use of material would result if the design were revised to supply only the *required* plastic moment for each span.

Moment-Balancing Method

When there are more than two or three redundants in the structure at failure, then a pseudo moment-distribution process may be used.[3.6] The plastic moment-distribution method was applied to the particular problem of making a moment check in Ref. 1.9 and is most suitable for cases in which the methods of moment distribution would otherwise be appropriate for an elastic analysis. Although it would not be used for such a simple example, the method of "moment-balancing" will be illustrated with the same continuous beam as was studied in the example on page 85.

EXAMPLE Three-Span Continuous Beam (Fig. 3.17)

For the continuous beam shown in Fig. 3.17*a* assume that the analysis has been completed and the ultimate load obtained as $P_u = 3M_p/L$. In the "moment balancing" process one starts with a computation of simple span moments M_s for all spans containing unknown moments. The resulting moment diagram is plotted in Fig. 3.17*b* and the M_s-values are tabulated in line 1 of (*c*), in terms of M_p.

* The equilibrium equations and sign convention are described and discussed in Procedure 8 of Table 7.3.

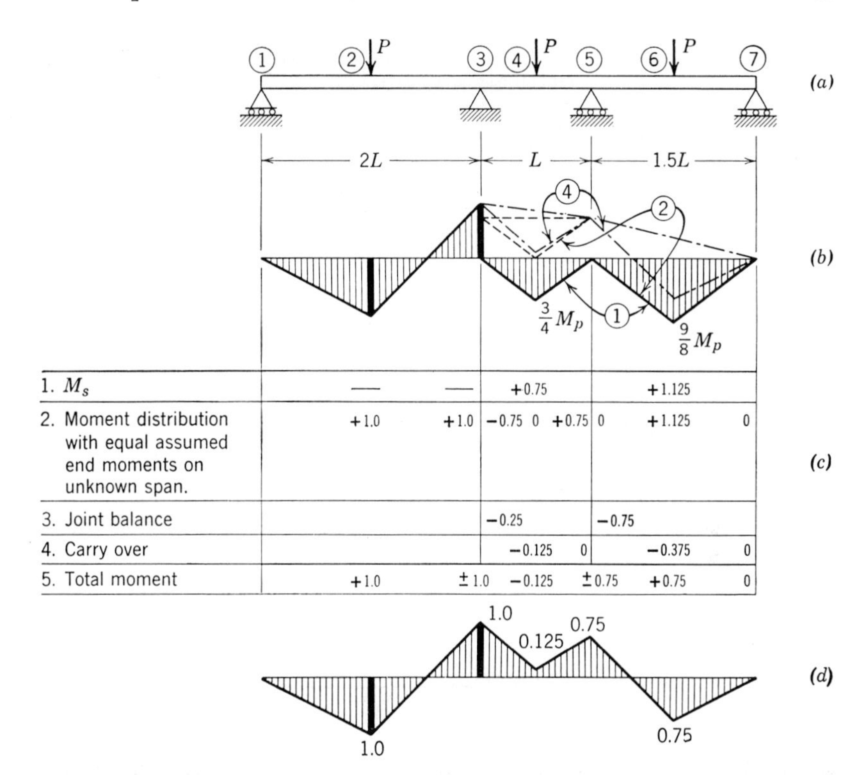

1. M_s	—	—		+0.75			+1.125	
2. Moment distribution with equal assumed end moments on unknown span.	+1.0	+1.0	−0.75	0	+0.75	0	+1.125	0
3. Joint balance			−0.25			−0.75		
4. Carry over				−0.125	0		−0.375	0
5. Total moment	+1.0	±1.0		−0.125	±0.75		+0.75	0

Fig. 3.17. Moment check performed for a continuous beam by the moment-balancing method

The sign convention is that clockwise end moments are positive. Within the beam, positive moments produce tension on the bottom fiber.

Next, a starting moment-distribution is selected on the basis that the moments at the end of a given span are equal in magnitude, keeping in mind the known end-conditions since this will shorten the calculations. For instance at section 3, $M = M_p$; at 7, $M = 0$. The resulting distributions are shown in line 2. In span 3–5 the end moments were assumed to be $0.75M_p$ with a resulting moment of zero at section 4. In span 5–7, the end moments were assumed to be zero.

The structure is now in equilibrium except at the joints. So, a moment balance is carried out to restore equilibrium, as shown in step 3. At section 3, for example, a moment of -0.25 must be applied to balance the moment M_p.

Since the application of the joint-balancing end moments upsets the equilibrium of the beam spans, a carry-over operation is necessary. This is done according to Table 3.1, being performed in such a way that moments greater than M_p do not result.

The function of Table 3.1 is to show the effect of a unit change in moment at one point in a span upon the moment at another point. For instance, in case (*a*) a unit change in ΔM_L, while ΔM_R is unchanged,

TABLE 3.1. CARRY-OVER FACTORS FOR USE IN THE MOMENT-BALANCING METHOD FOR OBTAINING A MOMENT CHECK

Case	ΔM_L	$\Delta M_{℄}$	ΔM_R	Incremental Moment Diagram
(a)	1	½	0	L ℄ R; +1, +½
(b)	1	0	1	+1, 0, +1
(c)	0	−½	1	−½, +1

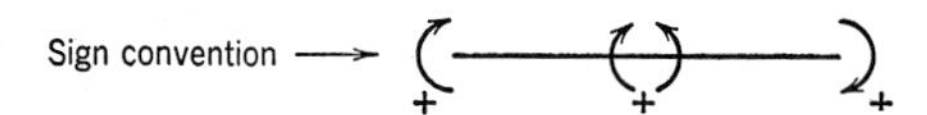

produces an increment of $+\frac{1}{2}$ at the centerline. The incremental moment diagram shows these changes for the three cases. If it were desired to hold the center-line moment unchanged (case b), then a unit change at one end would produce a unit change in the opposite end of the same sign. It is essential to keep in mind that *the continuity condition does not apply.* What is being sought is a *possible* statical moment diagram, not the *exact* one. For this reason the process being described only resembles moment distribution.

Starting from joint 3 of Fig. 3.17 and working to the right (line 4), for an applied moment at 3 of -0.25, and a zero change in moment at 5, the carry-over moment at 4 is -0.125 according to case (a) of Table 3.1. In span 5–7, for an applied end-moment increment of -0.75 at 5 and a zero change at 7, the moment increment at 6 is -0.375.

The final operation (line 5) is to sum the moments and check that M is less than M_p at all sections. The resulting moment diagram is shown in sketch (d) of Fig. 3.17. Although the diagram is somewhat different from that given in Fig. 3.16, it is still a solution that satisfies the *equilibrium* and *plastic moment* conditions and therefore represents a lower bound to the correct load. Thus the answer $P_u = 3M_p/L$ is correct.

It is of interest to note that the effect of this moment-balancing process is to shift the "simple-span" moment diagrams to different positions with respect to the base line in such a way that they are always within the positive and negative values of M_p. This is illustrated in Fig. 3.17*b* where the various steps are indicated by circled numbers.

EXAMPLE Rectangular Fixed-Base Frame (Fig. 3.18)

To illustrate the procedure, this structure will be checked by the moment-balancing method even though it is a case in which the trial and error method would be more suitable.

The ultimate load P_u, corresponding to the mechanism selected as a possible correct answer, is $P_u = 8M_p/L$. No determinate moments are written down for the beam since the final moments are known at sections 2, 3, and 4. Column moments are zero for step 1 because there is no load between the column ends. The starting column moments (step 2) may be determined from the sway equilibrium equation,

$$M_1 + M_2 + M_4 + M_5 + \frac{PL}{2} = 0 \tag{3.25}$$

On the basis of equal end moments,

$$M_1 = M_2 = M_4 = M_5 = -\frac{PL}{8} = -M_p$$

Therefore the value -1 is written down at each end of each column. The known beam moment values are recorded with proper sign. Step 3 is the joint balancing operation. Joint 4 is already in equilibrium. Equilibrium of joint 2 is accomplished by applying a moment of $+2.0M_p$. The carry-over moment at section 1 is zero according to case (*a*) of Table 3.1. The joints are in equilibrium as well as the beam 2–4; but the joint 2 balance moment of $+2.0M_p$ upsets equilibrium of the columns (Eq. 3.25). Thus adjustments are required in M_1 and M_5 in order that the sum of all changes $= 0$, and this correction would amount to -1.0 at each column base. But such a correction cannot be applied because the moments there are already at the maximum possible value. Further there is no other way to distribute this unbalance. Step 6 (total moment) shows that the plastic moment condition is violated and the chosen mechanism is thus incorrect.

It is emphasized that the procedure illustrated would never be used ordinarily for such a frame. Instead, alternate mechanisms (*a*) and (*b*) of Fig. 3.18 would be tried and the lowest load selected. The moment check in that case would be simple, involving only the equations of static equilibrium.

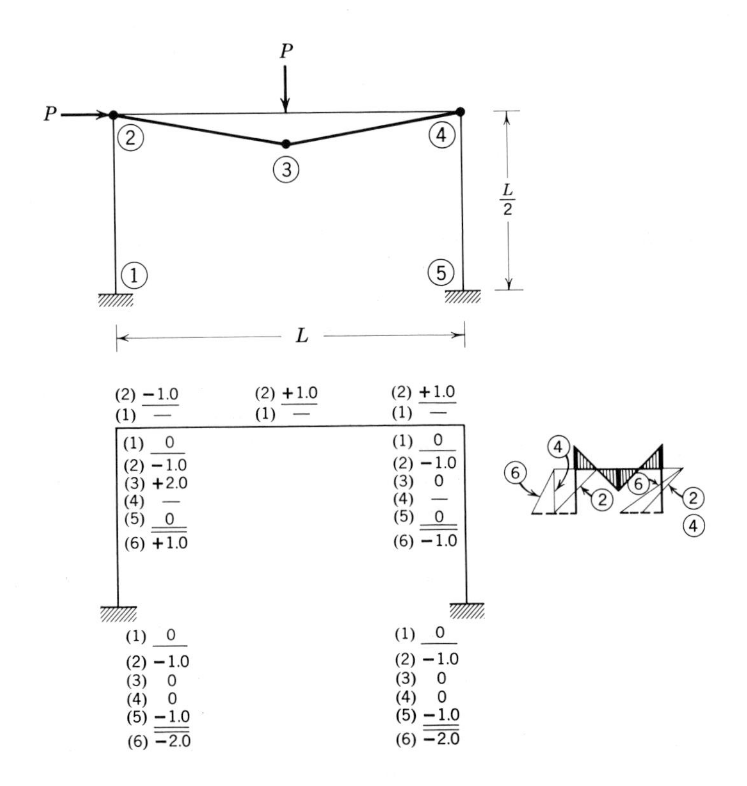

Step 1: Determinate (simple-span) moment
Step 2: Moment distribution with equal end moments on unknown span
Step 3: Joint balance
Step 4: Carry-over
Step 5: Column adjustment
Step 6: Total moment

Alternate mechanisms:

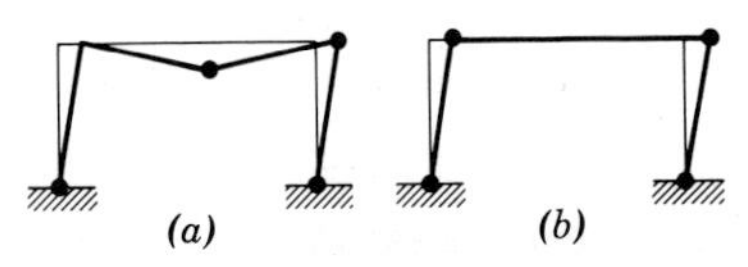

Fig. 3.18. Moment check for a fixed-base rectangular frame using the moment-balancing method.

EXAMPLE Analysis and Moment Check of Two-Span Frame (Fig. 3.19)

As a final illustration of moment-balancing, the problem shown in Fig. 3.19*a* will be solved. Suppose after trying four mechanisms it is decided that mechanism 1 gives the lowest load, $P_u = \frac{16}{5}(M_p/L)$ (see tabulation in sketch *b*). The frame is redundant at failure and from Eq. 3.23, $I = X - (M - 1) = 6 - (3 - 1) = 4$. So the moment-balancing method will be appropriate to make the moment check. The process is shown in sketch (*c*) of Fig. 3.19 and the final moment diagram is shown in sketch (*d*). The check is described according to the following steps:

Step 1: M_s at section 8 $= (2P)(L/4) = (PL/2) = \frac{8}{5}M_p = 1.60M_p$.

Step 2: Moments are known in span 2–4. For span 5–9 an arbitrary division is made, keeping the end moments equal. $M = 1.0$ assigned to section 8, $M = 0.6$ to ends. For the columns, since member 1–2 is twice the strength of the other columns, the sway moments will be distributed in the ratio, $M_1 = M_2 = 2M_6 = 2M_7 = 2M_9 = 2M_{10}$ and from the sway equilibrium equation (similar to Eq. 3.25), $M_1 = -PL/8 = -0.4M_p$.

Step 3: Joint 2 is balanced by applying $+2.40$ to the column. The unbalance of -1.20 on joint 4–5–6 is divided between 5 and 6, with both being loaded to their full M_p-value. At joint 9, $M = -0.40\ M_p$ is assigned to the column.

The application of moment to the column tops throws these members out of equilibrium. For use in later corrections, a "table of column unbalance" is kept as shown in sketch (*c*).

Step 4: In span 5–9, keeping zero change at 9, Table 3.1 case (*a*) shows that application of an increment of -0.40 at joint 5 produces a change of -0.20 at section 8.

Step 5: Column moment corrections to adjust for unbalance during step 3 are assigned to section 1, 7, and 10.

Step 6: Totals taken at all joints and $M \leq M_p$ at each section. Thus the trial mechanism was the correct one.

Semi-Graphical Method

It was pointed out at the end of the example on p. 86 (Fig. 3.17) that the effect of the moment-balancing process was to shift the "simple span" moment diagrams to different positions with respect to the base line, always keeping them within the positive and negative values of

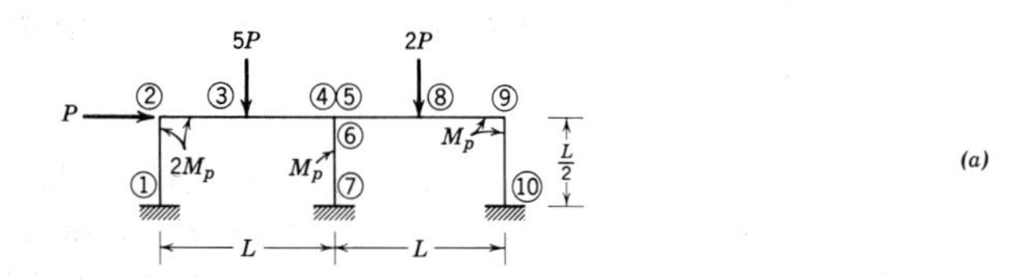

	Mechanism	$W_E/PL\theta$	$W_I/M_p\theta$	P_uL/M_p
1	② ③ ④ (1, 2, 1)	$(5)(\frac{1}{2}) = \frac{5}{2}$	$2 + 4 + 2 = 8$	3.2 ← P_u (trial)
2	⑤ ⑧ ⑨ (1, 2, 1)	$(2)(\frac{1}{2}) = 1$	$1 + 2 + 1 = 4$	4.0
3	(1, 1, 1, 1, 1, 1)	$(1)(\frac{1}{2}) = \frac{1}{2}$	$2 + 2 + 1 + 1 + 1 + 1 = 8$	16.0
4	(1, 1, 1, 1, 2, 1, 1, 2)	$(1)(\frac{1}{2}) + (2)(\frac{1}{2}) = \frac{3}{2}$	$2 + 2 + 1 + 1 + 1 + 2 + 2 + 1 = 12$	8.0

(b)

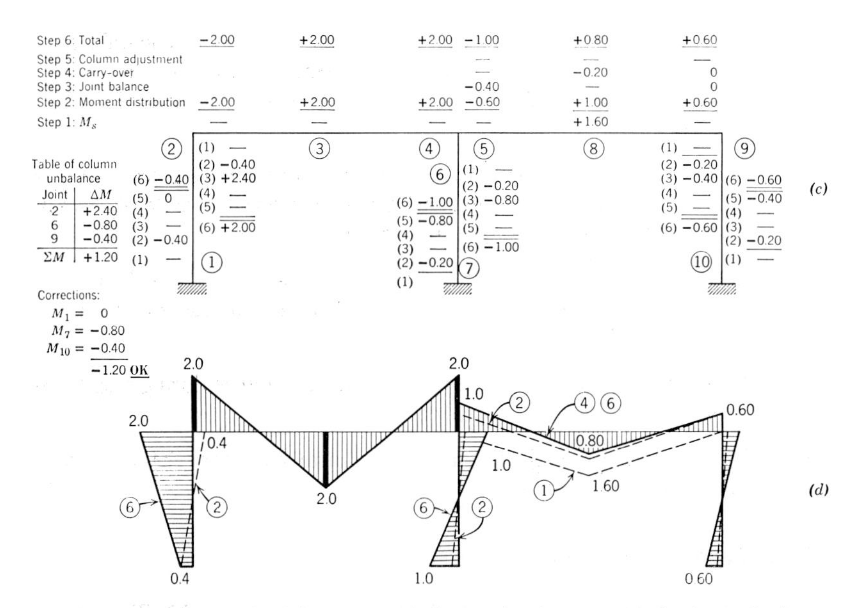

Fig. 3.19. Moment check for a two-span frame using the moment-balancing method.

M_p. This fact may be used effectively to make a moment check without the necessity of carrying forward all of the computations of the previous two methods.

In the example of Fig. 3.17, for instance, after the simple span moment diagrams are drawn for spans 3–5 and 5–7 (shaded in sketch *b*), the diagram for span 3–5 may be shifted upward until M_3 equals M_5 equals M_p. Then the base line for the simple span moment diagram of 5–7 may be shifted, holding 7 fixed until M_5 equals M_p. Since the plastic moment condition is not violated, the moment check has been completed.

EXAMPLE Two-Span Rectangular Frame (Fig. 3.20)

This structure is the same as the example on page 91 except the problem is worked by the "semi-graphical" method. The moments in span 2–4 are known

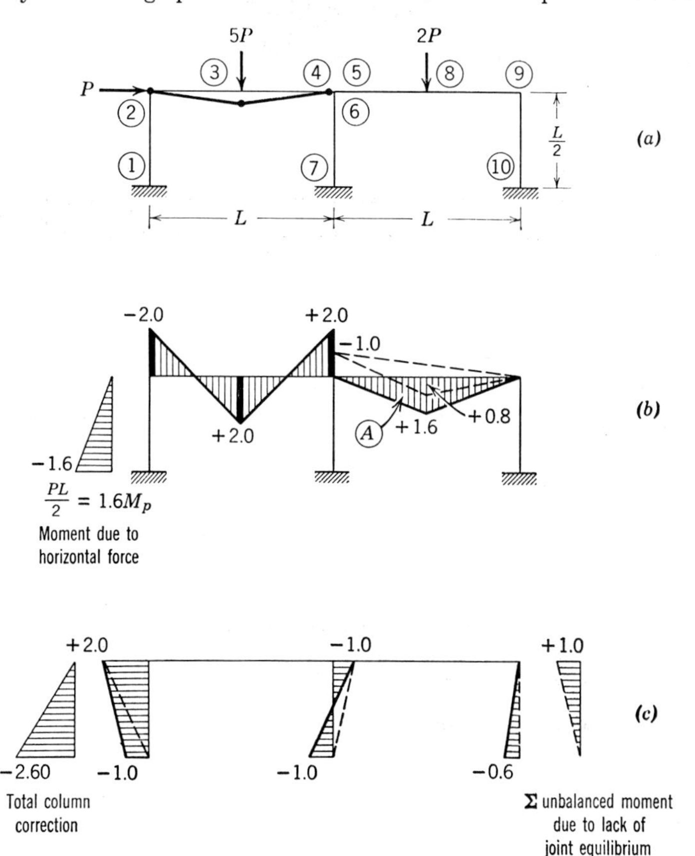

Fig. 3.20. Moment check for a two-span frame performed by the semi-graphical method.

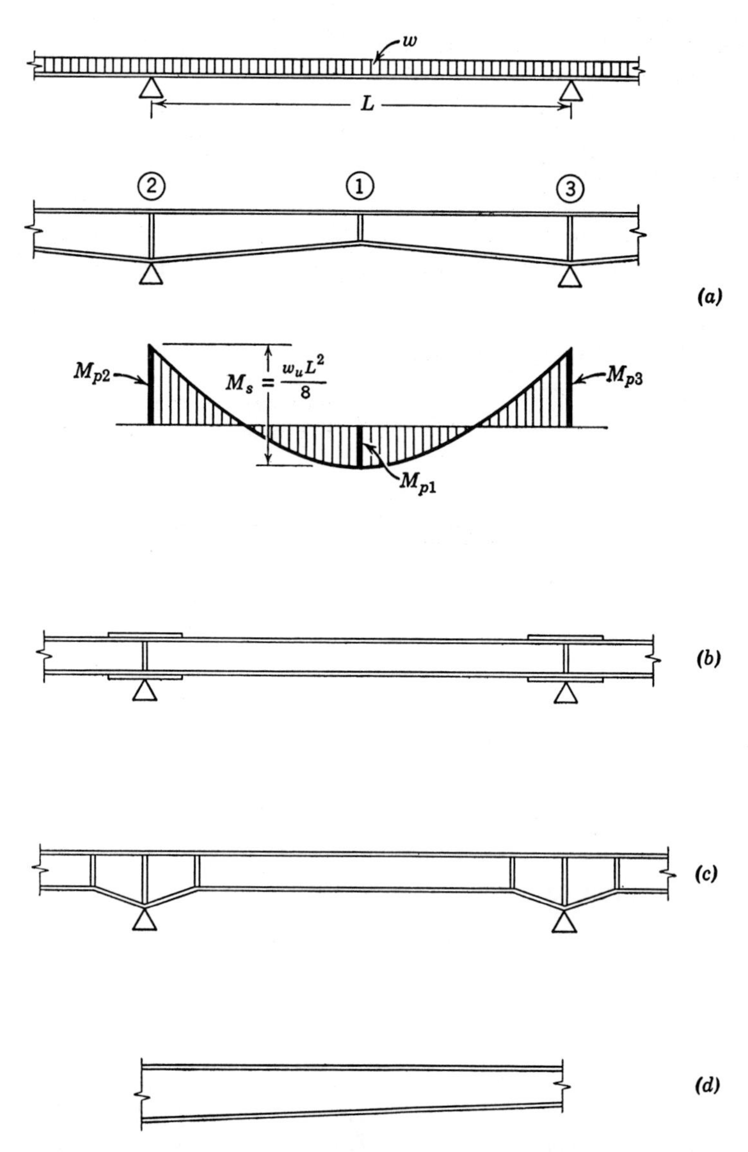

Fig. 3.21. Members of nonuniform cross section and their analysis.

and these are laid off in sketch (*b*). Next, the simple-span moment diagram is laid off for span 5–9. The moment at section 8 is $(2P)(L/4)$ equals $\frac{8}{5}M_p$. Since the plastic moment condition is violated at that section, element A must be shifted upward until M_8 is less than or equal to M_p. If M_5 is made equal to M_p, then in addition to satisfying the plastic moment condition at 8, the maximum possible restraint will have been supplied by the beam to balance the moment of -2.0 originally applied at joint 4–5–6.

Consider the action of the horizontal force P applied at the top of column 1–2. The moment due to this force must be resisted by the column end-moments and at the left of sketch (*b*) is a moment diagram representing the total moment about the column bases ($M = (PL/2) = 1.60M_p$). The structure is now "loaded" in sketch (*b*) with all the external forces that will act. Since the beams are in equilibrium without violating the plastic moment condition, only the columns need be investigated further and this is the next step.

Since the joints of the frame are not in equilibrium, the necessary balancing moments must be applied to the tops of the columns. The resulting moment diagrams are shown as dashed lines in sketch (*c*). The total unbalanced moment from this cause is $+1.0M_p$. A correction moment of opposite sign must be applied to maintain side-sway equilibrium, and thus the total moment that must be applied to column ends 1, 7, and 10 to establish equilibrium is equal to $-2.60M_p$, the moment diagram being shown to the left of Fig. 3.20*c*. This value may be apportioned to the various members in any ratio so long as the plastic moment value is not exceeded. The values, $M_1 = -1.0$, $M_7 = -1.0$, and $M_{10} = -0.6$ are selected, and the final column moment diagrams are obtained by adding the corresponding moment distributions to those already introduced by the joint-balancing moments. The final moment diagram is shown by the shaded portions of sketch (*c*).

The moment check is thus completed. Although the result differs from that obtained by the moment-balancing method, the equilibrium and plastic moment conditions are satisfied, and the ultimate load is thus correct.

3.9 MEMBERS OF NONUNIFORM CROSS SECTION

Since the elastic distribution of moments throughout a structure results in peak values that are usually of different magnitude, elastic design procedures often make use of haunches, cover plates, and tapered members to obtain a more economical structure. Plastic design, on the other hand, achieves its economy without the necessity for varying the cross section of the members. However, if there are instances where tapering, etc., might be desirable for other reasons, then such a structure can be analyzed without difficulty by the plastic methods.

Some of the possible methods by which the cross section of a beam may be varied are shown in Fig. 3.21. In all of these cases except sketch

(d), the beam may be analyzed for the case of uniform load by using the appropriate plastic moment values for the hinge which forms at the center and for the hinges which form at the ends. By the statical method for a uniformly distributed load w_u, the maximum strength of the beam may be determined according to the equation,

$$\frac{w_u L^2}{8} = M_{p1} + M_{p2} \tag{3.26}$$

where M_{p1} is the plastic moment value at the center and M_{p2} is the plastic moment value at the ends. For built-up H shapes, Eq. 2.24 may be used to compute the value of Z. When two equal cover plates are used on top and bottom flanges (sketch b) then

$$Z = Z_{WF} + A_p(d + t_p)$$

where A_p is the area of one plate, d is the depth of the rolled beam, and t_p is the thickness of the cover plates. If the built-up section is unsymmetrical, then Eq. 2.27 would be used to obtain Z.

Sketch (d) in Fig. 3.21 illustrates an exception to the rule that plastic hinges form at points of zero shear in beams. In all of the other cases, as long as the hinge forms at the support or in the uniform prismatic portion of the beam, or at the center of the beam for case (a), then the shear force is truly zero at the hinge. In the case typified by sketch (d), a shear force may be present at the plastic hinge which forms in the beam. Consider Fig. 3.22 and the tapered beam sketched in (a). The moment diagram due to the determinate loading is shown in sketch (b); the moment diagram due to loading by the redundant end moments is shown in sketch (c). How much load will the beam carry and where is the position of the plastic hinge in the beam? The end moments at 1 and 5 are known, and these may be used to commence construction of the composite moment diagram (sketch d) by laying off distances 1–a and 5–b. Line a–b then represents the base line of the determinate moment diagram. In order to determine the position of the hinge, the variation of moment capacity with distance along the beam must be considered. The curve of plastic moment capacity is shown by the two dashed lines a–c–b and a'–c'–b' in sketch (d). If x is selected as the distance from the smaller end to any section of the beam, then the moment capacity of the beam (and the equation of the curve) is given by

$$M_{px} = \sigma_y Z_x = \sigma_y \left\{ bt \left[d_5 + \frac{x}{L}(d_1 - d_5) - t \right] + \frac{w}{4} \left[d_5 + \frac{x}{L}(d_1 - d_5) - 2t \right]^2 \right\} \tag{3.27}$$

where the subscripts denote the dimensions at the appropriate sections in Fig. 3.22. Since the composite moment diagram must never fall outside the regions bounded by the two dashed curves (otherwise the

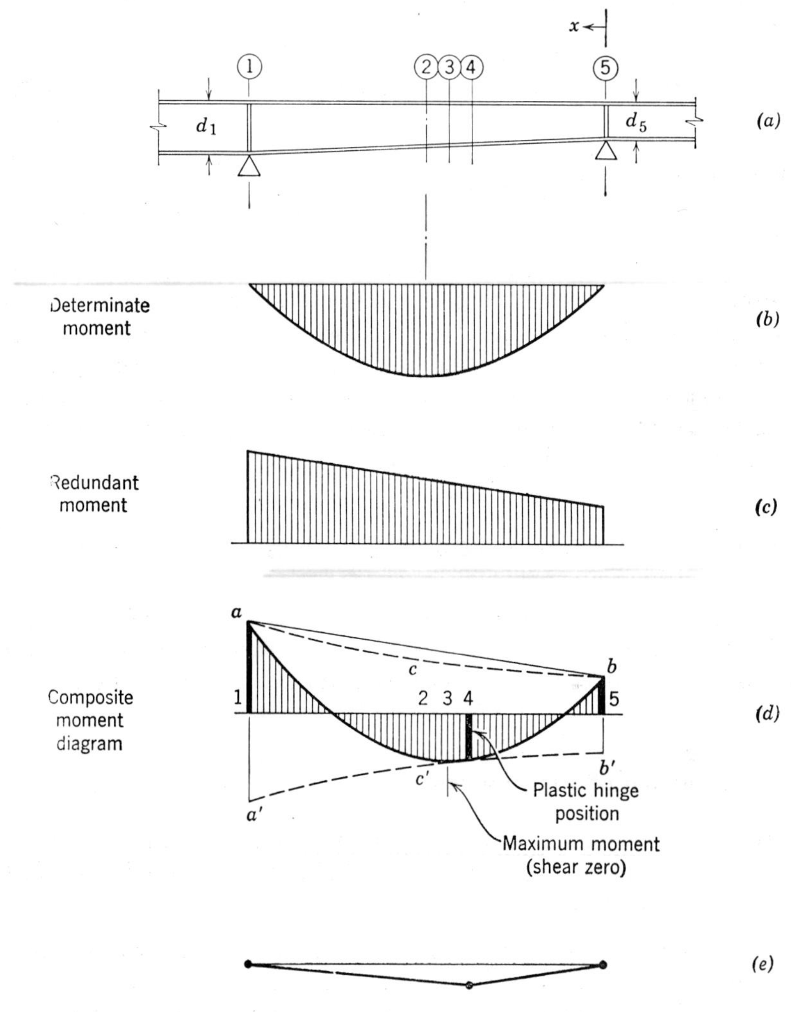

Fig. 3.22. Plastic analysis of a tapered member.

moment capacity would be exceeded), then the plastic hinge in the beam will form at section 4 where the statical moment diagram becomes tangent to the dotted curve representing moment capacity of the beam. The mechanism is shown in sketch (*e*). Notice that the hinge does not form at the point of maximum moment (zero shear) shown at section 3, because the moment capacity there is greater than required by the moment diagram.

If the magnitude of the plastic moment capacity of the beam is written as a function of the distance from section 5, there is obtained

$$M_{px} = \frac{w_u L}{2} x - \frac{w_u x^2}{2} - M_{p5} - \frac{x}{L}(M_{p1} - M_{p5}) \tag{3.28}$$

Solving Eqs. 3.27 and 3.28 for x will give the position of the plastic hinge at section 4; and with this information the ultimate load may be computed.

An example of the design of a tapered beam is given in Chapter 8.

Although in most cases the plastic hinge locations will correspond to positions of maximum moment, this discussion has shown that there are some exceptions. Thus, in structures with tapered members, part of the "moment check" requires an examination of the variation in moment capacity along the members to make sure that the plastic moment condition has not been violated.

3.10 TESTS OF CONTINUOUS STRUCTURES

In the previous articles it was shown how methods of structural analysis could be developed from the plastic theory. In this article will be shown the extent to which structural behavior confirms that theory.

Recapitulation of Assumptions

Six important assumptions are made with regard to the plastic behavior of structures. These are indicated diagrammatically in Fig. 3.23 and are as follows:

(1) The material is ductile. It has the capacity of absorbing plastic deformation without the danger of fracture.

(2) Each beam has a maximum moment of resistance (the plastic moment, M_p), a moment that is attained through plastic yield of the entire cross section (plastification).

(3) Due to the ductility of steel, rotation at relatively constant moment will occur through a considerable angle—resulting in the formation of a plastic hinge.

(4) Connections will transmit the desired plastic moment and provide for hinge action.

(5) As a result of the formation of plastic hinges at connections and other points of maximum moment, redistribution of moment will occur, allowing the formation of plastic hinges at points that are otherwise less highly stressed in the elastic region.

(6) The ultimate load may be computed with accuracy on the basis that a sufficient number of plastic hinges have formed to create a mechanism.

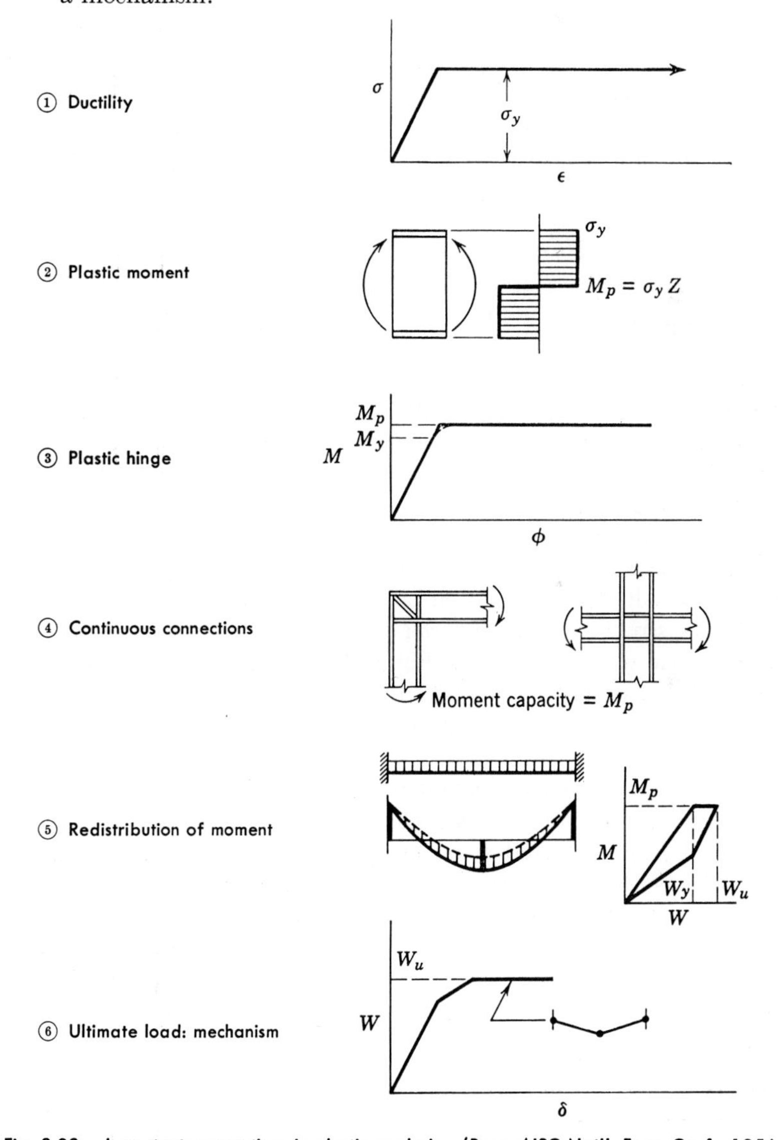

Fig. 3.23. Important assumptions in plastic analysis. (*Proc. AISC Nat'l. Engr. Conf.*, 1956.)

The experimental confirmation of assumptions 1–5 is given in Chapters 1, 2, and 5. It is desired to show here that continuous beams and rigid frames *will* attain the ultimate load computed on the basis of the above assumptions.

Strength of Continuous Structures

Typical of the structures tested both in this country and abroad was that shown in Chapter 1, Figs. 1.10 and 1.11. The structure carried the predicted ultimate load. Of 40-ft span, this welded structure was fabricated from 12WF36 shapes.

Further examples are shown in Figs. 3.24 and 3.25. They show in tabular form a number of these tests in which the members were fabricated from rolled sections. The structure and loading are shown to scale at the left. Next, the size of member (or members) is indicated. To the right is a bar graph on which is plotted the per cent of predicted ultimate strength exhibited by the test structure. (A test that reached 100%, reached the load predicted by the simple plastic theory.)

These two figures show that the actual strength of even the weakest structure was within 5% of its predicted ultimate load—an agreement much better than can be obtained at the so-called "elastic limit." Particularly remarkable among the continuous beam tests of Fig. 3.24 is the one conducted by Maier Leibnitz [1.2] (see the next to the last structure). In this experiment, prior to applying the vertical load, he raised the center support until the allowable stress was just reached, with the result that application of the first increment of external load caused the structure to be "overloaded." In spite of this, the computed ultimate load was attained. In Fig. 3.25 testing of the fourth frame was interrupted in order that the fifth test might be carried out on the same structure but with a different proportion of horizontal to vertical load.

Figure 3.26 shows tests conducted at Cambridge, England by Prof. Baker and his associates. I-shapes were used in these tests that include frames with pinned and fixed bases, and with flat, saw tooth, and gabled roofs. In each case the ultimate load computed by the plastic theory was reached and in numerous cases it was exceeded.

Tests conducted at the University of California[3.13] on a series of continuous beams with cover plates showed quite clearly that plastic hinges will form and subsequent redistribution of moment occur in such nonuniform members. All of these beams were thereby able to attain the computed ultimate load.

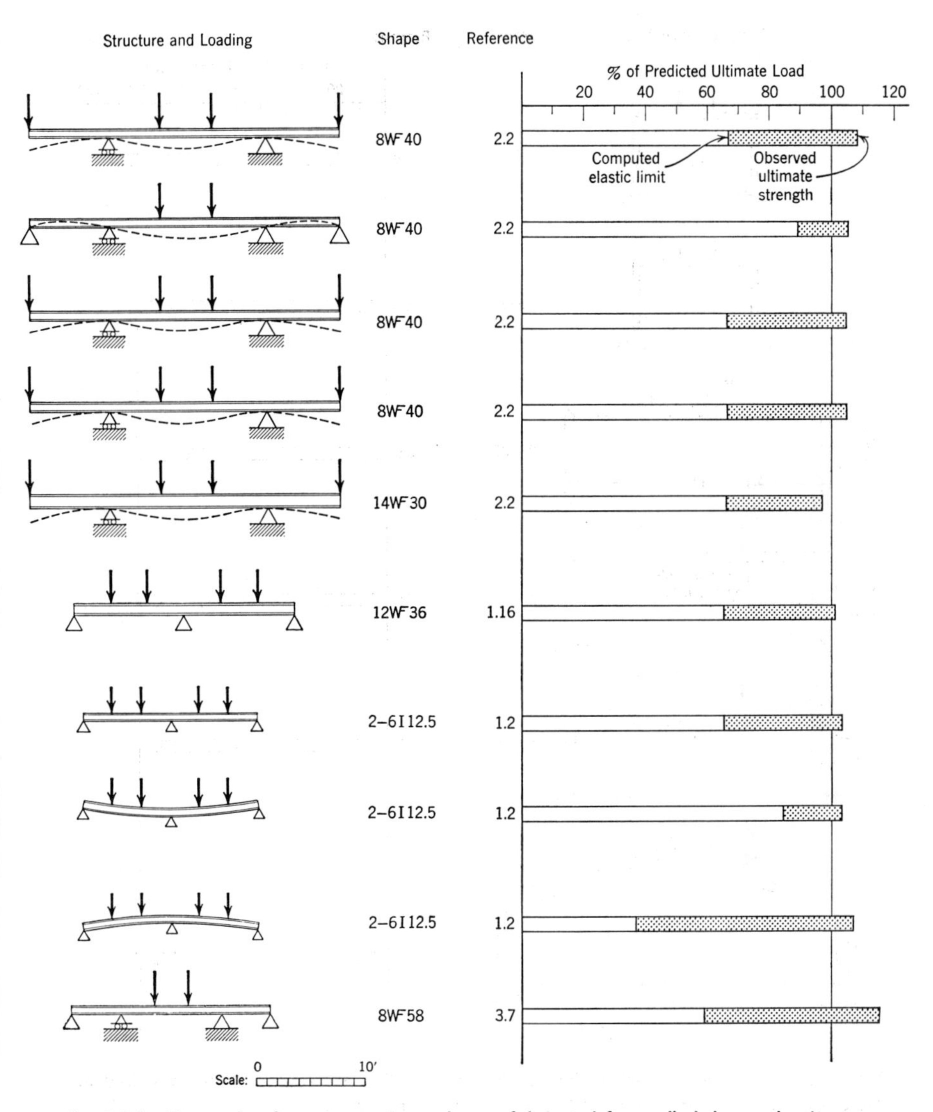

Fig. 3.24. The results of tests on continuous beams fabricated from rolled shapes showing correlation with predictions of plastic theory. (*Proc. AISC Nat'l. Engr. Conf.*, 1956.)

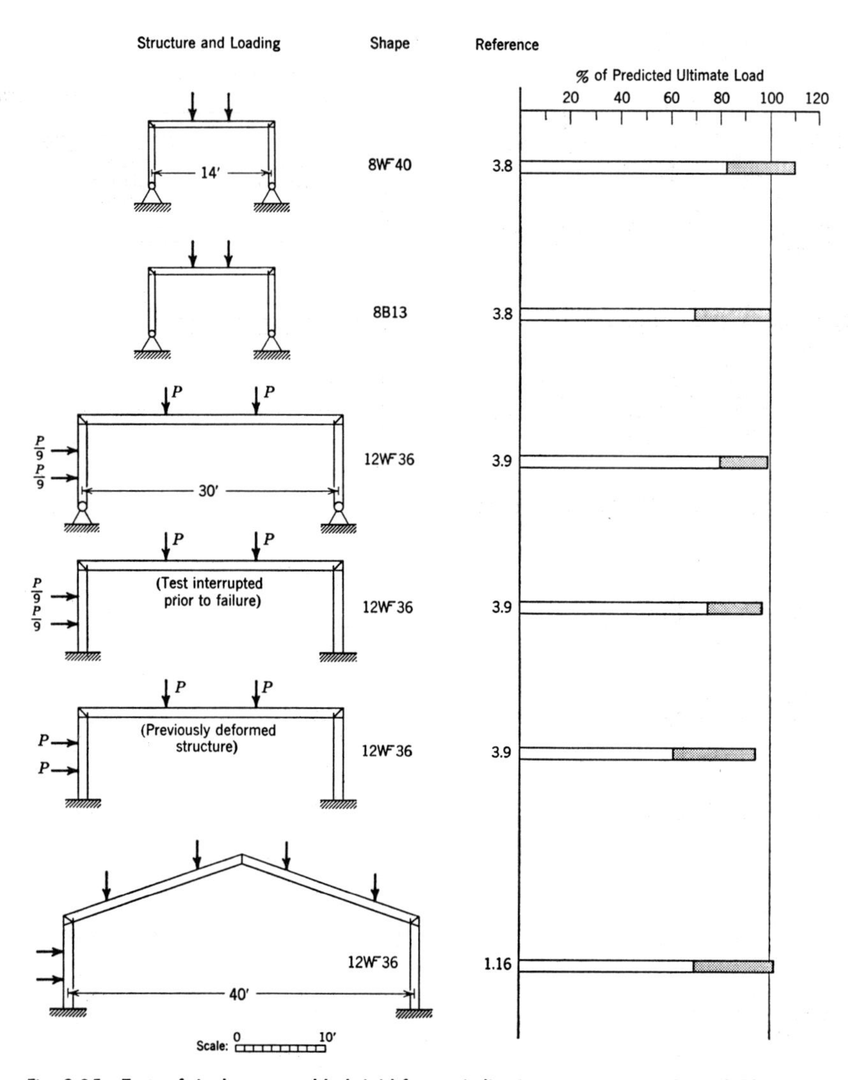

Fig. 3.25. Tests of single-span welded rigid frames indicating reserve strength available beyond the computed elastic limit. (*Proc. AISC Nat'l. Engr. Conf.,* 1956.)

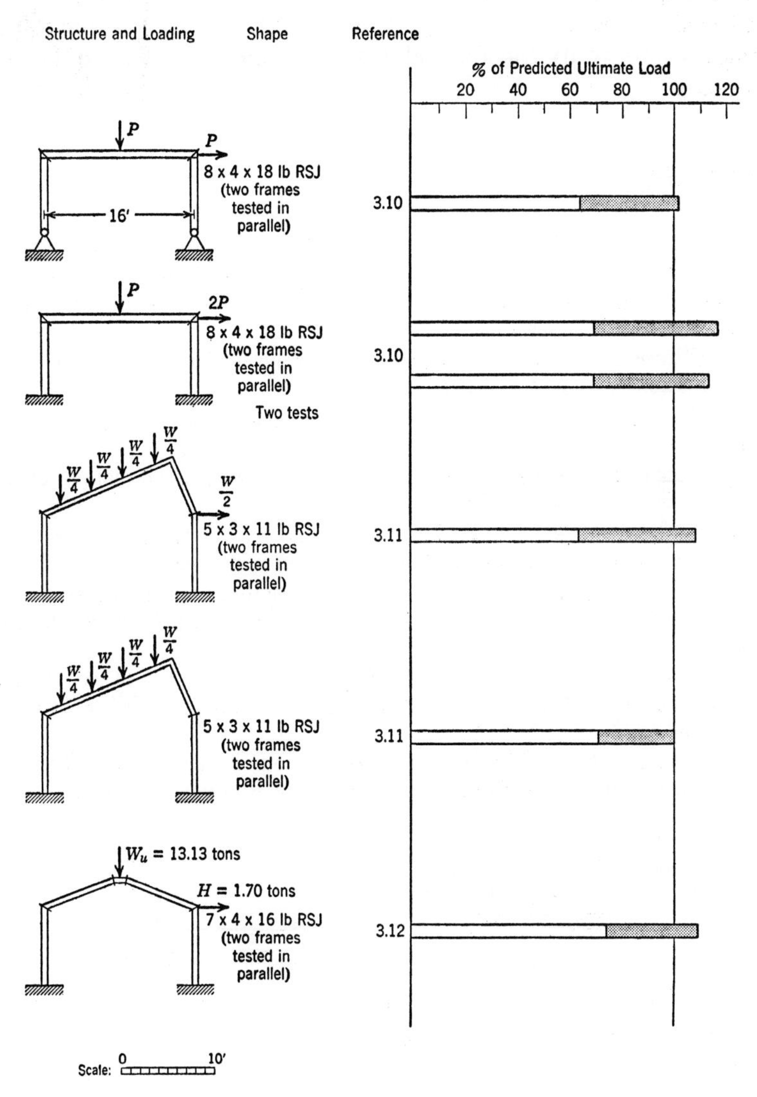

Fig. 3.26. Further tests of single span rigid frames (tests conducted in Europe). (*Proc. AISC Nat'l. Engr. Conf.,* 1956.)

The degree in which the basic assumptions of plastic analysis are confirmed by test is indeed remarkable and it is thus seen that experimental results verify the applicability of the plastic method of analysis to design.

PROBLEMS

3.1 What is the minimum required M_p value for the beam in problem 1.4?

3.2 Why is the continuity condition in elastic analysis not usually a necessary condition in plastic analysis? Under what circumstances must it be considered in plastic analysis?

3.3 Why is no internal strain-energy of the beams involved in the virtual work equations such as Eq. 3.10?

3.4 Why is the correct mechanism the one that corresponds to the lowest load?

3.5 How is the *equilibrium* condition satisfied in the statical method of analysis? In the mechanism method?

3.6 Explain why conventional elastic design is a "lower bound" solution? Under what circumstance is such a design also an upper bound solution?

3.7 Demonstrate the upper and lower bound theorems with the aid of a beam, loaded at the third points, fixed at one end but simply supported at the other.

3.8 A beam is fixed at one end and simply supported at the other. What is the maximum concentrated load that may be carried at the third-point closest to the fixed end?

3.9 Compute P_u for the three-span continuous beam shown in Fig. 3.27. What is the magnitude of P_3 and P_4 for complete failure of the beam?

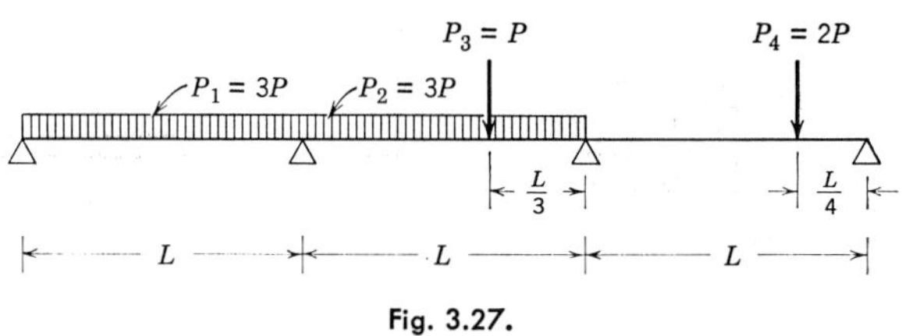

Fig. 3.27.

3.10 Suppose a single-span gabled frame with a total uniformly distributed load, W, has a plastic hinge at the crown and at the right hand eave. (A concentrated load P acts at the top of the left hand column towards the right.) Span length is L, column height is a, and rafter rise is b. Obtain the position of the instantaneous center of the right hand rafter. Obtain the mechanism angle at each hinge. Check by considering the deformed position of the frame and without using instantaneous centers.

3.11 A beam of length L has a moment capacity of $3M_p$ and is restrained at the ends by moments of $3M_p$ (left) and $2M_p$ (right). If the total distributed load is W, write the magnitude of that load when the plastic hinge forms in the beam. What is the location of that hinge?

3.12 Solve problem 3.11 by assuming that the plastic hinge occurs at midspan. Solve it by replacing the distributed load by 1, 2, and three equivalent concentrated loads.

3.13 A single-span rectangular frame with pinned base of span L and column height $L/2$ has a uniformly-distributed load of W and a concentrated side load of $W/3$ applied at the top of the column. Where are the plastic hinges located? What is W_u?

3.14 A rectangular single-span frame is loaded with a center concentrated load P and an equal horizontal load applied at the top of the left column acting to the right. Column bases are fixed. The beam strength is 1.5 times the column strength. Use the trial and error method to determine whether or not the beam mechanism is the correct one. What is the correct P_u?

3.15 A fixed-ended 16WF40 beam, 20 ft in length is reinforced at the ends with cover plates welded to top and bottom flanges. These plates are 5-in. wide and $\frac{3}{8}$ in. thick. How much load in kips will the beam support?

3.16 How long must the plates be in problem 3.15 in order that this ultimate load be reached.

3.17 Suggest another mode of failure (other than flexure) that might prevent a similar beam from reaching the computed ultimate load.

3.18 Compute P_u for the problem shown in Fig. 3.28. Make the moment check.

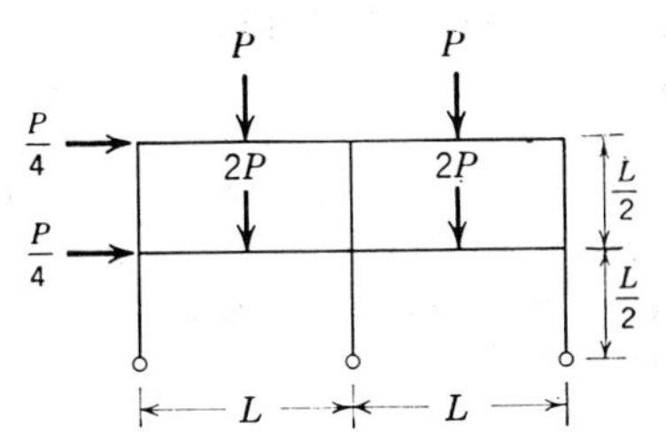

Fig. 3.28.

Secondary Design Problems

4.1 INTRODUCTION

In all the tests presented in Art. 3.10 (Figs. 3.24, 3.25, 3.26), the results confirm in satisfactory manner the predictions of the "simple plastic theory." This theory neglects such things as axial force, shear, and buckling, and yet the engineer knows they are present in most structures and he is accustomed to taking them into account in conventional elastic design.

Although it turned out in the tests just mentioned that such factors of secondary importance did not prevent the structure from reaching the desired load, methods must be available for accounting for these additional factors in order to have a design method that is generally applicable. Those factors that are neglected or are not included in the "simple" theory (and for which revision of that theory is sometimes needed) are the following:

Axial force and shear force (reduction in the plastic moment)
Instability (local buckling, lateral buckling, column buckling)
Brittle fracture
Repeated loading.

In addition, proper proportions of connections are required in order that the plastic moment will be developed. Furthermore, the deflection at working load must be tolerable. These two subjects are treated in Chapters 5 and 6.

In this chapter the effect and characteristics of the factors tabulated above will be indicated. Where appropriate, the results of theoretical analysis and of tests will be presented. This is followed by procedures to serve as a guide for checking the suitability of the original design. It should be kept in mind that the necessity for considering these additional factors is no different in principle from elastic design procedures. The design must always be checked for axial force, shear, and so on. It simply means that modifications or limitations in the form of "rules of design" are necessary as a guide to the suitability of a design based on the simple theory that neglects these factors.

In Chapter 7 the appropriate procedures are summarized in a form suitable for design reference.

4.2 INFLUENCE OF AXIAL FORCE ON THE PLASTIC MOMENT

In addition to causing column instability (treated in Art. 4.6) the presence of axial force tends to reduce the magnitude of the plastic moment. However, the design procedure may be modified easily to account for its influence because the important "plastic hinge" characteristic is still retained in the presence of this force even though the

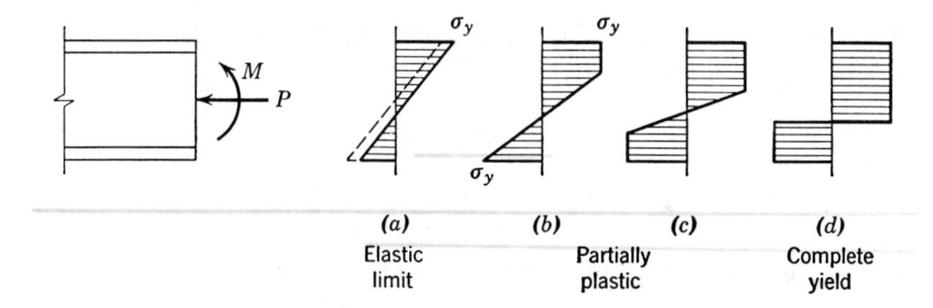

Fig. 4.1. Distribution of stress at various stages of yielding for a member subjected to bending and axial force.

moment capacity is reduced. The effect is small in the case of small axial load, and therefore in ordinary portal frame columns any reduction in hinge moment usually may be ignored. However, in the case of multi-story structures, the resisting moment of the columns in the lower floors would be reduced by axial load and evaluation of the ultimate load must then include such consideration.

Figure 4.1 shows the stress-distribution in a beam at various stages of deformation caused by thrust and moment. Due to the axial force, yielding on the compression side precedes that on the tension side. Eventually plastification occurs, but since part of the area resists the axial force the stress block no longer contains equal compression and tension areas (as was the case with pure moment). Thus, as shown in Fig. 4.2, the total stress-distribution (sketch *a*) may be divided into two parts: a part that is associated with the axial load (sketch *b*) and a part that corresponds with the bending moment (sketch *c*).

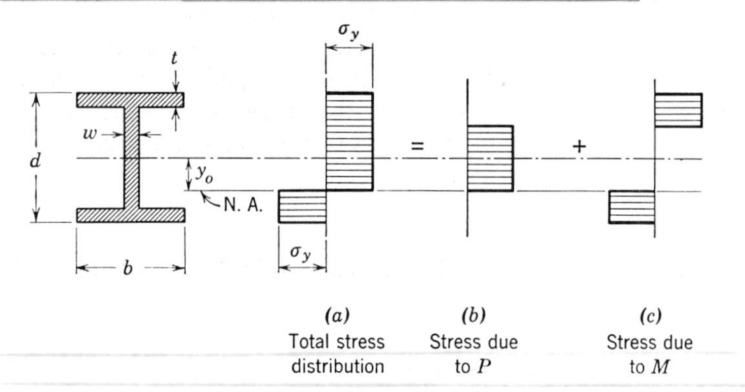

Fig. 4.2. Representation of (*b*) stress due to axial force and (*c*) stress due to bending moment for a completely plastic cross section subjected to bending and axial force. (*Welding J.*, 31(12), p. 607-*s*, 1952.)

For the condition shown in Fig. 4.2 with the neutral axis in the web and using the assumptions of Art. 2.2, the axial force P is given by

$$P = 2\sigma_y y_o w \tag{4.1}$$

where σ_y is the yield stress, y_o is the distance from the midheight to the neutral axis, and w is the web thickness. The corresponding bending moment M_{pc} is given by the following expression and represents plastic hinge moment modified to include the effect of axial compression:

$$M_{pc} = \sigma_y(Z - w{y_o}^2) \tag{4.2}$$

where Z is the plastic modulus. By substituting the value of y_o obtained from Eq. 4.1 into Eq. 4.2, the bending moment may be expressed as a function of the axial force P, or

$$M_{pc} = M_p - \frac{P^2}{4\sigma_y w}, \qquad [P \leq \sigma_y w(d - 2t)] \tag{4.3}$$

This relationship may be obtained in nondimensional form by dividing both sides of Eq. 4.3 by $M_p = \sigma_y Z$. Thus

$$\frac{M_{pc}}{M_p} = 1 - \frac{A^2}{4wZ}\left(\frac{P}{P_y}\right)^2, \qquad \left(0 < \frac{P}{P_y} < \frac{w(d - 2t)}{A}\right) \tag{4.4}$$

By the same process, an expression for M_{pc} as a function of P could be determined when the neutral axis is in the flange instead of the web. The resulting equations are:

$$P = \sigma_y[A - b(d - 2y_o)] \tag{4.5}$$

$$M_{pc} = \sigma_y b\left(\frac{d^2}{4} - y_o^2\right) \tag{4.6}$$

from which

$$M_{pc} = \frac{\sigma_y}{2}\left[d\left(A - \frac{P}{\sigma_y}\right) - \frac{1}{2b}\left(A - \frac{P}{\sigma_y}\right)^2\right],$$

$$(\sigma_y w(d - 2t) < P < P_y) \tag{4.7}$$

and in nondimensional form

$$\frac{M_{pc}}{M_p} = \frac{A}{2Z}\left[d\left(1 - \frac{P}{P_y}\right) - \frac{A}{2b}\left(1 - \frac{P}{P_y}\right)^2\right], \quad \left(\frac{w(d - 2t)}{A} < \frac{P}{P_y} < 1\right) \tag{4.8}$$

Figure 4.3 shows the "interaction" curve for an 8WF31 shape that results from this analysis (Eqs. 4.3 and 4.7). When the axial force is

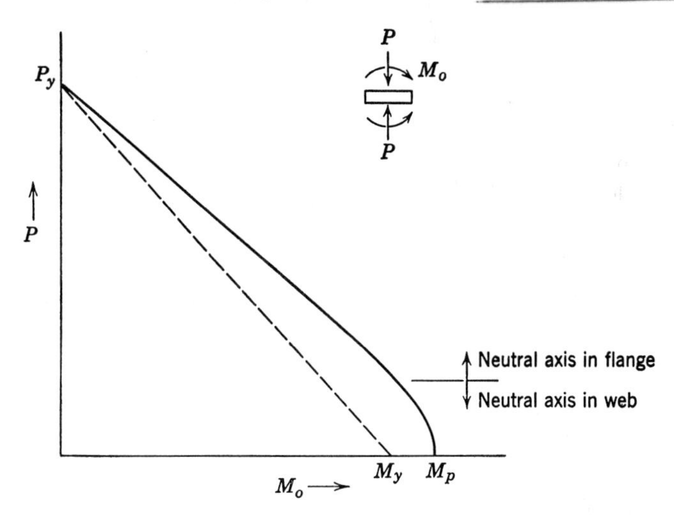

Fig. 4.3. Interaction curve for an 8WF31 cross section. (*Welding J.*, 31(12), p. 607-s, 1952.)

zero, $M = M_p$. When the axial force reaches the value $P = P_y = \sigma_y A$, then the moment capacity is zero. Between these limits the relationship is computed as described and the desired influence of axial force on the plastic moment has thus been obtained.

If, now, other WF shapes of different proportions were examined by this analysis, the relationship shown in Fig. 4.4 would be obtained. Here

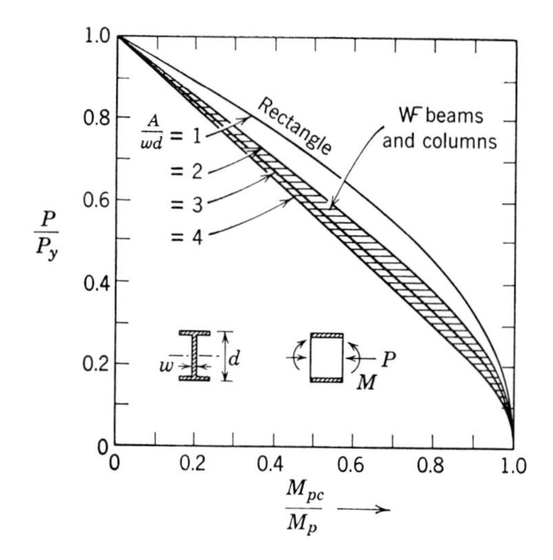

Fig. 4.4. Nondimensional interaction curve for various WF shapes.

the curve has been nondimensionalized and Eqs. 4.4 and 4.8 are applicable after some approximations have been made for certain geometrical ratios. The solution for the rectangle is also shown in Fig. 4.4.

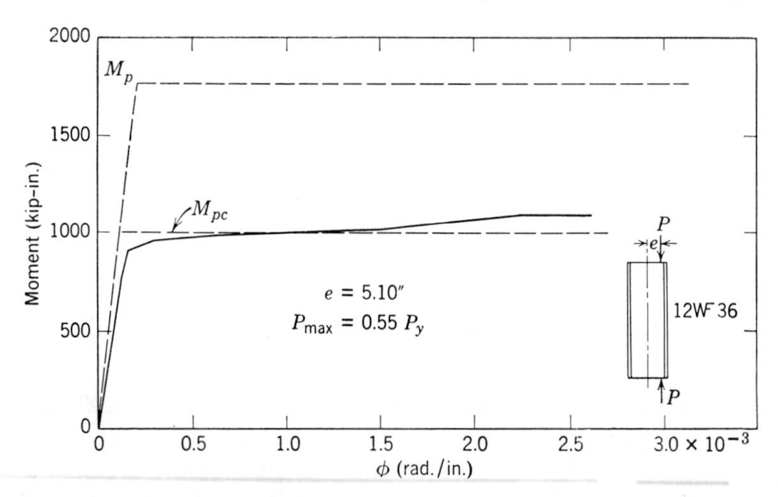

Fig. 4.5. Correlation between theory and test for an eccentrically loaded stub column. (*Proc. AISC Nat'l. Engr. Conf.*, 1956.)

Figure 4.5 shows the correlation of experiment with theory. The dotted lines represent the theoretical M–ϕ relationships for a certain 12WF36 member. The upper one would be for pure bending whereas the lower horizontal line is the theoretical action of a stub column loaded eccentrically as shown in the sketch. The solid line represents the

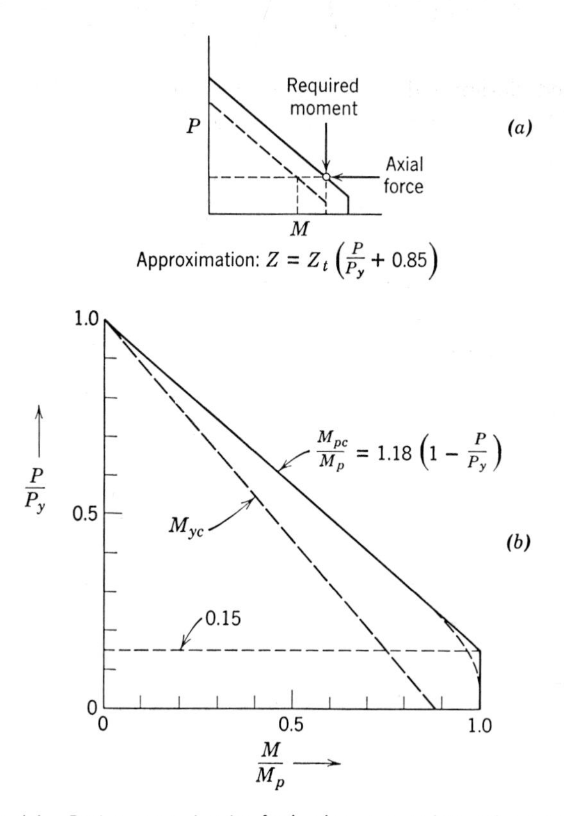

Fig. 4.6. Design approximation for load-vs.-moment interaction curve.

result of a test on this member. The "hinge" characteristic is retained, and the magnitude of the actual reduced plastic moment agrees with the value predicted from Eq. 4.7.

In order to account for the influence of direct stress in design, either curves such as Fig. 4.4 could be used, or, since most WF shapes have a similar curve when plotted on a nondimensional basis (see shaded portion of Fig. 4.4), the simple approximation of Fig. 4.6*b* could be used. With an error of less than 5%, axial load can be neglected up to $P/P_y =$ 0.15. For $P > 0.15P_y$ the reduction in moment capacity is given by the straight line equation shown as M_{pc}/M_p in Fig. 4.6*b*. Thus the moment capacity of a section in the presence of axial force is given by

$$\left.\begin{aligned} M_{pc} &= M_p, && \left(0 < \frac{P}{P_y} \leq 0.15\right) \\ M_{pc} &= 1.18\left(1 - \frac{P}{P_y}\right) M_p, && \left(0.15 < \frac{P}{P_y} \leq 1\right) \end{aligned}\right\} \tag{4.9}$$

The required design value of the plastic modulus Z for a member is determined by multiplying the trial value Z_t found in the initial design by the ratio M_p/M_{pc} or

$$Z = \frac{0.85 Z_t}{1 - P/P_y} \tag{4.10}$$

Actually Eq. 4.10 gives a value of the plastic modulus that is too great. As illustrated by Fig. 4.6*a* the P/P_y ratio will be less in the redesign and thus the reduction in M_p will be less than first computed. The equation

$$Z = Z_t(P/P_y + 0.85), \qquad \left(0.15 < \frac{P}{P_y} < 1.0\right) \tag{4.11}$$

is an approximation to account for this effect, P/P_y being the ratio obtained in the first design. It assumes that the area A is a linear function of the plastic modulus Z. The final selection should be checked by use of Eq. 4.9.

EXAMPLE Influence of Axial Force

An example illustrating the influence of axial force and the corresponding revision in design is as follows. Suppose a member must transmit a moment of 2400 kip-in. and an axial force of 100 kips. In the presence of this force, what shape will transmit the required hinge moment?

1. The required plastic modulus is computed neglecting axial force:

$$Z_t = \frac{M_p}{\sigma_y} = \frac{2400}{33} = 72.8 \text{ in.}^3$$

Try 16WF40
$Z = 72.7$ in.^{3}

2. The corresponding axial load ratio is determined:

$$\frac{P}{P_y} = \frac{100}{\sigma_y A} = \frac{100.0}{(33)(11.77)} = 0.258 > 0.15$$

3. Since $P/P_y > 0.15$ a redesign is made using Eq. 4.11:

$$Z = Z_t(P/P_y + 0.85) = 72.8(0.26 + 0.85) = 80.7 \text{ in.}^3$$

Use 16WF45
$Z = 82.0$ in.^{3}

4. Checking the selection according to Eq. 4.9, the plastic moment actually delivered is given by:

$$M_{pc} = 1.18(1 - P/P_y)\sigma_y Z$$

$$M_{pc} = 1.18\left(1 - \frac{100}{(33)(13.24)}\right)(33)(82.0) = 2460 \text{ kip-ins} > 2400 \qquad \underline{\text{OK}}$$

The fact that the effect of axial force can be neglected for $P/P_y < 0.15$ is of considerable benefit in the design of many single story structures, since it is usually found that the axial force is relatively small in the members. Therefore the design procedure is simplified since the axial force is neglected.

The relationships described in this article have been derived for flexure about the strong (x–x) axis because columns intended to develop considerable plastic moment would naturally be placed in this most advantageous position. In the event that the reduced plastic moment is desired for the weak (y–y) axis, two approaches are possible. Either the shape may be idealized to consist of two rectangles and the upper curve of Fig. 4.4 used to compute M_{pc}, or the relationship may be derived following the same procedure as that which led to Eqs. 4.4 and 4.8.

4.3 THE INFLUENCE OF SHEAR FORCE

If high shear exists in the web of a structural beam over a considerable length, as in the case of a short beam centrally loaded, or a long beam with a concentrated load near the end, the beam may yield rather generally in shear thus causing larger deflections than might otherwise be expected. The effect of shear force, therefore, is somewhat similar to that of axial force—it may limit the load on the beam to a value that is less than that which corresponds to the full plastic moment.

In order to arrive at a basis for assuring that the full plastic moment will be reached, a cantilever beam in bending will be considered as shown in Fig. 4.7. Of interest are the stress-distributions at various sections (A, B, and C) along the beam because they form a means of evaluating performance under different shear-to-moment ratios. At section A the beam is elastic and the flexure and shear stresses may be determined

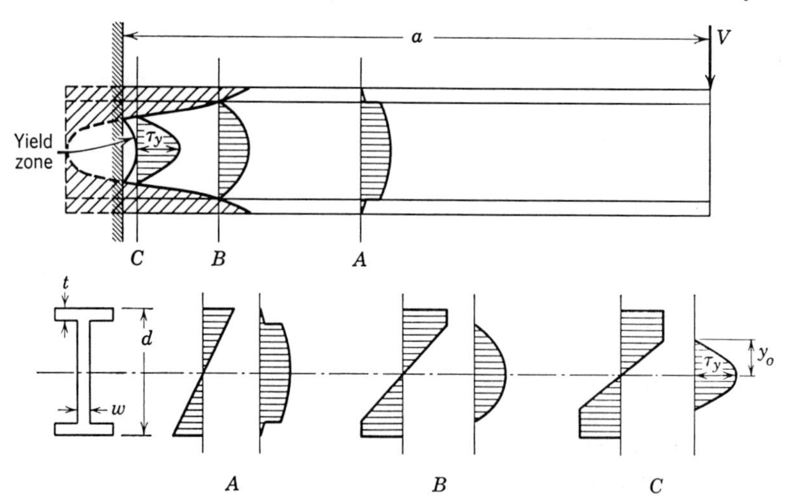

Fig. 4.7. Shear and flexural stress distributions in cantilever beam that is partially yielded in bending.

easily. At section B the flange is completely yielded due to flexure and therefore all of the shear stresses must be carried in the elastic core; thus the distribution of shear stresses is parabolic. At section C the shear stress has reached its yield value at the centerline.

Two possibilities of premature "failure" due to the presence of shear therefore exist:

(1) General "shear yield" of the web may occur in the presence of high shear-to-moment ratios (sections A and B of Fig. 4.7).

(2) After the beam has become partially plastic at a critical section due to flexural yielding, the intensity of shear stress at the centerline may reach the yield condition. (Section C of Fig. 4.7.)[1.9, 4.1]

Consider, first, case 1. The maximum possible shear according to this criterion is given by

$$V = \tau_y w(d - 2t) \tag{4.12}$$

If

$$\tau_y = \frac{\sigma_y}{\sqrt{3}}$$

then

$$V = \frac{\sigma_y}{\sqrt{3}} w(d - 2t) \tag{4.13}$$

The solution for case 2 is shown in Fig. 4.8 by three curving lines representing three different wide-flange proportions. It is obtained,

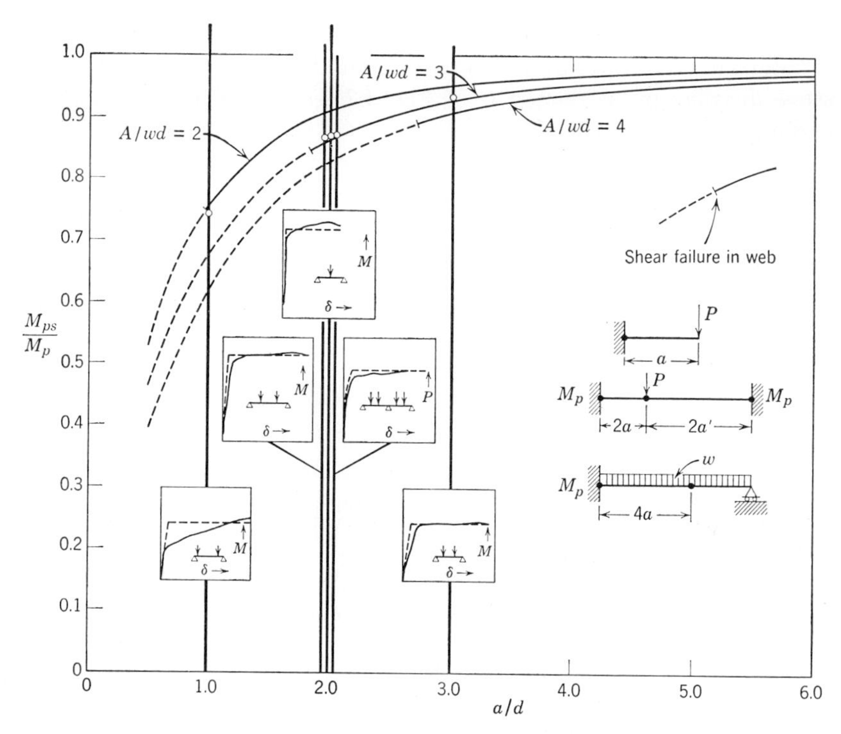

Fig. 4.8. Theoretical influence of shear force on the plastic moment and a comparison with tests showing that strain hardening counteracts this influence.

approximately, by considering the shear and flexural stresses of distribution C, Fig. 4.7. The maximum shear stress for a parabolic distribution is $\frac{3}{2}$ times the average value or

$$\tau_{\max} = \frac{3}{2}\frac{V}{w(2y_0)} = \frac{3}{4}\frac{V}{wy_o} \tag{4.14}$$

where w is the web thickness and y_0 is the distance from midheight to the elastic-plastic boundary. The maximum value of the shear stress is taken as the yield value ($\tau_{\max} = \tau_y = \sigma_y/\sqrt{3}$). If a is selected as the distance from the end of the cantilever to the "critical" section C, then the shear force V may be expressed as $V = M_{ps}/a$. With these substitutions, Eq. 4.14 becomes

$$\frac{\sigma_y}{\sqrt{3}} = \frac{3}{4}\frac{M_{ps}}{awy_0} \tag{4.15}$$

where M_{ps} signifies the magnitude of the bending moment at which the "limiting" condition of yielding due to shear force at the centerline is obtained. Now M_{ps} may also be computed directly from the flexural stress-distribution at section C of Fig. 4.7 by the principles described earlier. (See Eq. 2.12.) Then

$$M_{ps} = M_p - \sigma_y \frac{w y_o^2}{3} \tag{4.16}$$

If y_o is eliminated between Eqs. 4.15 and 4.16, and if both sides of the resulting expression are divided by $M_p = \sigma_y Z$, the following relationship is obtained which gives the influence of shear on the plastic moment:

$$\frac{9Z}{16a^2 w}\left(\frac{M_{ps}}{M_p}\right)^2 + \frac{M_{ps}}{M_p} - 1 = 0 \tag{4.17}$$

Although Eq. 4.17 may be solved for a given wide-flange shape, it is not in a sufficiently general form. From the approximate relationship *

$$Z = \frac{d}{2}\left(A - \frac{wd}{2}\right) \tag{2.25}$$

the ratio Z/wa^2 in Eq. 4.17 becomes

$$\frac{Z}{wa^2} = \frac{d^2}{a^2}\left(\frac{A}{2wd} - \frac{1}{4}\right) \tag{4.18}$$

Now the ratio A/wd represents the ratio of the total area to the area of the web. For WF beams it ranges from 2.0 to 3.0 and for columns it is about 4.0. Thus, for a given A/wd ratio, a solution may be obtained, giving M_{ps}/M_p as a function of a/d:

$$\frac{M_{ps}}{M_p} = \frac{-1 \pm \sqrt{1 + 4C\, d^2/a^2}}{2C\, d^2/a^2} \tag{4.19}$$

where $C = \frac{9}{16}[A/(2wd) - \frac{1}{4}] = \frac{27}{64}, \frac{45}{64}$, and $\frac{63}{64}$ for $A/wd = 2.0, 3.0$ and 4.0 respectively.

The solution of Eq. 4.19 for three values of A/wd (2.0, 3.0, 4.0) is shown in Fig. 4.8. The sketches in the inset show how these curves may be applied to other cases.

Below a certain value of a/d, the moment capacity would theoretically be limited because of complete yielding of the web (case 1, given by Eq. 4.13). Such a cut-off point is shown for each curve in Fig. 4.8.

* This relationship is approximate to about 5%, varying from +2%, to +10%.

The solid curve of Fig. 4.9 shows the behavior under test of a simply supported beam with the loads applied close to the supports. The long dashed lines give the relationship predicted by simple elastic and plastic theory, whereas the shorter dashed line reflects the theoretical influence of shear force in the elastic range. The maximum load that would

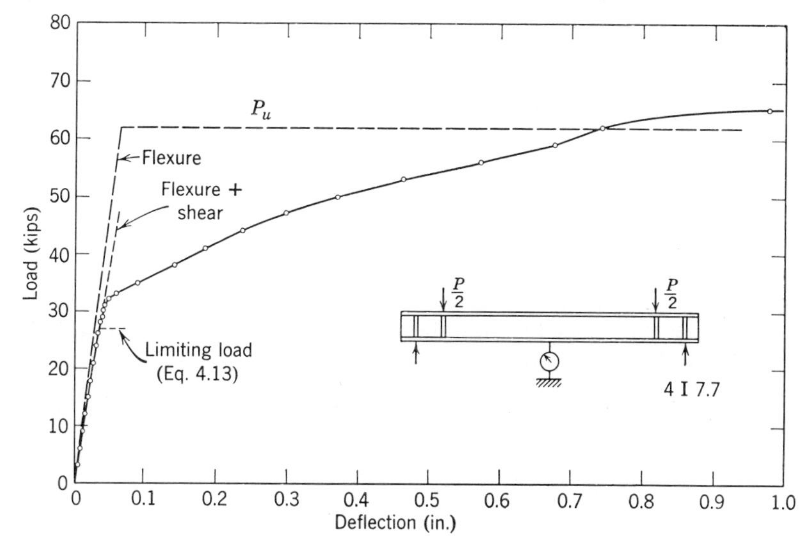

Fig. 4.9. Load–deflection curve of a beam subjected to shear.

be permitted by Eq. 4.13 is shown by the arrow on Fig. 4.9. (Eq. 4.17 does not apply in this case, yielding a higher load.) The limitations would thus prevent "failure" due to excessive shear deformations. The beam at the end of the test is shown in Fig. 4.10. Yielding occurred

Fig. 4.10. Photograph of the beam whose load–deflection curve is shown in Fig. 4.9.

first in the end panels which were subjected to the high shear force. After these panels strain-hardened, yielding occurred in the central portion under the action of pure moment.

Although Eq. 4.19 would call for a reduction in dependable M_p values in some instances, a consideration of the actual behavior of structures loaded in combined bending and shear shows that the implied limitation does not, in fact, exist for WF and I shapes. Since high shear and moment values occur in regions of localized yielding (steep moment gradient) the beneficial effects of strain-hardening usually enable such a beam to reach the full plastic moment. The experimental evidence for this is shown in Fig. 4.8 in the form of bar graphs, the top of which represent the ratio of maximum moment under test divided by M_p. The five test results all give maximum moment values equal to or greater than M_p. The corresponding moment- or load-deflection curves are shown in the small sketches.

Because of the effect of strain-hardening, application of Eq. 4.17 to design would be too conservative. Therefore a design guide to account for the influence of shear force may be obtained by a consideration only of Eq. 4.13. Since for WF shapes $d/(d - 2t) = 1.07$, and using $\sigma_y =$ 33,000 psi for structural steel, then the maximum allowable shear force in a beam at ultimate load (expressed in kips) may be computed * from

$$V_{max} = 18.0wd \tag{4.20}$$

where w = web thickness (in.) and d is the section depth (in.).

This equation is based on the assumption that all of the shear force is carried by the web. Further, the limit of usefulness is assumed to be the load at which the full web depth is stressed to the value τ_y. As shown by Fig. 4.9, this is a conservative assumption, since increased deformation is accompanied by an increase in load. Thus in design the problem of shear is primarily one of deflections. Even though there are theoretical reductions in load capacity, these may be relaxed if deflection is not a factor and if web buckling is not a problem.

EXAMPLE Shear Force in an Indeterminate Beam (Fig. 4.11)

A load $P = 125$ kips is applied three feet from the fixed end of this beam. The problem is to determine whether or not shear is a limitation and if so, what remedies might be taken.

* As discussed in Art. 7.4, it is open to question as to whether or not the maximum allowable shear force should be controlled by $\tau_y = \sigma_y/\sqrt{3} = 19{,}000$ psi or by the quantity, $\tau_w\sigma_y/\sigma_w$, where the subscript w refers to an elastic allowable working stress. In the latter case, $\tau_{max} = 13{,}000(\frac{33}{20}) = 21{,}500$ psi and Eq. 4.20 becomes

$$V_{max} = 20.0wd$$

(1) A beam is selected neglecting the influence of shear. By the mechanism method using the angles of sketch *b*

$$M_p = \frac{(P)(\frac{3}{4}L)}{(4+3)} = \frac{(125)(\frac{3}{4})(12)}{7} = 161 \text{ k}$$

$$Z = (161)(0.364) = 58.5 \text{ in.}^3$$

Try 16W 36
$Z = 63.9$ in.³

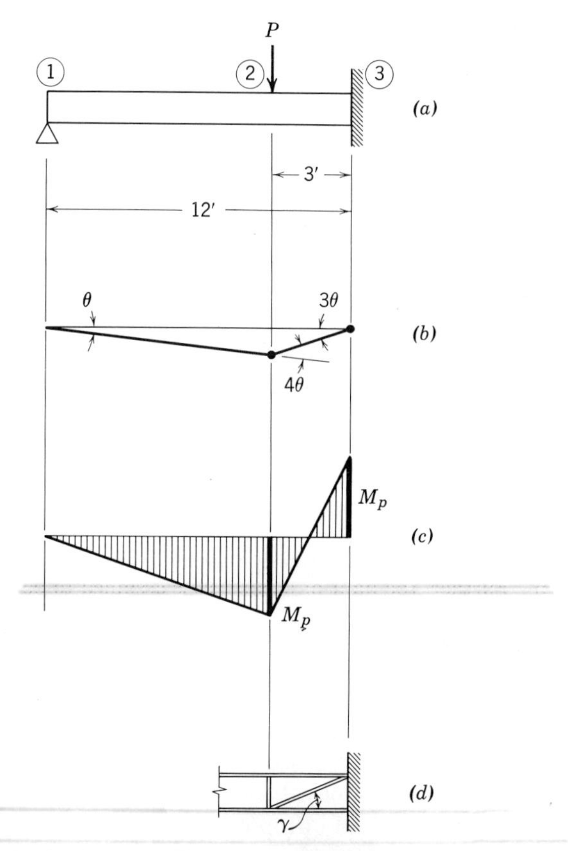

Fig. 4.11. Example of a design modification to account for the influence of shear force.

(2) The maximum shear in the beam is next computed.

$$V_{max} = \text{Shear in span 2–3} = \frac{2M_p}{3} = \frac{(2)(161)}{3} = 107 \text{ kips}$$

(3) Allowable shear is next computed according to Eq. 4.20

$$V_{allowable} = (18.0)(wd) = (18.0)(0.299)(15.85) = 85 \text{ kips} < 107 \text{ kips}$$

(4) Since there is insufficient web area to support the shear load without general yielding of the web, modification is required. Three possible solutions are as follows:

(*a*) Increase of section to provide sufficient web area:

$$\text{Area required} = wd = \frac{107}{18} = 5.95 \text{ in.}^2$$

$$\text{For 16WF45, } wd = (0.346)(16.12) = 5.60 \text{ in.}^2$$

This is considered adequate area in view of the conservative nature of this design check.

(*b*) Use of diagonal web reinforcement material:
The web is stiffened with a diagonal stiffener (tension-type) plate. Stiffener area computed on basis of truss web diagonal.

$$\text{Shear force} = 107 \text{ kips} - 85 \text{ kips} = 22 \text{ kips}$$

$$A_s = \frac{F_s}{\sigma_y} = \frac{\Delta V/\sin\gamma}{\sigma_y} = \frac{22}{(33)(0.25)} = 2.66 \text{ in.}^2$$

Use 2 plates 3″ x $\frac{1}{2}$″ as diagonal stiffeners on both sides of web.

(*c*) Use of web doubler:

$$w_d = w_r - w_a = \frac{107}{(18)(15.85)} - 0.299 = 0.376 - 0.299 = 0.077''$$

Use $\frac{1}{4}$″ plate doubler over web between 2–3 with slot welds along beam centerline.

4.4 LOCAL BUCKLING OF FLANGES AND WEBS

As a wide-flange beam is strained beyond the elastic limit, eventually the flange or the web will buckle. Figure 4.12 is typical of this action. Although stocky sections could be expected to retain their cross-sectional form through considerable plastic strain, with thin sections local buckling might occur soon after the plastic moment was first reached. Due to failure of a beam to retain its cross-sectional shape, the moment capacity would drop off; thus local buckling would prevent the section from sustaining the plastic moment until hinges are formed elsewhere. Therefore, in order to meet the requirements of deformation capacity (adequate rotation at the M_p value) compression elements must have width-thickness ratios adequate to insure against premature plastic buckling.*

A solution to this complicated plate buckling problem has been achieved [4.2,4.3] by requiring that the section will exhibit a rotation

* The problem of rotation capacity is treated in Chapter 6.

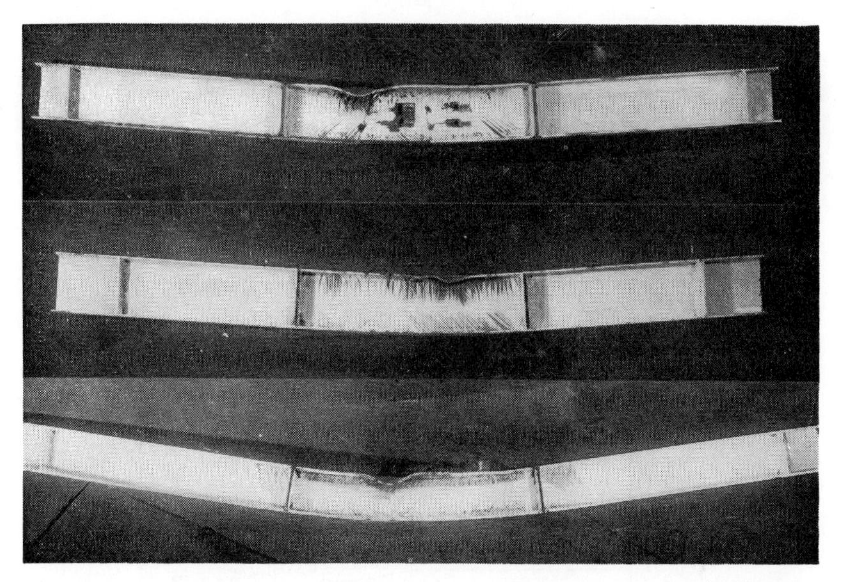

Fig. 4.12. Photographs of flange buckling in three beams with different b/t ratios.

capacity that corresponds to a compression strain equal to the strain-hardening value ϵ_{st} (Fig. 2.1). At this point the material properties may be more accurately and specifically defined than in the region between ϵ_y and ϵ_{st}.

The result of this analysis for flanges of WF shapes is shown in Fig. 4.13, together with the results of tests.[4.3] In the upper portion is shown the criterion for determining "adequate performance." When a flange buckles and the moment falls below the M_p value at a strain less than ϵ_{st} as shown by the heavy dashed curve, then the performance is said to be unsatisfactory. On the other hand when flange buckling results in eventual reduction in moment beyond ϵ_{st}, then the performance is deemed adequate. The results show that in the plastic range, the web exerts very little restraint on the flanges and that when the ratio of the total flange width to thickness is less than about 17, then "failure" will not occur until $\epsilon > \epsilon_{st}$.

It is evident that the number 17 is not sharply defined. Therefore for rolled shapes some upward variation in b/t ratio could be allowed. However, built-up members should be designed with 17 as the limit. From these results and from similar relationships established for webs, design guides may be established to assure that the compressive strains may reach ϵ_{st} without flange or web buckling.

These relationships have been derived on the basis that no stress

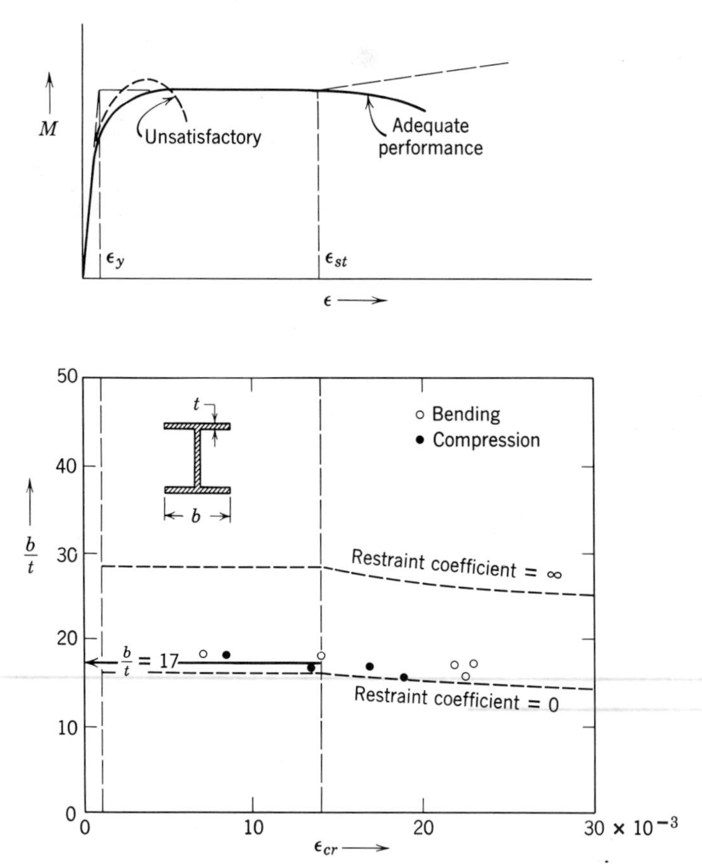

Fig. 4.13. Results of tests and theory showing a criterion for assuring that the hinge moment will be maintained until the strain reaches ϵ_{st}.

gradients exist along the member and that, in the case of flexure, no axial force exists. Since shear and axial forces also are usually present in beams, some consideration should be given to the influence of these factors.

Neither shear nor axial force will significantly modify the requirements of flange width–thickness ratios. Insofar as axial force is concerned the derivation has already assumed that the uniform strain across the flange must reach the strain-hardening value ϵ_{st}. There would be a small variation in the rotation angle at which the critical buckling condition would be reached.

The effect of axial force on web buckling is significant and this would be expected because the effect of the axial force is to increase the depth of the member that is yielded in compression (see, for example, Fig.

4.2a). Once the web is entirely yielded it would be expected that any further increase in P/P_y would not affect web failure. The curves, derived in Ref. 4.3 show the theoretical influence of this factor.

Summarizing, in order to assure that the compressive strains may reach ϵ_{st} without buckling, the cross-section proportions should be checked to see that flanges and webs of beams and columns comply with the following:

$$\left.\begin{aligned} &b/t \leq 17 \\ &d/w \leq 43 \quad \text{(columns in direct compression)} \\ &d/w \leq 55 \quad \text{(beams)} \end{aligned}\right\} \tag{4.21}$$

Fortunately, most existing WF and I shapes meet these requirements. The flanges of all I shapes are satisfactory and the flanges of WF shapes used as beams are all adequate except some of the smaller members. With regard to d/w ratios, practically all I-beams are satisfactory and most WF-beams meet the requirement. The d/w ratio of 55 is applicable for an axial thrust of 15% P_y. For $d/w > 55$ or for $0.15 < P/P_y < 0.27$, the following equation gives the maximum allowable d/w values.

$$d/w = 70 - 100P/P_y, \qquad (0 < P/P_y < 0.27) \tag{4.22}$$

which is an approximation to the theoretical analysis of Ref. 4.3.

Shapes with proportions that do not meet the requirements of Eq. 4.21 may be stiffened locally in the region of the plastic hinge. Figure 4.14 suggests some methods by which this may be accomplished. (The

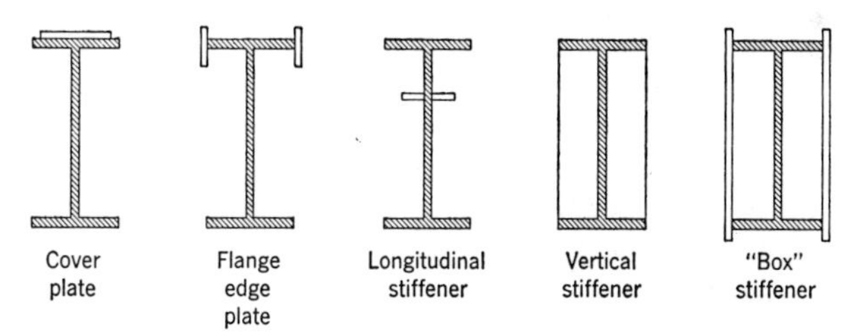

Fig. 4.14. Methods of stiffening a WF shape to prevent local instability.

width–thickness ratio of any compression or load-bearing stiffeners for this purpose should not be greater than 8.5.) Of course, such devices add to the expense of the structure, and they should be used only when it is reasonably certain that choice of another shape will not solve the problem. In this connection, it is noted that the "box" type has the advantage of a somewhat higher shape factor and greatly improved lateral buckling resistance.

4.5 LATERAL BUCKLING

The effect of lateral buckling is much like that of local buckling. In fact, in many tests the two frequently occur simultaneously.

Figure 4.15 shows a moment–curvature relationship that is undesirable; the moment does not remain at near-constant value through a sufficient angle change. This result was obtained in a test of a simply supported beam with a span length intentionally made so long that

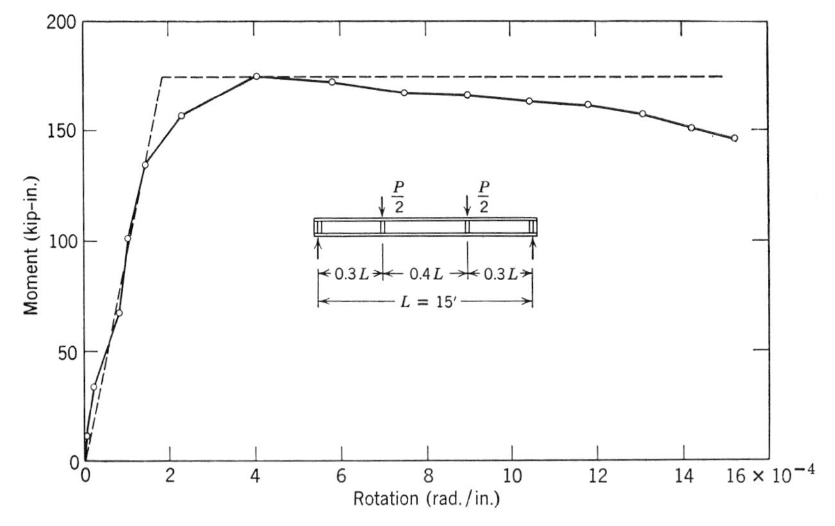

Fig. 4.15. Moment rotation curve showing the effect of lateral buckling.[1.16]

premature buckling was inevitable.[1.16] A photograph of three beams that have buckled laterally is shown in Fig. 4.16. The center beam corresponds with Fig. 4.15.

The problem of specifying the critical length of a beam such that premature lateral buckling will be prevented has not been completely solved. Theoretical studies have been made somewhat along the lines of those which proved to be successful in the case of local buckling. The problem to be solved is to determine the critical length of an otherwise unsupported member whose flanges may be strained to the point of strain hardening while still remaining in the plane of the bending moment.

Fig. 4.16. Photograph of three beams that have buckled laterally.

If a WF shape is idealized to consist of two rectangles as shown in Fig. 4.17*a* then from Ref. 4.4 the basic equation for the critical moment is given by

$$M_{cr} = \frac{\pi}{L}\sqrt{EI_yGK}\sqrt{1 + \pi^2 \frac{E}{L^2GK}\left(\frac{I_yd^2}{4}\right)} \tag{4.23}$$

where the member is bent in pure moment and the beam is "pin-ended" in both directions (Fig. 4.17*b*). The critical moment is M_p. For the idealized section, the following may be written:

$$\left.\begin{aligned} M_p &= \sigma_y \frac{Ad}{2} \\ I_y &= Ar^2 \\ A &= 2bt \\ K &= \tfrac{2}{3}bt^3 = \tfrac{1}{3}At^2 \end{aligned}\right\} \tag{4.24}$$

If these substitutions are made, the following equation is obtained:[1.9]

$$\frac{(L/r)^2}{\sqrt{\left(\frac{L}{r}\right)^2 + \frac{3}{4}\pi^2 \frac{E}{G}\left(\frac{d}{t}\right)^2}} = \frac{2\pi\sqrt{EG/3}}{\sigma_y(d/t)} \tag{4.25}$$

which gives implicitly the critical slenderness ratio in terms of the material properties E and G and as a function of the geometry d/t. For

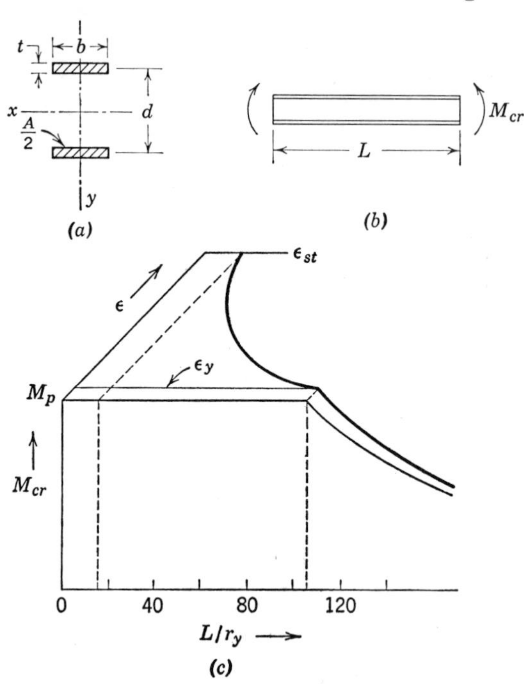

Fig. 4.17. (a) Idealized cross section; (b) beam under pure moment; and (c) diagrammatic representation of the moment–slenderness–strain relationship (idealized).

WF shapes used as beams the value of d/t is about 32, so there only remains the question of assigning values of E and G.

For the shape just to reach M_p without buckling, and assuming idealized behavior, the following elastic properties are used:

$$E = 30{,}000{,}000 \text{ psi}$$
$$G = 12{,}000{,}000 \text{ psi}$$
$$\sigma_y = 33{,}000 \text{ psi}$$

and the critical L/r becomes 106.

However the problem requires that the flanges strain to the point of strain-hardening *prior* to buckling. In order to obtain this solution, one would substitute into Eq. 4.25 the properties of the material at the point of strain-hardening, namely,

$$E_{st} = 700{,}000 \text{ psi}$$

$$G_{st} = 2{,}400{,}000 \text{ psi *}$$

$$\sigma_y = 33{,}000 \text{ psi}$$

and the critical slenderness ratio becomes about 15.

The result is shown in Fig. 4.17*c* in the form of a three-dimensional plot. It shows the strains that correspond to the two points just computed on the basis of idealization of the action and properties of a WF shape.

Now in the case of local buckling the solution using the "strain-hardening constants" was such that it embraced nearly all of the rolled shapes and was therefore an economically feasible design guide. However, in the present instance, a slenderness ratio arrived at by similar reasoning (namely $L/r \cong 15$) would be too restrictive, requiring substantially more points of lateral support than at present, thus adding to the expense of the structure. Further, numerous tests show that this value would be too conservative.

There are two approaches for achieving a solution to this problem that will make possible a practical engineering design. The one is to consider whether or not "full rotation" to the point of strain-hardening is actually required. The other approach is to refine the theory, considering those factors that were neglected in the derivation. In the final solution, a combination of both approaches is made.

Consider the matter of required rotation. Obviously, certain hinges in a structure will be required to rotate more than others. The first hinge to form will rotate the most and the hinge to form last requires a very small rotation and indeed the moment must only *reach* M_p. In the fixed-ended beam of Fig. 3.4, for instance, the first hinges will form at the ends and the hinge at the center will be the last to form; therefore, rotation capacity is required at the ends and none is required at the center. The requirement that a section be capable of absorbing rotations to the point of strain-hardening was developed as a "maximum" requirement. In view of the fact that there is such a marked difference in L/r as between full rotation and none at all, then it would be unduly restrictive to apply it to all cases. Studies on this problem are showing

* Based on tests noted in Ref. 4.2.

how much rotation is required at a plastic hinge in different types of structures.[4.5]

The second approach is to consider the effect of factors that were ignored in the theory. Upward revisions in critical slenderness ratio are possible from a consideration of moment gradient, St. Venant's torsion, the extent of yielding (part of the beam elastic and part plastic), and the effect of end restraint provided by the adjoining beams. The effect of moment gradient and of end restraint are particularly significant.

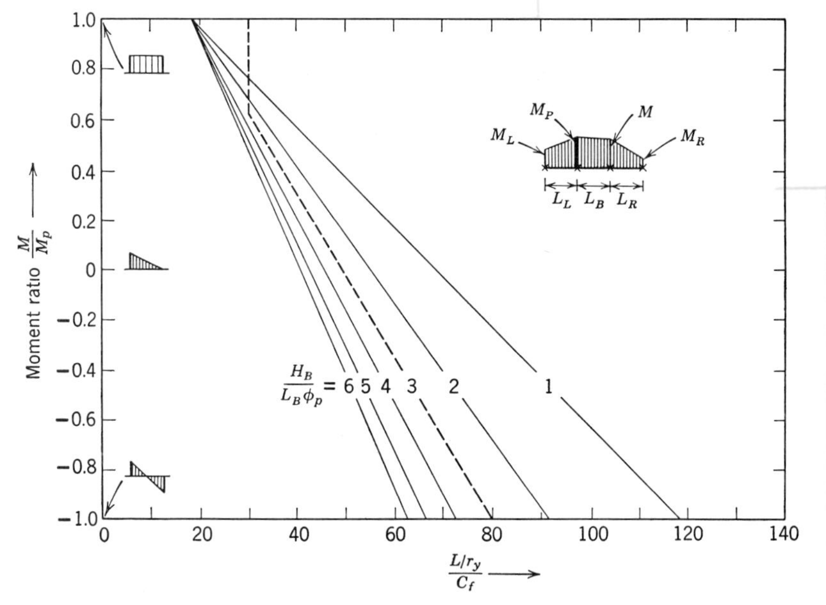

Fig. 4.18. Approximation to account for the influence of moment ratio, restraint coefficient, and hinge rotation on the critical buckling length of beams.

Figure 4.18 shows the result of the work done in Refs. 4.6 and 4.7 where these theoretical problems were considered. For a member containing a plastic hinge, the moment ratio (M/M_p) is the ratio of the moment at one end to the plastic moment at the other. The abscissa $\left(\frac{L}{r_y}\right)\left(\frac{1}{C_f}\right)$ is the slenderness ratio in the weak direction of a WF shape divided by a correction factor C_f to account for the fixity (or restraint) provided by the adjoining segments. The parameter in this figure is a function of the hinge angle H_B within the braced span L_B. (Dividing by ϕ_p makes the function nondimensional.) In other words, each of the diagonal lines which start from a "basic" value of 18 for uniform moment is drawn for a different magnitude of required hinge rotation.

In Ref. 4.5 values of $H/\phi_p L$ for different structures are presented, and Fig. 4.19 presents some of the limiting values obtained as a result of that study. It shows that the last hinge occurs in the rafter in most cases until the columns become unusually high with respect to the frame span.

Figure 4.18 shows a cut-off at $L/r_y C_f$ of 30; this is warranted from a consideration of test results. For M/M_p less than six-tenths, the relationship may be written in the form,

$$\left(\frac{1}{C_f}\right)\left(\frac{L}{r_y}\right) = 18 + a(1 - M/M_p) \tag{4.26}$$

and for $H_B/L_B\phi_p$ equal to 3.0, the quantity a is equal to 31. Thus Eq. 4.26 becomes

$$\frac{1}{C_f}\frac{L}{r_y} = 48 - 30\frac{M}{M_p}, \qquad \left(0.6 > \frac{M}{M_p} > -1.0\right) \tag{4.27}$$

and this is the equation of the dashed line shown in Fig. 4.18. A value of 3.0 for the hinge-angle parameter has been selected because it represents a reasonable upper limit.

Summarizing, since yielding markedly reduces the resistance of a member to lateral buckling, bracing will be required at those points at which plastic hinges are expected. Intermediate between these critical sections, conventional rules may be followed to protect against elastic lateral buckling. In the vicinity of a plastic hinge a guide to the spacing of lateral support is as follows:

The critical spacing is given by

$$\left.\begin{aligned} \frac{L}{r_y} &= 30, \qquad \text{for } M/M_p > 0.6 \\ \frac{L}{r_y} &= 48 - 30\frac{M}{M_p}, \qquad \text{for } M/M_p < 0.6 \end{aligned}\right\} \tag{4.28}$$

where L = distance between bracing points

r_y = radius of gyration in the "weak" direction

$\frac{M}{M_p}$ = moment ratio in the length L_B between bracing points.

If the unbraced length of the member is greater than this value, then additional bracing may be required. Appendix 1 outlines a procedure that may be used to check such a greater length; it involves an evaluation of the correction factor, C_f.

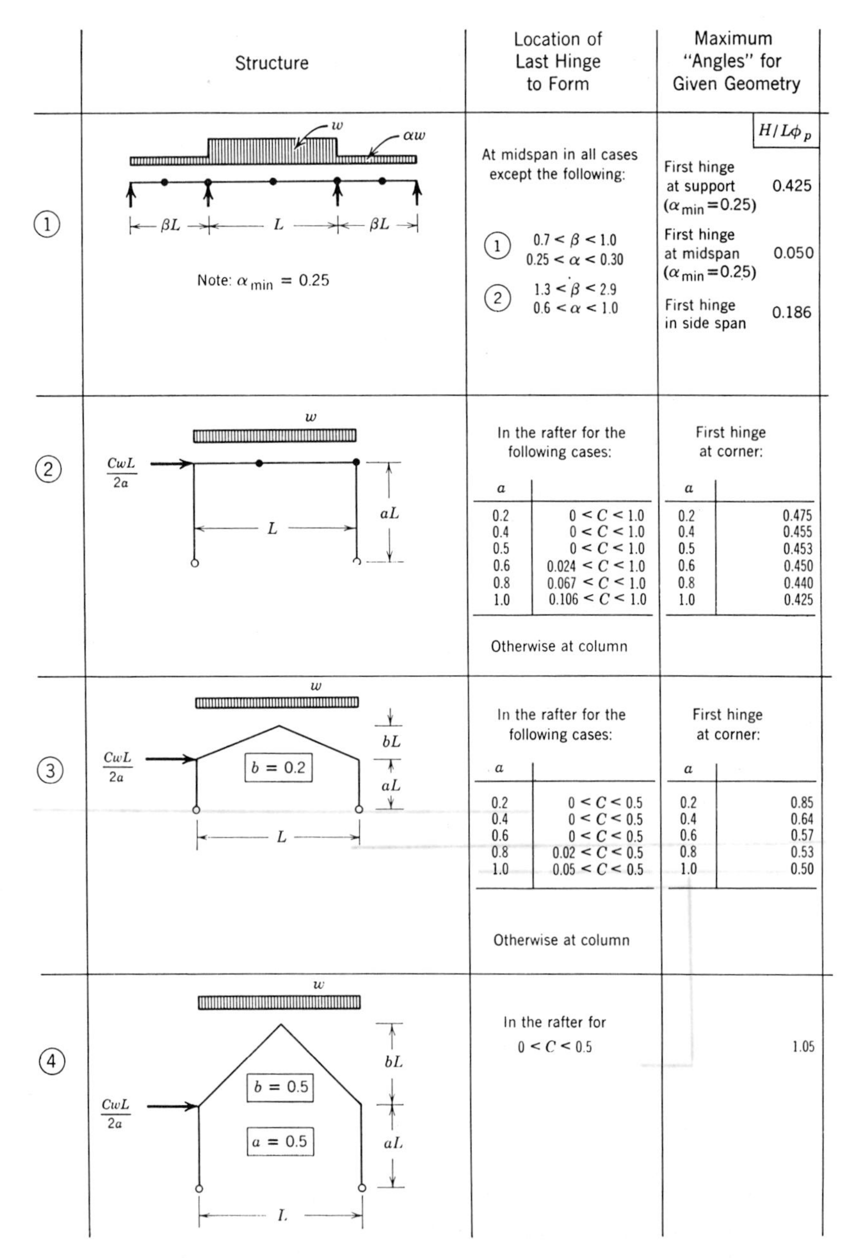

Fig. 4.19. Location of the last hinge to form in several practical structures; maximum values of the function $H/L\phi_p$ for the same structures.

Instead of computing C_f for each case, a simplified design guide could be obtained by assuming an average value for the correction factor. For practical structures C_f varies from 1.10 to 1.40. Using an average of 1.25, Eq. 4.27 becomes

$$\left.\begin{aligned} \frac{L}{r_y} &= 60 - 40\frac{M}{M_p}, &\quad \left(0.625 > \frac{M}{M_p} > -1.0\right) \\ \frac{L}{r_y} &= 35, &\quad \left(1.0 > \frac{M}{M_p} > 0.625\right) \end{aligned}\right\} \tag{4.29}$$

A somewhat more conservative value for C_f has been used for the "cut-off" at $L/r_y = 35$, reflecting the more critical nature of the flatter moment gradients.

The magnitude of the forces required to prevent lateral buckling is small,[4.8,3.9] and slenderness ratio requirements will normally govern. In tests the lateral force has seldom exceeded a value given by

$$T = 0.01\sigma_y A$$

where A is the area of member being braced. If this value is doubled to account for uncertain field conditions, then the required force, normal to plane of the frame, becomes

$$T = 0.02\sigma_y A$$

To provide stiff bracing, it is desirable to keep working stresses low in the bracing member. Thus

$$T_w = A_b\sigma_w = 0.02\sigma_y A$$

or

$$A_b \cong 0.04A$$

Bracing members (the purlins) must themselves be braced with respect to other parts of the frame such as by roof bracing. Both the compression and the tension flanges should be braced at changes of section. Although this same type of bracing is desirable at other points of support, the important requirement is to prevent lateral and torsional displacement.

An illustrative example is included in Appendix 1 and in Chapter 9 these procedures are applied in Plates V, VI, VIII, and XI.

4.6 COLUMN STABILITY

In this article, two kinds of columns will be considered briefly. The first is the centrally loaded, pin-ended column; the second is the column as it is found in a frame, bent or restrained (as the case may be) by other members that are rigidly joined to it. The former "ideal" loading condition is not too frequent in occurrence, but attention must be given to it because columns must be checked for failure about the weak axis. The latter "framed column" is by far the most common compression member that will be encountered in plastically designed structures.

Unfortunately the situation regarding our knowledge of the behavior of these different kinds of columns is just the reverse; a great deal is known about centrally loaded pin-ended members, but much is yet to be learned about framed columns. From the work of the Column Research Council, established to stimulate research on buckling problems, new knowledge is rapidly becoming available,[4.9] Bleich's treatise on buckling[4.4] representing the first result of the Council's activity.

Centrally Loaded Columns

Most columns used in steel building frames have a slenderness ratio sufficiently low that failure does not occur due to elastic buckling; instead the column buckles after yielding has occurred. Besides the yield stress level itself, the one factor that is of predominant influence on the failure of such members in the inelastic range is residual stress.

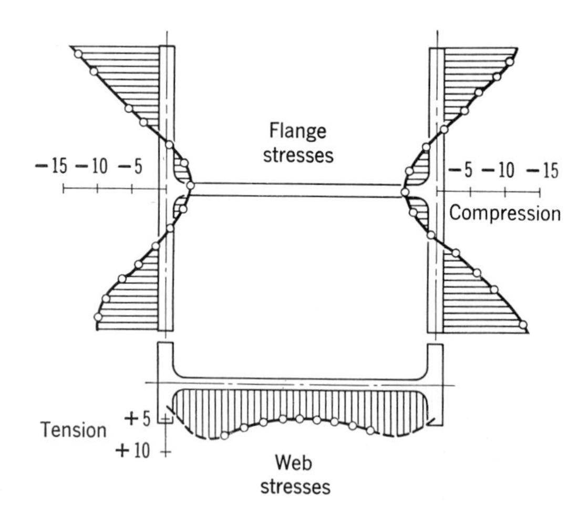

Fig. 4.20. Typical residual stress pattern in 8WF31 rolled shape.

The buckling strength of centrally loaded columns has been studied at Lehigh University for a number of years in a project to determine the behavior of columns containing residual stresses.[1.14, 2.2, 2.3] Such stresses remain in the member after cooling from the rolling temperature, after cold bending, or after welding. These studies also have shown that wide flange shapes used as columns may be expected to have compressive residual stresses in the flange tips amounting to 13,000 psi due to cooling after rolling. A typical pattern is shown in Fig. 4.20.

Columns of a material like structural steel with a definite yield stress level, will have a reduced buckling strength once the total stress (residual plus applied stress) reaches the yield stress level. When compared with values that otherwise would be predicted on the basis of coupon tests, these reductions are the largest in the region of slenderness ratio from 70 to 90. The studies previously mentioned also included a careful evaluation of the yield stress level for a complete range of WF columns of ASTM A7 structural steel. It showed that the average compressive yield stress level is about 34,000 psi, a value so close to the minimum specification value of 33,000 psi that the latter should be used.

The basic equation for the critical strength of a column containing residual stresses was derived in Ref. 2.2 and is given by

$$P_{cr} = \frac{\pi^2 E I_e}{L^2} \tag{4.30}$$

where I_e is the moment of inertia of the elastic portion of the cross section of a member that has been strained beyond the point of first yielding. In terms of the average critical stress, Eq. 4.30 may be written as

$$\sigma_{cr} = \frac{\pi^2 E I_e / I}{(L/r)^2} \tag{4.31}$$

and this equation is the basic equation for a column containing residual stresses.

In Ref. 1.14 two methods were presented for obtaining a solution to Eq. 4.31 for WF shapes with cooling residual stress. One method makes use of measured residual stresses and the other is based on the determination of the average stress-strain curve of a "stub column" that contains the residual stresses.* Both methods are based on a modified tangent-modulus concept of buckling.

* It has been shown in Ref. 4.10 that these same methods may be applied to H-shaped columns built up by welding, although the constants are different.

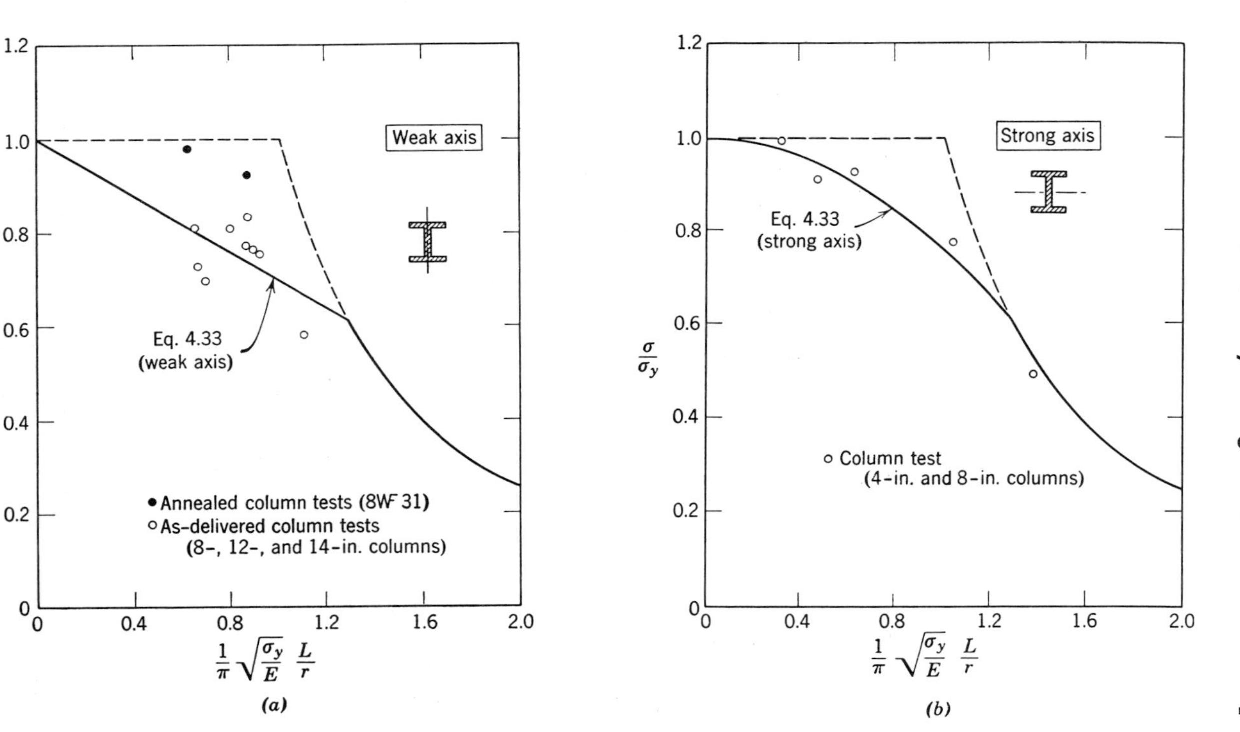

Fig. 4.21. Column test results correlated with predictions based upon the influence of residual stresses: (a) Columns bent about the weak axis and (b) Columns bent about the strong axis.

The following equations are suggested as approximations to the above method for determining the maximum strength of rolled WF columns

$$\left.\begin{aligned} \sigma_{x-x} &= \sigma_y - \frac{\sigma_p}{\pi^2 E}(\sigma_y - \sigma_p)\left(\frac{L}{r}\right)^2, & \frac{L}{r} &< \pi\sqrt{E/\sigma_p} \\ \sigma_{y-y} &= \sigma_y - \frac{(\sigma_y - \sigma_p)}{\pi}\sqrt{\frac{\sigma_p}{E}}\left(\frac{L}{r}\right), & \frac{L}{r} &< \pi\sqrt{E/\sigma_p} \\ \sigma &= \frac{\pi^2 E}{(L/r)^2}, & \frac{L}{r} &> \pi\sqrt{E/\sigma_p} \end{aligned}\right\} \tag{4.32}$$

where σ_{x-x} is the maximum stress for flexure about the strong axis, σ_{y-y} the corresponding value for failure about the weak axis.

Using $\sigma_y = 33{,}000$ psi, $E = 29{,}600{,}000$ psi, and $\sigma_p = \sigma_y - \sigma_{rc} = 20{,}000$ psi, these equations reduce, approximately to

$$\left.\begin{aligned} \sigma_{cr}(x\text{–}x) &= 33{,}000 - 0.9(L/r)^2, & (L/r &\leq 120) \\ \sigma_{cr}(y\text{–}y) &= 33{,}000 - 100L/r, & (L/r &\leq 120) \\ \sigma_{x-x} = \sigma_{y-y} &= \frac{292{,}000{,}000}{(L/r)^2}, & (L/r &> 120) \end{aligned}\right\} \tag{4.33}$$

Figure 4.21 shows the degree to which these equations are confirmed by tests, the curves being shown in nondimensional form.

For design use, these formulas are more conveniently written by dividing by the yield stress level or,

$$\left.\begin{aligned} P/P_y &= 1 - \frac{(L/r)^2}{35{,}600} & &\text{STRONG AXIS} & (L/r &\leq 120) \\ P/P_y &= 1 - \frac{L/r}{330} & &\text{WEAK AXIS} & (L/r &\leq 120) \\ P/P_y &= \frac{8850}{(L/r)^2} & &\text{BOTH AXES} & (L/r &> 120) \end{aligned}\right\} \tag{4.34}$$

For columns loaded in direct compression, Eqs. 4.34 would be used to limit the allowable axial load on the column.

For columns whose design is controlled by combined bending and axial loads Eqs. 4.34 would be useful for checking for failure of the column in the plane normal to the principle plane of bending. In this case, the equations would be written more conveniently,

$$\left.\begin{array}{ll}(L/r)_{x-x} < 189\sqrt{(1 - P/P_y)} & P/P_y > 0.625 \\ (L/r)_{y-y} < 330(1 - P/P_y) & P/P_y > 0.625 \\ L/r < \sqrt{\dfrac{8850}{P/P_y}} & P/P_y < 0.625\end{array}\right\} \quad (4.35)$$

Framed Columns

The plastic theory assumes that failure of the entire frame (in the sense that a mechanism forms) is not preceded by failure of the columns due to instability. Consequently, after the frame has been designed and the members selected, the columns must be checked to see that they meet the stated or implied design requirements. They must be stable under the applied axial force and end moments; further, any plastic hinges forming at the ends must have adequate rotation capacity.

Although the load at which an isolated column will fail when it is loaded with axial force and bending moment can be predicted with reasonable accuracy, the buckling problem becomes exceedingly complex when the column is a part of the framework.[4.11] A complete solution to this latter problem is not in hand, but conservative procedures are available for checking the suitability of columns selected for a plastically-designed structure. Currently under sponsorship by Column Research Council and others, work is going forward on the solutions to these problems; consequently, revisions are to be expected.[1.17]

As already implied, the failure load of a column and its ability to transmit plastic moments are dependent to a great extent upon the loading conditions. These are as follows:

(1) Double curvature with plastic hinges at both ends;
(2) Double curvature with a plastic hinge at one end and the opposite end intermediate between pinned and at plastic hinge value;
(3) Single curvature with one end pinned and moment applied at the opposite end;
(4) Single curvature with unequal end moments;
(5) Single curvature with equal end moments.

Reference 4.12 treats the important cases (3) and (5) and suggests formulas that should assist the designer. Figures 4.22 and 4.23 contain some of the results. The problem was approached on the basis that the criterion of failure is the maximum load the member will support considering inelastic action. Since the columns in structures to be designed by the plastic method may be thought of as bending members, the maximum carrying capacity is actually expressed in terms of a member of given slenderness supporting a certain axial thrust; and the infor-

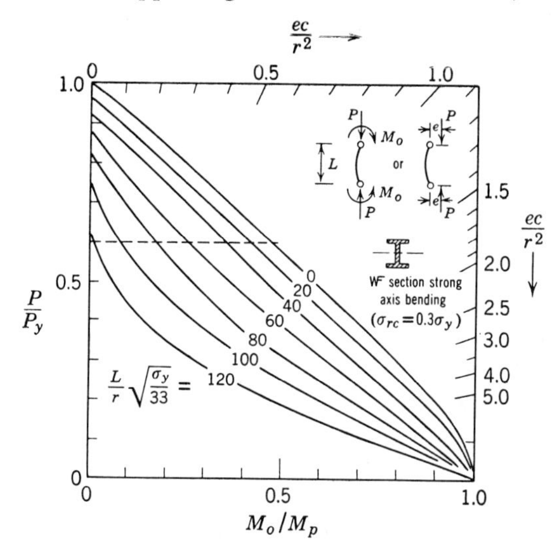

Fig. 4.22. Modified end-moment values for columns with equal and opposite end moments.[4.12]

mation to be supplied is the maximum moment that such a member can sustain.

The procedure then used in Ref. 4.12 was to determine the relationship between applied end moment and end rotation as the member is

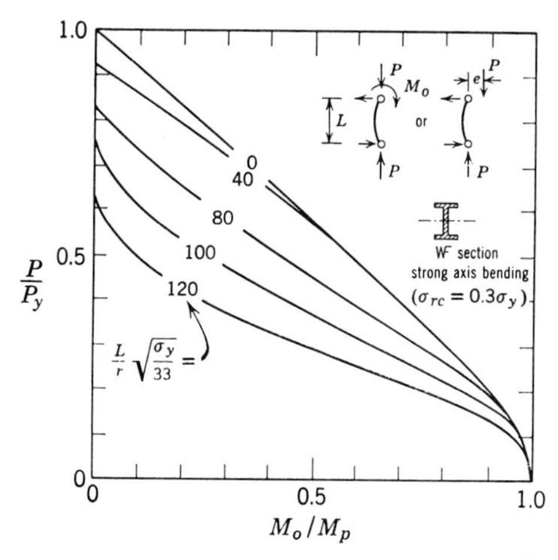

Fig. 4.23. Modified end-moment values for columns with moment applied at one end only.[4.12]

bent about the strong axis. When the resulting curve reaches a horizontal tangent as shown in Fig. 4.24 then the maximum moment for the given axial thrust and slenderness ratio has been reached. The procedure was carried out using a systematized numerical integration procedure.[4.13] M–ϕ curves (modified for axial force) that were used in the numerical integration procedure were presented in Ref. 4.14. Local and lateral-torsional instability were not considered. Local instability will not be a limitation as long as the requirements of Art. 4.4 are met.

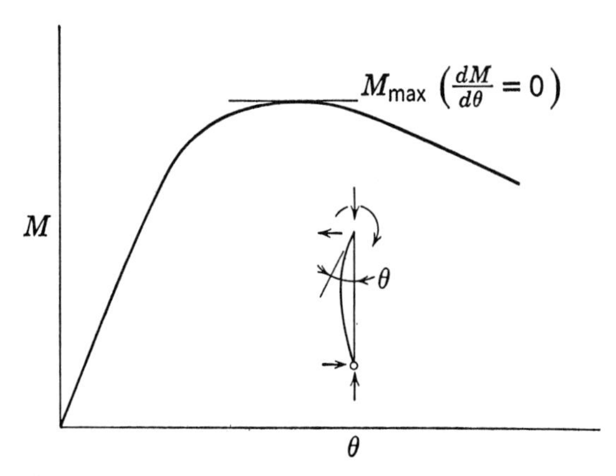

Fig. 4.24. Typical moment-vs.-rotation curve for a column.

In most practical cases, wall panelling and other bracing in the wall systems will supply torsional resistance; where this is true, these equations apply as well as when lateral restraint for this purpose is supplied.*

The results of this analysis for the most severe case ("single curvature") are shown in Fig. 4.22. For an adequately braced member it shows on a nondimensional basis how much end moment M_o may safely be applied at the ends of columns of different slenderness ratios and under different amounts of axial force. For example, if the axial force were 40% of the yield value ($P/P_y = 0.40$) then a column of slenderness ratio of 60 could resist an end moment equal to 48% of its M_p value. Accordingly, a proportionately larger section would be required. A similar chart developed for loading condition (3) is shown in Fig. 4.23. An important feature of Fig. 4.23 is that for relatively low axial loads, the effect of that axial force upon the moment capacity is so small that it may safely be neglected up to a limiting value. Such information will be useful in design.

* It is also assumed that wall panelling acting with floor systems, or other restraints, prevents instability due to side-sway of columns.

In design one could either use the curves of Figs. 4.22 and 4.23, or analytical expressions could be derived to fit the curves. This has been done in Ref. 4.12, the results being tabulated for design use with the limitation that P/P_y be less than 0.6.

Summarizing, the following design guides are suggested:

(1) Columns in industrial frames with low axial force (P/P_y less than 0.15) and with unequal end moments be proportioned without considering axial force.

(2) Columns loaded in combined axial load and bending moment should be such that P/P_y is less than 0.60 and L/r_x is less than 120. This limitation is recommended pending the results of further research on framed columns.

(3) All columns with P/P_y greater than 0.15 and all columns bent in single curvature with equal and opposite end moments would be designed according to Fig. 4.22 or 4.23, depending on the loading condition. Such members must have sufficient wall bracing or other support.

EXAMPLE Portal Frame Column

The following example is given to illustrate the procedure for checking a design of a portal frame to see if the columns will transmit the required moment. It presumes adequate torsional bracing. The structure is a pinned-base rectangular portal frame with a column height of 16 ft and the plastic moment at the top of the column is 2000 kip-in. The vertical reaction at the base of the column is 80 kips.

(1) Select trial member size neglecting effect of axial force

$$Z_t = \frac{M_p}{\sigma_y} = \frac{2000}{33} = 60.7 \text{ in.}^3$$

Use 16WF36
$Z = 63.9$ in.^{3}

(2) Determine axial force on the member from frame reactions. Compute P/P_y ratio.

$$P = 80 \text{ kips}$$

$$P/P_y = \frac{80}{(33)(10.59)} = 0.229$$

(3) For the loading condition on the column, enter the appropriate graph (Fig. 4.22 or 4.23) with the above P/P_y value and given L/r and find allowable M/M_p value.

Loading condition: pinned at base, moment at top (use Fig. 4.23).

$$L/r_x = \frac{(16.0)(12)}{(6.49)} = 29.7$$

$$M/M_p = 0.88$$

(4) Compute required plastic modulus.

$$Z = \left(Z_t \frac{M_p}{M}\right)$$

$$= \frac{(60.7)}{(0.88)} = 69.0 \text{ in.}^3$$

(5) Select a new section for the column.

Use 16WF40 (Z = 72.7 in.^{3})

(6) Check the weak axis according to Eq. 4.35.

$$L/r_y < \sqrt{\frac{8850}{P/P_y}} = \sqrt{\frac{8850}{80/(33)(11.77)}} = 204$$

$$L/r_y = \frac{(16)(12)}{1.50} = 128 < 204 \qquad \underline{\text{OK}}$$

4.7 BRITTLE FRACTURE

Since brittle fracture would prevent the formation of a plastic hinge, it is important to assure that such failure does not occur. This is an equally important aspect of conventional elastic design when applied to fully welded continuous structures, and, as already pointed out in previous chapters, the assumption of ductility is important in elastic design with numerous design assumptions relying upon it.

In past years the failures of ships and pressure vessels have focused attention on the importance of this problem, and although hundreds of articles have been published on the problem of brittle fracture, no single easy rule is available to the designer. Nonetheless, an examination of the conditions that have led to brittle failures in the past should be helpful as a background for formulating good practice.

Brittle fractures are caused by a combination of adverse circumstances that may include several of the following:

(1) Local stress concentrations and residual stresses
(2) Poor welding
(3) Notch sensitive steel

(4) Shock loading
(5) Low temperature
(6) Strain aging
(7) Triaxial tension state of stress.

In plastic design the engineer should be guided by the same principles that govern the proper design of an all-welded structure designed by conventional methods, since the problem of brittle failure is of equal importance to both. First, the proper material must be specified to meet the appropriate service conditions. Secondly, fabrication and workmanship must meet high standards. In this regard, punched holes in tension zones and the use of sheared edges are not permitted. Such severe cold working exhausts the ductility of the material. Thirdly, design details should be such that the material is as free as possible to deform. The geometry should be examined so that triaxial states of tensile stress will be minimized.[4.15]

Reference 4.16 contains an excellent review of the role played by *material*, by *fabrication*, and by *design* upon the brittle fracture problem. In addition, although prepared primarily for plate structures, it suggests practical procedures to be followed whenever the engineer suspects that brittle fracture might be a problem in a particular situation.

The effect of various edge conditions on the brittle failure of steel have been studied in Ref. 4.17. The loss in ductility due to sheared edges was positively demonstrated; however the authors suggest that under certain circumstances this undesirable effect can be overcome by a subsequent flame-softening treatment.

How may one be sure that brittle fracture will not be a problem even if the above suggestions are followed? Although no positive guarantee is possible, experience with tests of rolled members under normal loading conditions (but with many "adverse circumstances" present that might be expected to lead to failures) has not revealed premature brittle fractures of steel beams. Further, the use of fully continuous welded construction in actual practice today has not resulted in failures, even though factors otherwise neglected in design have most certainly caused plastic deformations in many parts of such structures.

Summarizing, the following guides are suggested for assuring structural ductility:

(1) Ordinary structural grade steel for bridges and buildings may be used with modifications, when needed, to insure weldability and ductility at lowest service temperature.

(2) Fabrication processes should be such as to promote ductility. Sheared edges and punched holes in tension flanges are not per-

mitted. Punched and reamed holes for connecting devices would be permitted if the reaming removes the cold-worked material.

(3) In design, triaxial states of tensile stress set up by geometrical restraints should be avoided.

4.8 REPEATED LOADING

Up to this point the tacit assumption has been made that the ultimate load is independent of the sequence in which the various loads are applied to the structure. Thus one would also suppose that a certain degree of fluctuation in the magnitude of the different loads would be tolerable as long as the number of cycles did not approach values normally associated with fatigue, in which case the design would be controlled by an endurance limit and not by maximum plastic strength.

The importance of fatigue or repeated load as a factor in design of beams in the plastic range is apparant to a reader of the summary report on University of Illinois fatigue tests conducted by Professor W. M. Wilson and his associates.[4.18] There is, nevertheless, some ground for encouragement on the part of the designer of any structure if 100,000 cycles of load or less are expected in which the load goes from a minimum stress of zero to a maximum value. In this range, beam sections fabricated with uniform cross sections and with continuous welds had fatigue strengths in which the stresses were above the initial yield point of the material. On the other hand, built-up beams with partial length cover plates (such as are used in elastic design procedures to distribute the material and utilize it at as high a stress as possible) had lower fatigue strengths than the beams with constant cross section. According to this report[4.18] "any plain rolled beam without attachments or flange holes will have a greater fatigue strength than any cover plated beam or any built-up beam of the same or somewhat greater section modulus." Since the application of plastic design is best suited to the use of uniform beam and column sections, the relatively good fatigue strength of such members is of advantage. Thus, the mere existence of repetitive loading does not necessarily exclude the application of plastic analysis to design.

The uncertain fatigue life of "knee" connections such as are used in portal frames presents a limitation to both the "elastic" and "plastic" design, and more fatigue tests of these connections are greatly needed. Simple and economical portal frame connections of the square knee type almost always include a connection of the fillet welded tee type at a region of maximum moment. Fatigue failure in such a connection is likely after a few thousand cycles of load but there is a lack of sufficient information to form definite conclusions, and until further test work is carried out it seems essential that the application of plastic design to such structures be limited to relatively few repetitions of the major loads.

For ordinary building design no further consideration of variation in loads is warranted. However, if the major part of the loading may be completely removed from the structure and reapplied at frequent intervals, it may be shown theoretically that a different mode of "failure" may occur. It is characterized by loss of deflection stability in the sense that under repeated applications of a certain sequence of load, an increment of plastic deformation in the same sense may occur during each

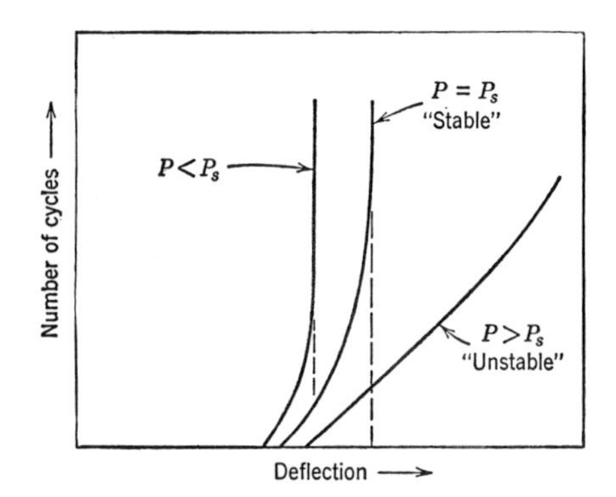

Fig. 4.25. Diagrammatic representation of deflection stability ("shakedown").

cycle of loading. The question is, does the progressive deflection stop after a few cycles (does it "shake down") or does the deflection continue indefinitely? If it continues, the structure is "unstable" from a deflection point of view, even though it sustains each application of load.

Loss of deflection stability by progressive deformation is characterized by the behavior shown by the line $P > P_s$ in Fig. 4.25. If the load is variable and repeated and is greater than the stabilizing load P_s then the deflections tend to increase for each cycle. On the other hand, if the variable load is equal to or less than P_s, then, after a few cycles the deflections will stabilize at a constant maximum value and thereafter the behavior will be elastic. The line marked $P > P_s$ is typical of this case.

In the event that the unusual loading situation is encountered, methods are available for solving for the stabilizing load P_s and the design may be modified * accordingly.[1.9,4.15] As mentioned earlier, how-

* Another repeated loading effect is called "alternating plasticity" or "plastic fatigue." It is characterized by an actual reversal of stress of a magnitude sufficient to cause plastic deformation during each cycle. Unless the design criterion should be controlled by fatigue, the discussion in this section applies equally well to "plastic fatigue" as well as to "deflection stability."

ever, this modification will not be necessary in the large majority of cases. In the first place the ratio of live load to dead load must be very large in order that P_s be significantly less than P_u, and this situation is unusual. Secondly, the load factor of safety is made up of many factors other than possible increase in load (such as variation in material properties and dimensions, errors in fabrication and erection, etc.).

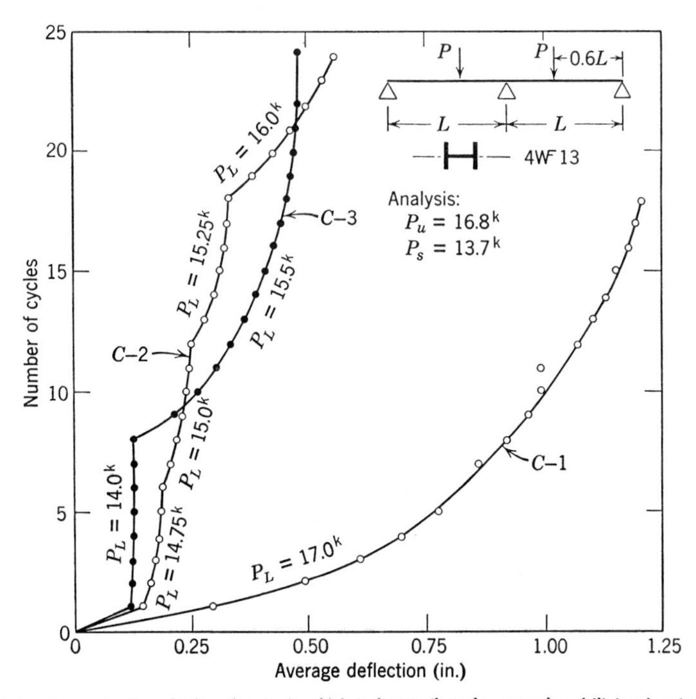

Fig. 4.26. A group of cyclic loading tests which indicate that the actual stabilizing load is higher than theoretically predicted.

Variation in live load, alone, could not be assumed to exhaust the full value of the factor of safety; and thus the live plus dead load would probably never reach P_s. Further, as discussed in Ref. 4.19, failure in this sense is accompanied by a very definite warning that loss of deflection stability is imminent. This implies that a lower load factor would be appropriate as regards P_s than as regards P_u.

Figure 4.26 lends further assurance that the problem is largely academic for ordinary building construction. It represents the action of a two-span continuous beam with an off-center concentrated load in each span.[4.20] Even in this extreme case in which the *full* value of the separate loads was removed and then re-applied, the actual reduction in load capacity was only about 8% instead of the 20% predicted by the theory of deflection stability.

In summary, plastic design is intended for cases normally considered as "static" loading. For such cases the problem of repeated loading may be disregarded. Where the full magnitude of the principal load(s) is expected to vary, the ultimate load may be modified according to analysis of deflection stability.

PROBLEMS

4.1 Derive the axial force-vs.-moment capacity interaction equation for a WF shape for the condition that the neutral axis is in the web.

4.2 Draw the P-vs.-M interaction curve for a 14WF426 member and for a 36WF300 beam.

4.3 Draw the P–M interaction curve for an 8WF31 column bent about the weak axis. Compare with strong axis curve.

4.4 Why is the effect of axial force on moment capacity neglected when P/P_y is less than 0.15?

4.5 For an 8WF67 column draw the theoretical M–P–ϕ relationship at the following P/P_y values: 0, 0.15, 0.25, 0.50, 0.75. Use a common system of non-dimensional coordinates and neglect the curving portion of the M–ϕ diagram.

4.6 Select a member whose cross section will transmit a moment of 3000 kip-in. in the presence of an axial force of 150 kips.

4.7 For a span of 20 ft, how much load would a 16WF40 beam support without failure due to "uniform web shear?"

4.8 A load of 1500 kips is applied 4 ft from the end of a fixed-ended beam, 16 ft long. Select a member. Check its adequacy for shear and modify the design if necessary.

4.9 Comment upon the modification due to shear force of a design of a uniformly loaded, fixed-ended beam.

4.10 For the problem shown in Fig. 3.16 if the distance L is 30 ft, would a purlin spacing of 6.0 ft be adequate?

4.11 Compare Eqs. 4.33 with a "conventional" formula for centrally loaded columns and show how the factor of safety varies with L/r.

4.12 A column, loaded in single curvature with equal end moments (M_p = 1800 kip-in.) must carry an axial force of 70 kips. Select a section neglecting the effect of axial force and then check to see whether or not the design is adequate. Revise if necessary. $L = 16$ ft.

5 Connections

5.1 INTRODUCTION

Connections play a key role in determining whether or not a structure will reach its computed ultimate load, because plastic hinges usually form at the intersection of two or more members. Often the connection must transfer large shear forces; and since the connecting devices (welds, rivets, or bolts) are often located at points of maximum moment, they are subjected to the most severe loading conditions. Design procedures must, therefore, assure the performance that is assumed in design—namely, that the connection will develop and subsequently maintain the required plastic moment.

The ability of fabricators to successfully join members by welding has lent impetus in recent years to the application of plastic design methods; because by welding it is possible to join members with sufficient strength that the full plastic moment may be transmitted from one member to another. However, this is but one of the methods of fabrication for which plastic design is suitable. Plastic design is also applicable to structures with partially welded (top plate) or with riveted or bolted connections whenever demonstrated that they will permit the formation of plastic hinges.

The various types of connections that might be encountered in steel frame building structures are shown in Fig. 5.1 and are designated as follows:

(1) Corner connections (straight, haunched)
(2) Beam-column connections

(3) Beam-to-girder connections
(4) Splices (beam, column, roof)
(5) Column anchorages
(6) Miscellaneous connections (purlins, girts, bracing).

Primary attention is given in this chapter to design considerations appropriate to corner connections and to beam-column connections.

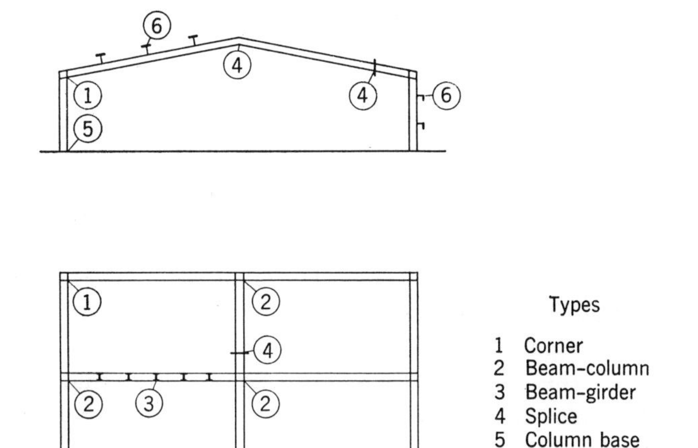

Fig. 5.1. Types of connections in building frames, classified according to their function.

5.2 REQUIREMENTS FOR CONNECTIONS

The design requirements are introduced by considering the general behavior of different types of corner connections as observed under load. A straight connection is shown in Fig. 5.2 in which two WF shapes

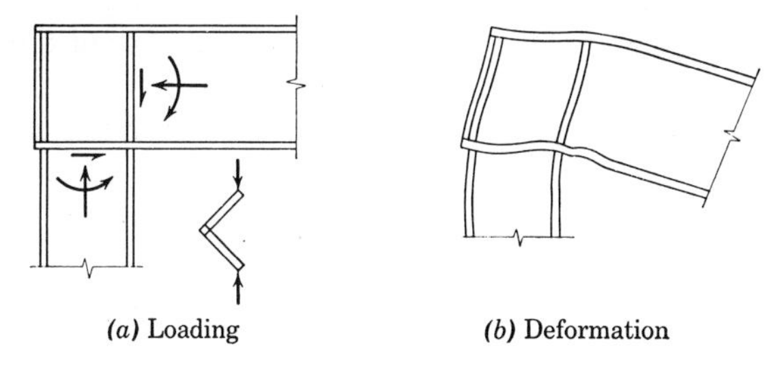

(a) Loading *(b)* Deformation

Fig. 5.2. Typical loading and resulting deformation of a straight corner connection without stiffening of the web.

are joined without additional stiffening. The behavior of such a connection is shown by curve A in Fig. 5.3. Because of insufficient web thickness, yielding due to shear force commences at a relatively low load. The connection rotates beyond the required hinge rotation (as indicated by the dotted vertical line) but the plastic moment assumed

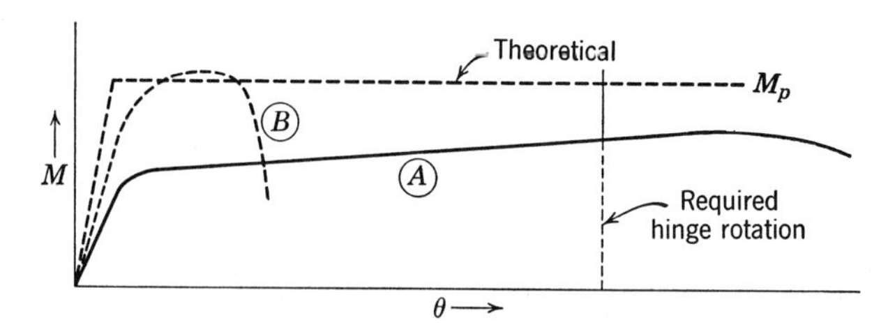

Fig. 5.3. Moment vs. angle change relationship for two "unsatisfactory" corner connections.

in the design is not developed. Also the elastic deformation is considerably greater than the value assumed.

Still another connection might behave as shown by curve B in Fig. 5.3. Although the elastic stiffness and maximum strength are satisfactory, the connection buckles prior to realizing the needed hinge rotation. (See Art. 6.4.)

The inadequacies of these connections make it possible to formulate four principal design requirements—requirements that in principle are common to all connections. These are: (1) strength, (2) stiffness, (3) rotation capacity, and (4) economy. Requirements (1)–(3) are now discussed in the light of the behavior of corner and interior connections. Obviously extra connecting materials must be kept to a minimum, else there may be loss of over-all economy.

Strength

The connection must be designed in such a way that it will develop the plastic moment M_p of the members (or the weaker of the two members).

For straight connections the critical or "hinge" section is assumed at point H (Fig. 5.4a). As will be seen below, for haunched connections, the critical sections are assumed at R_1 and R_2, Fig. 5.4b.

Stiffness

Although it is not essential to the development of adequate strength of the completed structure, it is desirable that the average unit rotation

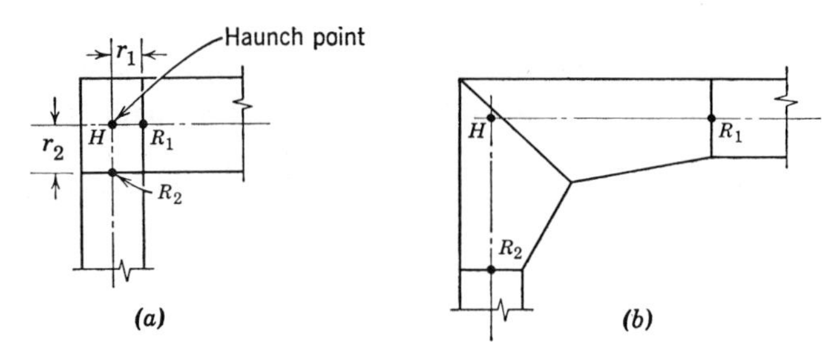

Fig. 5.4. Designation of critical sections in straight and haunched connections.

of the connecting zone not exceed that of an equivalent length of the members being joined. The equivalent length ΔL is the length of the connection or haunch measured along the frame line. Thus, in Fig. 5.4*a*,

$$\Delta L = r_1 + r_2 \tag{5.1}$$

This requirement therefore reduces to the following:

$$\theta_h \leq \frac{M_h}{EI} \Delta L \tag{5.2}$$

which states that the change in angle between sections R_1 and R_2 as computed[4.8] should not be greater than the curvature (rotation per unit of length) times the equivalent length of the knee.

Normally an examination of the design to see if it meets the stiffness requirement will not be necessary. Compared with the total length of the frame line, the length of the connection is small. Therefore if it is slightly more flexible than the beams which it joins, the general over-all effect cannot be very great.

Rotation Capacity

Of greater importance than adequate rigidity in the elastic range is an adequate reserve of ductility after the plastic moment value has been reached. This rotation is necessary to assure that all required plastic hinges will form throughout the structure. Thus all connections must be proportioned to develop adequate rotation at plastic hinges. This subject is discussed in further detail in Art. 6.4.

Ordinarily the loading tends to push the two arms of a straight connection together as shown in Fig. 5.2*a*. Somewhat different behavior may be observed if the connection is loaded so that the adjoining arms open outward, placing the reentrant corner in tension.[5.1,5.2] The severe high stress concentration and the general restraint are conducive to fractures, and such failure would limit connection strength or rotation capacity. However, two facts make this possible limitation of not much concern. In the first place, in ordinary building frames it is highly unlikely that the loading will be such as to place the corners in "tension." And if such loading does occur, the "tension" hinges are usually the last to form. Secondly, the only premature fractures that have been observed in tests have been associated with welds of poor quality. No premature failures have occurred under tension loading where the welding was sound.

5.3 STRAIGHT CORNER CONNECTIONS

Straight corner connections will now be considered in the light of the requirements outlined above. The plain connection without extra stiffening devices will be analyzed first in order to determine when it can be used without supplementary strengthening. Then an analysis follows which leads to suggestions for stiffening an otherwise inadequate design. Brief mention is made of connection stiffness.

Corner Connections without Stiffening

The connection and loading are shown in Fig. 5.5. The design objective is to avoid the behavior shown by curve A in Fig. 5.3, which

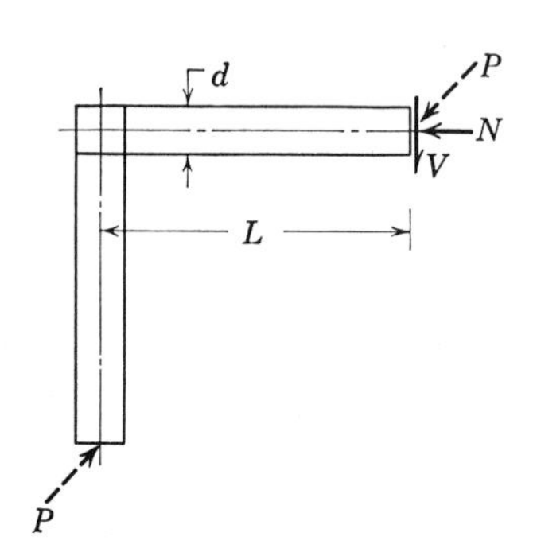

Fig. 5.5. Idealized loading on straight corner connection.

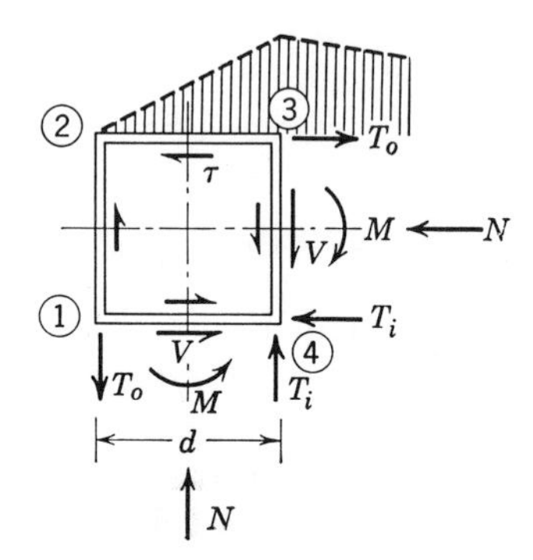

Fig. 5.6. Forces and stresses assumed to act on unstiffened straight corner connection.

reflects inadequate strength to resist shear force. The following assumptions are made:

(*a*) The shear stress at yield is given by Mises' yield criterion:

$$\tau_y = \sigma_y/\sqrt{3} \tag{5.3}$$

(*b*) The shear stress is uniformly distributed in the web of the knee.
(*c*) The web carries shear stress, the flanges carry flexural stresses.

With these assumptions, it will be possible to obtain a value of the haunch moment at "shear yield" $M_{h(\tau)}$, which can then be equated to M_p.

The resulting forces and stresses are shown in Fig. 5.6. The haunch moment at shear yield, $M_{h(\tau)}$, is derived as follows:

$$\tau = T_o/wd$$

$$T_o = \frac{M}{d} - \frac{N}{2}$$

Both the moment M and the direct force N may be expressed in terms of V which in turn may be expressed in terms of the moment at the haunch M_h. Thus,

$$T_o = \frac{M_h}{d}\left(1 - \frac{d}{L}\right) \tag{5.4}$$

and therefore

$$\tau = \frac{M_h}{wd^2}\left(1 - \frac{d}{L}\right)$$

Using assumption (*a*),

$$M_{h(\tau)} = \frac{wd^2\sigma_y}{\sqrt{3}\left(1 - \frac{d}{L}\right)} \tag{5.5}$$

Equation 5.5 gives the moment that may be carried by an unstiffened corner connection of the type shown in Fig. 5.5.

Now, the desired flexural strength is $M_p = \sigma_y Z$. Equating this to Eq. 5.5 the required web thickness becomes

$$w = \frac{\sqrt{3}\,fS}{d^2}\left(1 - \frac{d}{L}\right) \tag{5.6}$$

Since the product $f\left(1 - \frac{d}{L}\right)$ is approximately 1.0, the required web thickness is given by

$$w \geq \sqrt{3}\,\frac{S}{d^2} \tag{5.7}$$

In the event that two dissimilar sections are joined, then Ref. 4.8 suggests that the value of S for the weaker of the two members be used and the product of the section depths employed instead of the term d^2.

If substitutions are made in Eq. 5.7 using Eqs. 2.15 and 2.11, one obtains the equation

$$w \geq \frac{\sqrt{3}\, M_p}{f \sigma_y d^2}$$

If M_p is expressed in kip-ft and the dimensions in inches, then this equation is approximately equivalent to Eq. 5.49 when the term ΔM in the latter is replaced by M_p.

Examination of the commercially available rolled WF shapes using Eq. 5.7 shows that many of them require stiffening to realize the design objective for straight connections. Several alternates are possible when such stiffening is required. Web reinforcement plates ("doublers") might be used, and these could be proportioned according to the equation

$$w_d = \frac{\sqrt{3}\, S}{d^2} - w \tag{5.8}$$

In many cases a diagonal stiffener will be more suitable, and suggestions for proportioning them are contained in the following section.

Bounding the knee proper, the flanges of one of the rolled sections may be allowed to continue on through the joint. Insert plates of area

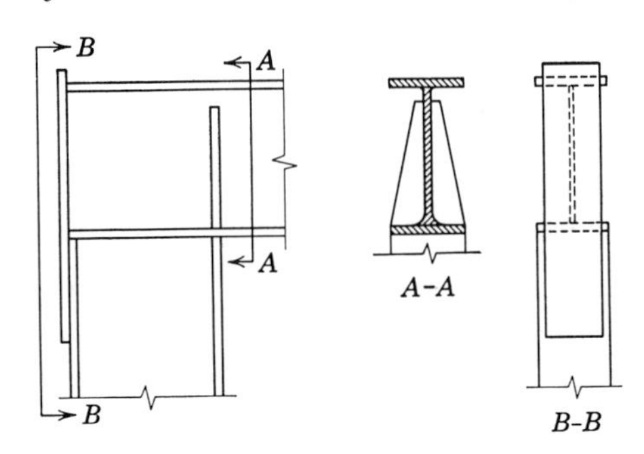

Fig. 5.7. Typical details for straight corner connection using fillet welding.

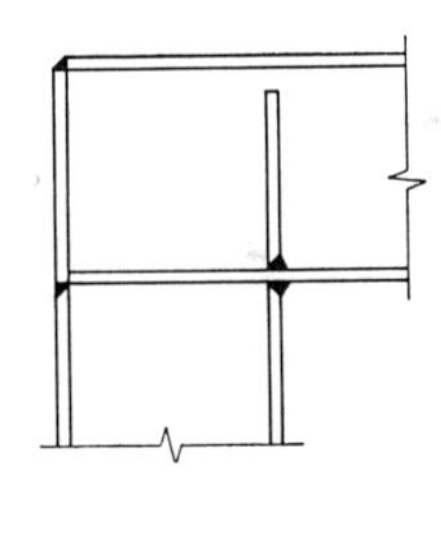

Fig. 5.8. Typical details for straight corner connection using fillet and butt welding.

equal to the flange area may then be added to act as extensions to flanges of the other member. Figure 5.7 shows one design that is suitable for the lighter shapes where fillet welding is practicable. The sniped stiffener at section A–A eliminates the need for careful fitting of plates—

a detail that might be desirable for the sake of economy. Also, accurate fitting of the end plate is not needed in the sense that it must be cut to fit the precise dimensions of a given rolled beam. The sniped bracket need not extend the full depth of the member; all that is needed is a sufficient length to transfer the column flange thrust into the beam web. In fact, many connections have shown satisfactory characteristics under test and have had only "half-depth" stiffeners. The length of the end plate that laps over the rolled beam would be governed by the amount of weld required to develop its strength.

Figure 5.8 shows a typical layout for a connection joining members whose flanges are so thick that it would be uneconomical to attempt to develop their strength with fillet welds.

Strength of Stiffened Connections

Assuming, now, that the knee web is deficient as regards its ability to resist the shear force, a diagonal stiffener may be used. As suggested in Ref. 5.3, a plastic approach may be used to analyze such a connection (Fig. 5.9). The force T_o is made up of two parts, a force carried by the web in shear and a force transmitted at the end by the diagonal stiffener, or

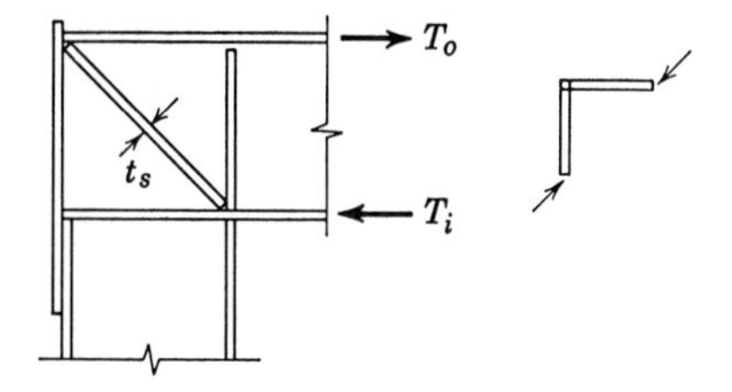

Fig. 5.9. Corner connection with diagonal stiffener.

$$T_o = T_{\text{web}} + T_{\text{stiffener}}$$

The maximum force that may be carried by the web equals the area of the web times the average yield stress in shear, or

$$T_w = \tau_y wd = \frac{\sigma_y}{\sqrt{3}} wd$$

The maximum component of the force in the diagonal in the direction of T is T_s, and

$$T_s = \frac{A_s\sigma_y}{\sqrt{2}} = \frac{b_s t_s \sigma_y}{\sqrt{2}}$$

where A_s, b_s, and t_s represent the area, width, and thickness of stiffener, respectively. Thus, when both the web and the diagonal stiffener have reached the yield condition,

$$T_o = \frac{\sigma_y wd}{\sqrt{3}} + \frac{\sigma_y b t_s}{\sqrt{2}} \tag{5.9}$$

From Eqs. 5.4 and 5.9 therefore, the available moment capacity of the type of connection shown in Fig. 5.9 is given by

$$M_h = \frac{\sigma_y d}{\left(1 - \frac{d}{L}\right)} \left(\frac{wd}{\sqrt{3}} + \frac{bt_s}{\sqrt{2}}\right) \tag{5.10}$$

In design it is desirable to determine the required thickness of stiffener t_s such that $M_h = M_p$. Using $M_p = \sigma_y Z$ in Eq. 5.10 there is obtained,

$$\sigma_y \frac{wd}{\sqrt{3}} + \frac{\sigma_y bt_s}{\sqrt{2}} = \frac{\sigma_y fS}{d}\left(1 - \frac{d}{L}\right) \tag{5.11}$$

Assuming that $f\left(1 - \frac{d}{L}\right)$ equals 1.0, Eq. 5.11 becomes

$$\frac{t_s b}{\sqrt{2}} = \frac{S}{d} - \frac{wd}{\sqrt{3}}$$

Thus a guide to the required thickness of diagonal stiffener in a corner connection that would otherwise be deficient in shear resistance is

$$t_s = \frac{\sqrt{2}}{b}\left(\frac{S}{d} - \frac{wd}{\sqrt{3}}\right) \tag{5.12}$$

Examples to illustrate the use of this equation will be found in Chapter 9. The use of a diagonal stiffener with a thickness equal to that of the rolled section flange normally will be adequate, although the thickness will be greater than required.

The remaining portions of the connection would be proportioned to have adequate plastic strength as described for the unstiffened knee above. Vertical stiffeners are still required but their role is modified somewhat. Even though a portion of the flange force is transmitted to the diagonal stiffener, the remaining portion must be transmitted to the web. Further, such a stiffener is needed to prevent crippling of the web.

A vertical stiffener formed of material whose thickness is equal to that of the beam flange will have adequate strength. If it is desirable to use the least possible thickness, then a relationship may be obtained by equating the force carried by the column flange to the total resisting forces supplied by the diagonal stiffener, the vertical stiffener, and the beam web.

Further details for designing connections with diagonal stiffeners will be found in Ref. 5.4.

In arriving at the above expressions, it was assumed that although the stiffener had yielded, it would remain straight throughout the entire range of connection deformation. Consistent with the procedures suggested in Art. 4.4, therefore, the width of the outstanding flange of stiffener plates loaded in compression should not be greater than about $8\frac{1}{2}$ times its thickness in order that inelastic local buckling not limit its effectiveness. Frequently this requirement would necessitate the use of a stiffener with thickness greater than computed according to Eq. 5.12 if the width of stiffener is made equal to the flange width of the adjoining member. There is no restriction against using a narrower stiffener plate as long as its area is equal to that implied by Eq. 5.12. For illustration see the design of connection 10 in Plate XI, p. 323.

Stiffness Considerations

Equation 5.2 formulates a desirable characteristic with regard to connection stiffness. If for some reason it were necessary to make the connection at least as stiff as the members joined by it, then the following equation gives the required web thickness to assure the needed rigidity [4.8]

$$w \geq \frac{2.6S}{d^2} \tag{5.13}$$

Should it be desired to predict the actual rotation of a connection, Ref. 5.16 may be consulted.

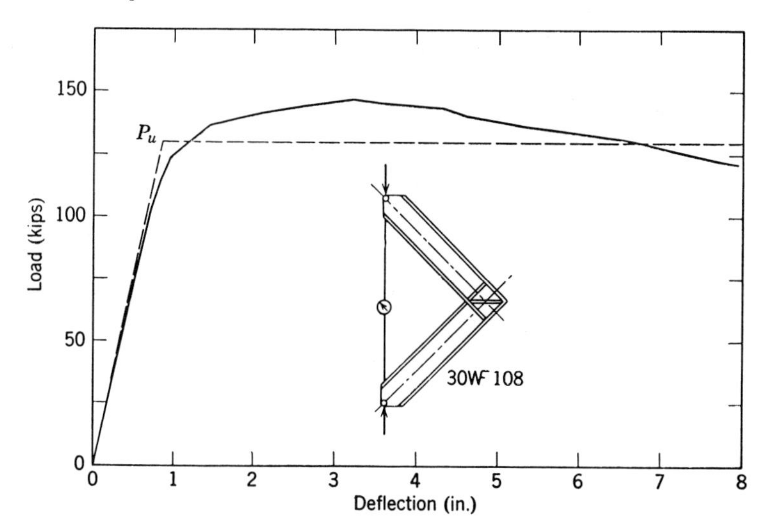

Fig. 5.10. Behavior of a 30-in. welded corner connection showing the development of a plastic hinge.[1.16]

Experiments

Figure 5.10 shows the efficiency with which a properly proportioned straight connection will perform. The load corresponding to the plastic moment is realized and there is an adequate amount of rotation at this load to allow the necessary redistribution of moment to other portions of a structure of which the connection might be a part. References 4.8, 5.1, 5.2, and 5.4 may be consulted for further experimental verification of the procedures outlined in this article.

5.4 HAUNCHED CONNECTIONS

In order to improve the esthetic appearance of a structure it is often the practice to put haunches at the connections. Frequently they also have an economic value in conventional elastic design. By placing them at regions of maximum moment at the eaves, the size of rolled shapes required to support the loads is reduced.

Actually, haunched connections are a product of the elastic design concept whereby material is placed in conformity with the moment

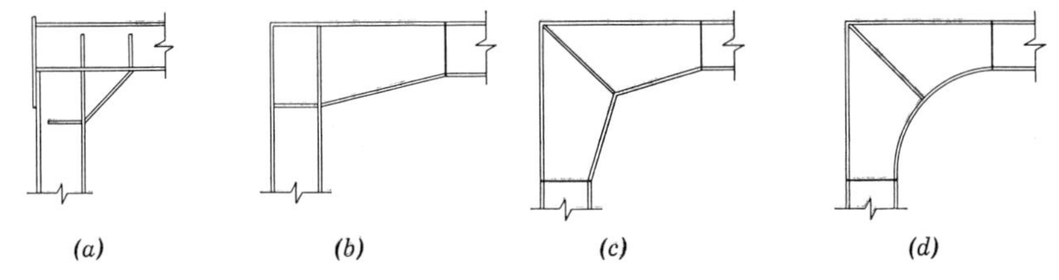

Fig. 5.11. Typical haunched corner connections.

diagram to achieve greatest possible economy. In plastic design, on the other hand, it is frequently possible to achieve the same saving in weight without necessity for the use of a haunch, and a further saving in cost because an expensive fabrication detail has been eliminated.

If a haunch is to be specified for architectural considerations in a structure to be proportioned by plastic methods, then the designer might just as well realize the additional savings in material by considering its action in the analysis of the structure. It is the purpose of this article to consider procedures for proportioning such connections so that the structure will behave as assumed in the design.

Four types of haunched connections are shown in Fig. 5.11. Analysis and test have shown all of them to be suitable in design, although more frequent demand may be found for (*c*) and (*d*). In a sense they are all special cases, since the rolled shapes are shown joining at right angles. Industrial frames will in many cases have gabled or curved roofs (see Fig. 5.1); but connections which meet the requirements for "right angle" orientation will be suitable for the gabled case as well.

The design requirements generally will be quite similar to those stated in Art. 5.2, except for one additional feature; haunched knees may exhibit poor rotation capacity[4.8]. Due either to insufficient bracing or to inadequate proportions of the haunch itself, the knee may buckle laterally before the desired design conditions have been reached. The tendency for this mode of failure is greater than in the straight connections because in haunched knees the stress distribution is more nearly uniform along the compression flange, it cannot be supported laterally along the full length, and therefore a large amount of energy can be released by buckling.

Therefore, the design of a haunched connection for use in a plastically designed structure must embody both strength and stability considerations. The haunch is proportioned with sufficient strength so that a plastic hinge is formed at the end of the haunch (at the point where it joins the rolled member). This facilitates the analysis of the frame and assures adequate rotation capacity. Secondly, the haunch is proportioned in a manner that will provide adequate resistance to lateral buckling, supplemented by bracing to prevent such deformation.

Tests of haunched connections have been conducted and reported upon in Refs. 4.8, 5.1, 5.5, 5.6, 5.7, and 5.8. A number of studies have been made resulting in design procedures that have proved satisfactory in the past for conventional elastic design.[5.9, 5.10, 5.11, 5.12] Reference 5.13, on the other hand, explored the plastic behavior of the haunch itself. The strength of tapered and curved connections was analyzed by the plastic methods and the stability of the haunch was checked according to a consideration of its behavior in the inelastic range. The procedures that follow are based in part on these later studies.

Tapered Haunches

Referring to Fig. 5.12, the following assumptions are made:

(1) The moment diagram is linear from the point of inflection in the beam (point O) to the haunch point, H.

(2) A plastic hinge forms at the end of the rolled shape (section R).

(3) The length O-R is taken as approximately $3d$ and represents about as severe a condition as might be encountered in practice.

(4) Lateral support will be provided at the extremities and at the common intersection points of the haunch.

(5) The width of the haunch flange is assumed equal to that of the adjoining rolled shape.

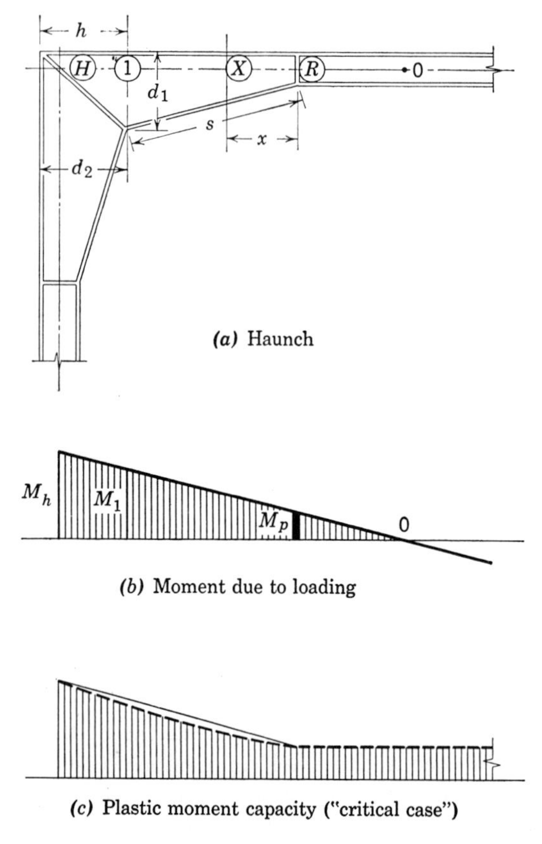

(a) Haunch

(b) Moment due to loading

(c) Plastic moment capacity ("critical case")

Fig. 5.12. A tapered haunched connection loaded in the "critical" condition.

Strength of the haunch. Consider first the strength of the haunch at section R where the rolled section and the haunch are joined as shown in Fig. 5.13*a*. In sketch *b* the two parts at section R are separated and the stress distribution corresponding to the plastic moment applied by the beam is shown to the right. If the thickness of the haunch tension flange is equal to that of the rolled section ($t_t = t$) and if the web thickness and flange width are the same, the yield stress above the neutral axis will simply persist across section R, as shown in the stress distribution to the left in sketch *b*. This will also be true for the web below the neutral axis. Since the maximum stress in the haunch compression flange is σ_y acting along the flange direction, the horizontal component of the corresponding force available to resist the compression in the lower flange of the beam is

$$T_c = \sigma_y A_c \cos \beta \tag{5.14}$$

where A_c is the area of the haunch flange and β is the angle between the two flanges as shown in Fig. 5.13*b*. Of course, this force is insufficient

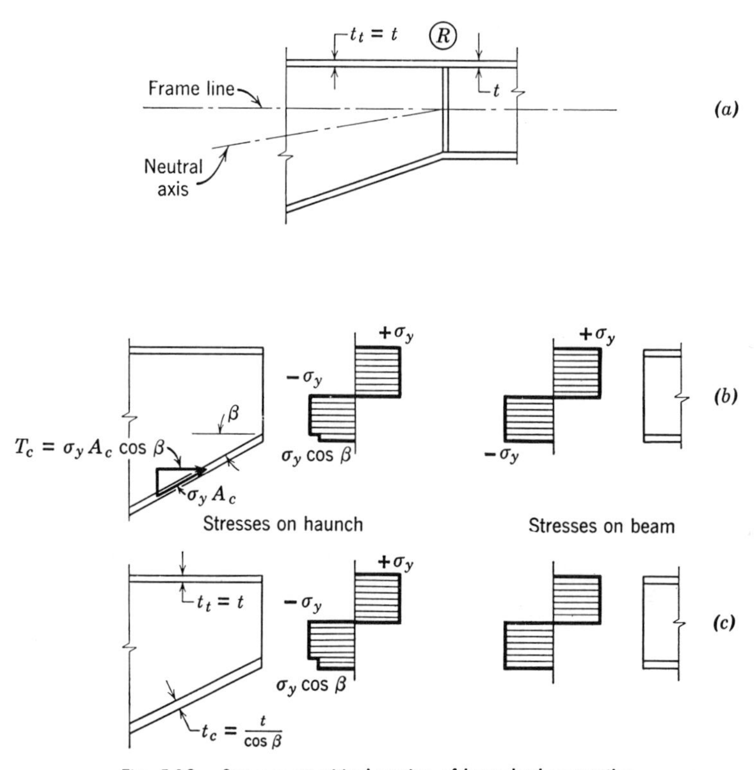

Fig. 5.13. Stresses at critical section of haunched connection.

to maintain equilibrium; the resulting stress distribution is shown to the left in sketch *b*. If, now, the thickness of the lower flange is increased in the ratio $t/\cos \beta$, the horizontal force T_c will be precisely equal to the yield force delivered by the lower flange of the rolled shape. The neutral axis of the haunch will remain essentially at the mid-depth between flanges. Thus the moment capacity will equal that of the rolled beam if the lower flange is proportioned according to

$$t_c = \frac{t}{\cos \beta} \tag{5.15}$$

Fig. 5.13*c* shows the resulting stress-distribution.

The next section that might be investigated in the haunch of Fig. 5.12 is section 1 at the common intersection point of the two trapezoids that make up the connection. The plastic modulus Z_1 of a vertical section normal to the outside flange at section 1 is given by Eq. 2.24 if the value of the flange thickness is taken as that of the outer (tension) flange and if the inner compression flange is increased in thickness by Eq. 5.15. Thus

$$Z_1 = bt(d_1 - t) + \frac{w}{4}(d_1 - 2t)^2 \tag{5.16}$$

If the moment due to the external loading on this section is given by M_1 (Fig. 5.12*b*), then the necessary condition to meet design assumption (2) is that

$$M_1 \leq \sigma_y Z_1 = \sigma_y \left\{ bt(d_1 - t) + \frac{w}{4}(d_1 - 2t)^2 \right\} \tag{5.17}$$

This condition may be met either by varying the depth d_1 while holding the other dimensions constant or by varying the flange thickness t while holding the depth constant. Two different studies have shown [5.5, 5.13] that if the angle β is greater than about 12° for the assumed loading condition, then the critical section will always be at section R. If β is less than 12° then the haunch proportions must be modified so that section 1 will have a strength that corresponds at least to that at section R.

For a section X which is located intermediate between sections R and 1, the plastic modulus is obtained from Eq. 2.24:

$$Z_x = bt(d_x - t) + \frac{w}{4}(d_x - 2t)^2 \tag{5.18}$$

where $d_x = d + x \tan \beta$. If a haunch were proportioned such that it were just adequate to meet the moment requirements at sections R and 1, then the moment capacity would be somewhat less than required at a section between the two ends since Z_x is a function of the square of the depth. This is shown diagrammatically in sketch (*c*) of Fig. 5.12. For the usual proportion of tapered haunches, the "error" amounts to not more than about 3% and would be encountered only under the "critical" condition for which $\beta \cong 12°$. This small variation is neglected.

The effect of variation in β as it influences the location of the critical section is shown in Fig. 5.14. This figure shows the moment capacity that actually would be supplied for three designs that would all meet the requirement that at no point should the haunch strength be less than that necessary to develop M_p of the rolled beam. For β considerably greater than $\beta_{cr} = 12°$ (about 24°) the depth at section 1 is considerably greater than required and the critical section is at R. If $\beta = \beta_{cr}$ the compression flange is under a state of nearly uniform stress. Finally, for the very flat haunch ($\beta < \beta_{cr}$) the critical zone is at section 1 and an appropriate increase in flange thickness would be necessary according to Eq. 5.17.

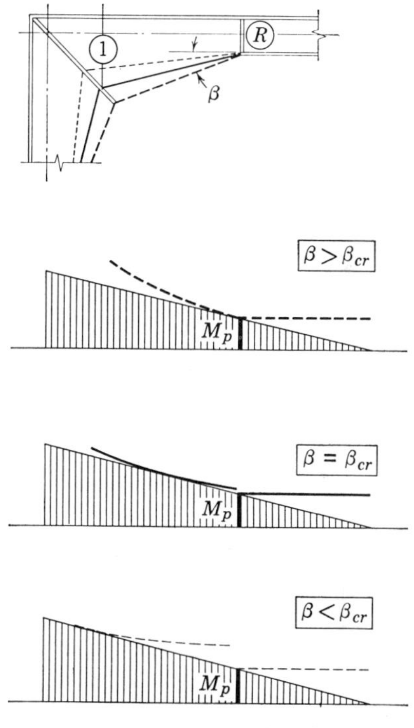

Fig. 5.14. Variation of haunch moment capacity as a function of angle between flanges (diagrammatic).

The influence of shear force and of axial force on the tapered portion of the haunch will be less severe in the haunch than it would be in the rolled section. If the latter was adequate, then the haunch will be satisfactory as well.

Within the corner portion itself (that is, within dimensions d_1 and d_2 of Fig. 5.12) the strength of the member in shear should be examined. This will be treated below under "details."

Lateral stability. To assure that the connection does not buckle laterally, bracing is first of all supplied at the extremities and the common intersection points of the compression flange.

The next concern is with the haunch compression flange between the bracing points. A conservative assumption for checking this flange is to assume that it acts as a column, simply supported at the ends, the web providing restraint only in a direction normal to the flange. Consistent with the plastic stress distribution shown in Fig. 5.13, it is next assumed that the strain in the flange attains the strain-hardening point, ϵ_{st}. The buckling strength of the compression flange acting as a column [4.3] is then given by

$$\sigma_{cr} = \sigma_y = \frac{\pi^2 E_{st}}{(s/r)^2} \tag{5.19}$$

where s is the length of the compression flange of the haunch (Fig. 5.12). The critical slenderness ratio is thus

$$\frac{s}{r} = \pi \sqrt{\frac{E_{st}}{\sigma_y}} \tag{5.20}$$

For $\sigma_y = 33{,}000$ psi and $E_{st} = 700{,}000$ psi

$$\frac{s}{r} = 14.1 \tag{5.21}$$

For a rectangle $r = b/\sqrt{12}$. Therefore, Eq. 5.21 becomes, as an approximation,

$$s_{cr} = 4.1b \tag{5.22}$$

This means that the length of the compression flange from sections R to 1 may not be greater than 4 times the flange width b. If this value is exceeded, then intermediate bracing must be supplied or the value of r must be increased.

Equation 5.22 may be restrictive in many large haunches and it is therefore of value to see if some means may be found to eliminate the supplementary bracing that otherwise would be required. Now, it will be remembered from the discussion regarding local and lateral buckling (Arts. 4.4 and 4.5) that the critical length is a function of the maximum strain that the member is required to absorb. Equation 5.21 assumes that the strain must reach strain-hardening. If the strain was only required to reach the elastic limit, then the critical s/r value would be about 100 instead of 14 if there were no residual stresses or accidental eccentricities. Quite evidently, the critical length in the plastic range will be a function of the strain which the member must absorb.

One method of limiting the strain in the haunch is to increase its strength so that the large plastic deformations will occur instead in the rolled section. An expression relating the required increase in flange thickness to the s/b ratio may be derived assuming that the compression flange is a column, restrained to bend about its strong axis. Making use of the results of Ref. 4.10 for I-shaped welded assemblies, the haunch may be made sufficiently strong so that the average stress will not exceed that which would be permitted on a column of corresponding slenderness ratio. For s/b not greater than 17, the resulting increase in flange thickness Δt is given by

$$\Delta t = 0.1 \left(\frac{s}{b} - 4\right) t, \qquad \left(4 < \frac{s}{b} < 17\right) \tag{5.23}$$

Increasing the haunch flange thickness by the amount required by Eq. 5.23 assures lateral stability to the compression flange when bracing is supplied at the ends and at the common intersection point.

If the haunch depth can be increased substantially over that which is required, then indirectly, a similar result is achieved. Frequently, this will be a convenient alternative, particularly when the haunch flanges would otherwise become unusually thick. According to Ref. 5.13 when $\beta > 24°$ then the increase called for by Eq. 5.23 would not be required.

Details. The following summarizes the procedure to be used for certain details in the haunch.

b/t of flanges. This ratio is automatically limited to the same values that apply to rolled beams because the haunch flange could never be thinner than the beam flange and still meet the requirements posed above. Therefore,

$$\frac{b}{t_c} \leq 17 \tag{4.21}$$

Diagonal stiffener thickness. Employing an approach similar to that used for the straight connections, assuming that the tension flange is at the yield stress throughout at section 1, and making use of Eq. 5.9, we obtain the following relationship:

$$\sigma_y A_t = \frac{\sigma_y w d_1}{\sqrt{3}} + \frac{\sigma_y b t_s}{\sqrt{2}} \tag{5.24}$$

where A_t = area of tension flange = bt_t, d_1 = depth at section 1 (assumed equal to d_2, see Fig. 5.12), and t_s = thickness of diagonal stiffener. Equation 5.24 becomes

$$t_s = \sqrt{2}\, t_t - \frac{0.82 w d_1}{b} \tag{5.25}$$

Of course if the answer is negative, then no shear stiffening will be required. This equation may also be applied to haunches used in gable frames where the angle at the eaves is not a right angle.

The diagonal stiffener must also resist the component of thrust at the inner corner delivered by the compression flanges. This total force is given by:

$$T_s = 2T \sin (45° - \beta) \tag{5.26}$$

The quantity T_s is equal to $b_s t_s \sigma_y$ and the force T may be determined by using for T_c the quantities given by Eq. 5.15. Therefore,

$$bt_s\sigma_y = 2bt_c\sigma_y \sin(45° - \beta)$$

$$t_s = \frac{2t}{\cos\beta}\frac{1}{\sqrt{2}}(\cos\beta - \sin\beta)$$

$$t_s = \sqrt{2}\,(1 - \tan\beta)t \tag{5.27}$$

In no case should the thickness be less than $b/17$.

Transverse stiffeners. At the ends of the haunch (section R), transverse stiffeners are required for transmitting the compressive thrust. Referring to Fig. 5.13 the thrust against the stiffener plate is given by the quantity $\sigma_y A_c \sin\beta$, where β is the angle made by the haunch flange with the flange of the prismatic member. Thus the stiffener plate area may be determined from

$$\sigma_y A_s = \sigma_y A_c \sin\beta \tag{5.28}$$

and for equal width of plating,

$$t_{tr} = t_c \sin\beta \tag{5.29}$$

Usually the required thickness according to Eq. 5.29 will be less than the minimum value permitted according to $b/t < 17$.

Design procedures. Prior to summarizing the design procedure for a tapered haunch, it will be of value to review the variations in geometry that should be made to achieve a satisfactory connection. First, the compression flange is increased in the ratio $1/\cos\beta$ according to Eq. 5.15. If the angle β is at the critical value and $s/b > 4.0$, then in order to control the plastic strains in the haunch to prevent lateral buckling a further increase in flange thickness is suggested in Eq. 5.23. The net effect on a tapered haunch under "critical" loading is reflected in the following flange thickness requirements:

$$\left.\begin{aligned} &\textbf{Tension flange:}\; t_t = \left[1 + 0.1\left(\frac{s}{b} - 4\right)\right]t \qquad (4 < s/b < 17) \\ &\textbf{Compression flange:}\; t_c = \frac{t_t}{\cos\beta} \qquad (4 < s/b < 17) \end{aligned}\right\} \tag{5.30}$$

If the angle β were increased substantially beyond 12° ($\beta > 24°$) then all of the plastic action will occur near the end of the haunch and the increase in thickness to control the plastic strain is not required. The only increase needed is that given by Eq. 5.15. If architectural considerations will permit, this alternate approach should be investigated.

The design procedure for a tapered haunch is therefore as follows:

(1) Select the general proportions of the haunch—its length along the girder and column and the angle between the two flanges. If these are not specified from architectural considerations, then some adjustment is possible for greatest economy.

(2) Check the plastic modulus furnished at the common intersection point (section 1 of Fig. 5.12) making use of Eq. 5.17. If it is less than the value required from the moment diagram (M_1 in Fig. 5.12*b*) then a change in proportions must be made to increase it to this value, assuring that the hinge will be "forced" at the end of the haunch.

(3) If the moment at section 1 is close to the required value, and an increase in depth of haunch is not desirable, then increase the flange thicknesses according to Eq. 5.30. If $s/b > 4.0$, one alternative to this procedure is to proportion the flanges according to Eq. 5.15 and to supply intermediate bracing. Another alternative is to increase the depth at section 1 so that the plastic modulus is increased by a substantial amount. Then the tension flange may be proportioned equal to the rolled section flange and the compression flange according to Eq. 5.15.

(4) Lateral support is provided at the ends and the common intersection point.

(5) The web of the haunch is made equal to that of the rolled shape.

(6) Next, a diagonal stiffener in the corner is proportioned according to Eq. 5.25 and/or Eq. 5.27, but in no case should the stiffener projection from the web be greater than 8.5 times its thickness.

(7) Finally, transverse stiffeners are proportioned according to Eq. 5.29 subject to the same limitation in step 6 above.

EXAMPLE Tapered Haunch for Gabled Frame (Fig. 5.15)

The haunched connection shown in this figure will be designed for a gabled frame, the moment diagram being as indicated. The haunch will be proportioned with sufficient strength to force the formation of plastic hinges at sections C and G.

Column section:

$$Z = (0.364)(1650 \text{ kip-ft}) = 600 \text{ in.}^3$$

Use 36WF160
Z = 623.3 in.^{3}

$$\begin{cases} d = 36.00 \text{ in.} \\ b = 12.00 \text{ in.} \\ t = 1.02 \text{ in.} \\ w = 0.653 \text{ in.} \end{cases}$$

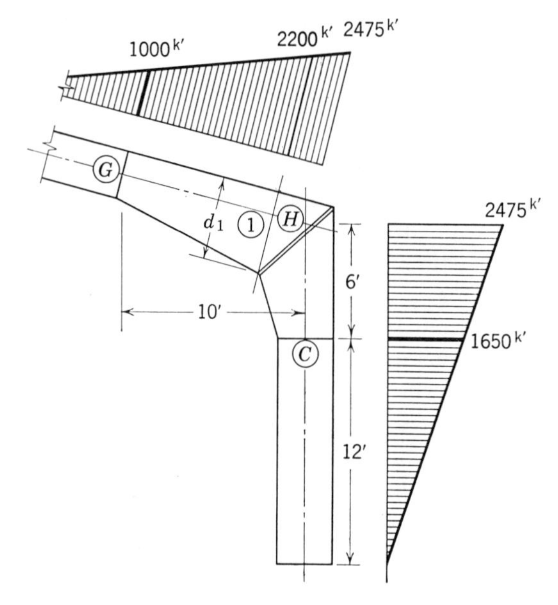

Fig. 5.15. Design example: tapered haunch.

Beam section:

$$Z = (0.364)(1000) = 364 \text{ in.}^3$$

Use 30WF116 $\left\{ \begin{array}{l} d = 30.00 \text{ in.} \\ b = 10.50 \text{ in.} \\ t = 0.85 \text{ in.} \\ w = 0.564 \text{ in.} \end{array} \right.$

$\mathbf{Z = 377.6 \text{ in.}^3}$

General haunch proportions:

The horizontal length is made equal to 10 ft, the vertical height equal to 6 ft. It is assumed that the depth d_1 was specified to be 50 in.

Strength at section 1:

$$M_1/\sigma_y = (0.364)(2200) = 800 \text{ in.}^3$$

$$\begin{aligned} Z_1 &= bt(d_1 - t) + \frac{w}{4}(d_1 - 2t)^2 \\ &= (10.5)(0.85)(49.15) + \frac{0.564}{4}(48.30)^2 \\ &= 770 \text{ in.}^3 < 800 \text{ in.}^3 \end{aligned}$$

Since the plastic modulus furnished is 3.75% less than required, an increase of 5% in flange thickness will be specified.*

* For the usual proportions of haunches, an increase of flange area is about 75% effective in increasing Z.

Flange thickness (girder portion):

From Eq. 5.30

$$t_{t1} = \left[1 + 0.1\left(\frac{105}{10.5} - 4\right)\right] t$$

$$= 1.60t$$

Total thickness is given by

$$t_t = (1.60)(1.05)t = (1.68)(0.85) = 1.43 \text{ in.}$$

$$t_c = \frac{1.43}{\cos 11°} = \frac{1.43}{0.98} = 1.46 \text{ in.}$$

Use $1\frac{1}{2}$-in. plates throughout

Flange thickness (column portion):

Since β is large the critical section is at C. Therefore

$$t_t = t = 1.02'' \qquad \textbf{Use 1-in. plate}$$

$$t_c = \frac{t}{\cos \beta} = \frac{1.02}{\cos 20°} = \frac{1.02}{0.94} = 1.09$$

Use $1\frac{1}{8}$-in. plate

NOTE: For ease of fabrication, **use $1\frac{1}{2}$-in. plate** throughout the haunch.

The flange width is selected as 12 in. from C to H and is tapered to 10.5 in. at section G.

Web thickness:

Use same thickness as WF column ($w = 0.653$). **Use $\frac{11}{16}$-in. plate.**

Diagonal stiffener:

Examine haunch for shear. From Eq. 5.25

$$t_s = \sqrt{2}\,(1.0) - \frac{(0.82)(0.688)(50)}{12}$$

$$= 1.414 - 2.35$$

Thus no stiffener is required for shear resistance.

From Eq. 5.27:

$$t_s = \sqrt{2}\,(1 - \tan 11°)(0.85)$$

$$t_s = 0.96 \text{ in.}$$

Use 1-in. plate stiffener

$$b/t = 10.5 < 17$$

OK

Transverse stiffeners:

A nominal thickness will be adequate.

$$t_{tr} = \frac{b}{17} = \frac{12.0}{17} = 0.71 \text{ in.}$$

Use $\frac{3}{4}$-in. plate

Curved Knees

The analysis of haunches with curved inner flanges follows a procedure that is quite similar to that used for tapered haunches. Consequently, only the resulting guides for proportioning the haunch will be presented, reference being made to earlier cited reports for further information.[5.13] A design procedure is outlined in the steps that follow.

The design of the curved knee starts with a selection of its over-all length CH and HG as shown in Fig. 5.16. These distances fix the

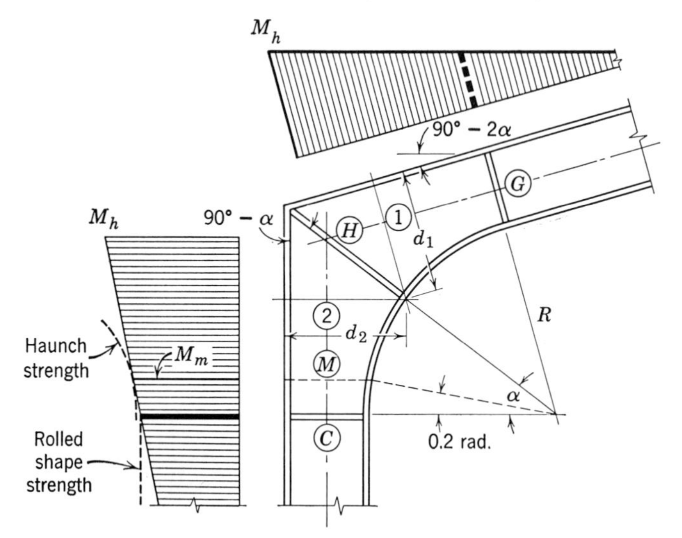

Fig. 5.16. Curved knee for gabled frame.

position of possible plastic hinges and influence the size of the resulting rolled shapes that are selected.* It is assumed in this discussion that the haunch is symmetrical and that the largest plastic moment is required in the column (section C).

The web thickness and flange width of the haunch are next proportioned to be the same as in the rolled shape.

When positive lateral support is provided to the compression flange at the points of tangency and at the midpoint (sections C, H, and G of Fig. 5.16) then by an analysis similar to that leading to Eq. 5.22 it may be shown that the maximum allowable radius is given by

$$R \leq \frac{4b}{\alpha} \tag{5.31}$$

* For illustration see Plate VIII, p. 288.

where α is the angle between sections C and H expressed in radians. If the actual radius is larger than the value given by this equation, then the haunch must be supplied with supplementary bracing or it must be strengthened to control the strain. As in the case of the tapered haunch, strengthening may be accomplished by an increase in flange thickness according to

$$\Delta t = 0.1 \left(\frac{R\alpha}{b} - 4 \right) t, \qquad \left(4 < \frac{R\alpha}{b} < 17 \right) \tag{5.32}$$

The required thickness of the compression and tension flanges may now be determined. It has been shown that the "critical" section of a curved haunch is about 0.2 radian from the point of tangency at section C. Since the corresponding angle β would be 0.2 radian, the correction in flange thickness due to non-parallel flanges is negligible. However to provide strength at the critical or "minimum" section "M" sufficient to develop M_p at section C (see dotted lines representing haunch strength in Fig. 5.16), an increase of flange thickness is necessary. From Eq. 2.24

$$t_t = t_c = \frac{d_m}{2} - \sqrt{\frac{d_m^2 b}{4(b - w)} - \frac{M_m/\sigma_y}{b - w}} \tag{5.33}$$

Where d_m is the depth at the critical section,

$$d_m = d + 0.02R \tag{5.34}$$

and M_m is the moment at the "minimum" section. The thickness of the tension and compression flanges must be increased according to Eq. 5.32 if the radius is greater than $\frac{4b}{\alpha}$.

The curved inner flange must next be checked for the "cross-bending" effect, a check that is necessary because the flange tends to bend across the web. If

$$t_c \geq \frac{b^2}{2R} \tag{5.35}$$

then the selected flange plate will be adequate.[5.13] Otherwise its thickness must be increased.

Next, the web of the connection should be examined for shear at the outer corner. Even though the flange stress at section 2 is considerably less than the yield value (the critical region is at section M), for simplicity it is assumed that it is transmitting its maximum capacity. Consequently, Eq. 5.25 will be a conservative guide to the adequacy of the web and the minimum thickness of diagonal stiffener at section H. Even when Eq. 5.25 does not require a stiffener, one should be supplied to stiffen the web against the general inward thrust introduced by the curved inner flange and to assist in supporting that flange against lateral buckling. A minimum stiffener permitted according to $b/t \leq 17$ will be adequate for this purpose.

Transverse stiffeners at sections C and G theoretically will resist no compressive thrust. Nonetheless, they will be supplied to provide lateral stability to the flange.

The influence of axial force and of shear force on the moment capacity of the haunch may be taken into account in the same way as is done for the rolled sections. If the rolled shape is adequate, then no modification of the haunch design will be required.

EXAMPLE Haunched Connection with Curved Inner Flange (Fig. 5.16)

The procedure just outlined will be illustrated on a curved haunch similar to that shown in Fig. 5.16. The given information includes a length of haunch of 6 ft measured along the frame line, and column and girder members of 24WF100 shapes. Lateral support is to be provided at sections C, H, and G. $M_c =$ 780 kip-ft, $M_h = 1170$ kip-ft.

Web thickness:

Use web equivalent to 24WF100 ($w = 0.468$ in.) cut web from $\frac{1}{2}$-in. plate.

Flange width:

Maintain width of 24WF100 ($b = 12.00$ in.). Use 12-in. plate width for tension and compression flange.

Haunch radius:

With lateral support at sections C, H, and G, maximum radius for "minimum" flange is given by Eq. 5.31.

$$\text{Trial radius} = 6.83 \text{ ft}$$

$$\frac{4b}{\alpha} = \frac{4(1.0)}{0.65} = 6.15 \text{ ft} < 6.83 \text{ ft}$$

Increase flange thickness by Eq. 5.32

$$\Delta t = 0.1\left[\frac{(6.83)(0.65)}{1.0} - 4\right] t = 0.046t \text{ in. (negligible)}$$

Flange thickness:

Supply sufficient strength at critical section M to develop M_p at C.

$$d_m = d + 0.02R = 24 \text{ in.} + (0.02)(6.83)(12) = 25.64 \text{ in.}$$

$$M_m = 780 + \frac{0.2(R)}{6}(1170 - 780) = 869 \text{ kip-ft}$$

$$t_t = t_c = \frac{25.64}{2} - \sqrt{\frac{(25.64)^2(12)}{4(12.00 - 0.47)} - \frac{(889)(0.364)}{(12.00 - 0.47)}} = 0.90 \text{ in.}$$

Check for cross-bending:

$$t_c \geq \frac{b^2}{2R} = \frac{(12)^2}{(2)(6.83)(12)} = 0.88 < 0.90 \qquad \underline{\text{O.K.}}$$

Use plate of $\frac{15}{16}$ in. thickness for tension and compression flange of haunch (compares to $t = 0.775$ in. for 24WF100).

Check for shear in corner (diagonal stiffener):

$$d_2 = 48 \text{ in.}$$

$$t_s = (\sqrt{2})(0.938) - \frac{(0.82)(0.50)(48)}{12} = 1.33 - 1.64 = -0.31$$

Although no stiffener is required for shear, supply a "minimum" stiffener for stability.

$$t_s = \frac{b/2}{8.5} = \frac{6.0}{8.5} = 0.70 \text{ in.}$$

Use $\frac{3}{4}$ x $5\frac{3}{4}$ in. plate stiffeners.

Transverse stiffeners:

Use $\frac{3}{4}$ x $5\frac{3}{4}$ in. plate stiffeners.

Experiments

Although a complete set of tests has not been performed on haunched connections, partial data are available and more are being collected.

Figure 5.17 shows the behavior of a connection with tapered haunch. It might be called a "marginal" test since it barely met the requirements of transmitting M_p of the 12WF36 shape; a considerable portion of the plastic action occurred in the haunch. With an s/b ratio of 10.0, Eq. 5.30 would have required flanges 60% thicker than the 12WF36 rolled shape. Actually an increase of 50% in thickness was used; thus on the basis of this test the procedure is conservative.

The curved connections tested in Ref. 4.8 were all designed to join two 8B13 shapes at right angles. None were proportioned with sufficient

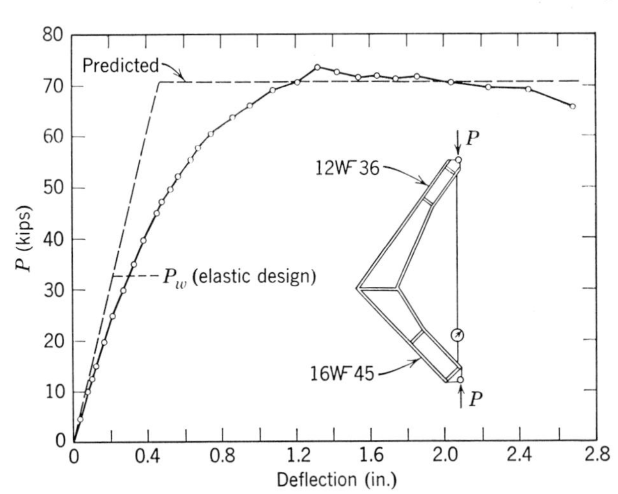

Fig. 5.17. Comparison between theory and test of a tapered haunch.

strength to meet the suggestions of the previous section and thus indirectly lend support to the suggestions made.

Effect of Haunches on Frame Analysis

Insofar as analysis of the frame is concerned, the effect of haunches is to increase the number of sections at which plastic hinges may form and to eliminate the possibility of a hinge forming precisely at the corner. Because the hinge location has been moved from the corner, then a frame with a haunch will carry more load than an identical frame in which "straight" connections are used. The following example is illustrative.

EXAMPLE Portal Frame with Haunches (Fig. 5.18)

This frame has symmetrical haunches that extend a distance equal to $L/10$ from the corner. The loads and dimensions are as shown in sketch (*a*) and the prismatic portion of both the beams and the columns is equal to M_p. The frame will be analyzed by the statical method. The determinate and redundant moment diagrams are laid out in sketches (*b*) and (*c*) and the composite moment diagram is drawn in sketch (*d*).

At the right hand connection there is a possibility of a hinge at sections 5 and 6. By laying out the moment diagram to scale it is evident in this problem that for equalization of moments at section 4 and at the haunch, the fixing line 1–*a*–*b*–7 must be drawn with the moment at 4 equal to the moment at 5. Else-

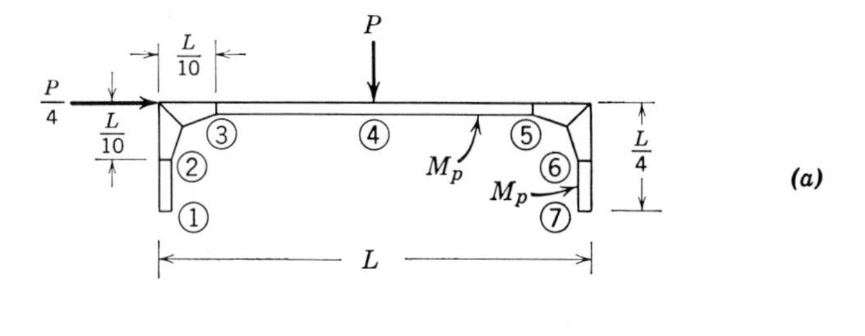

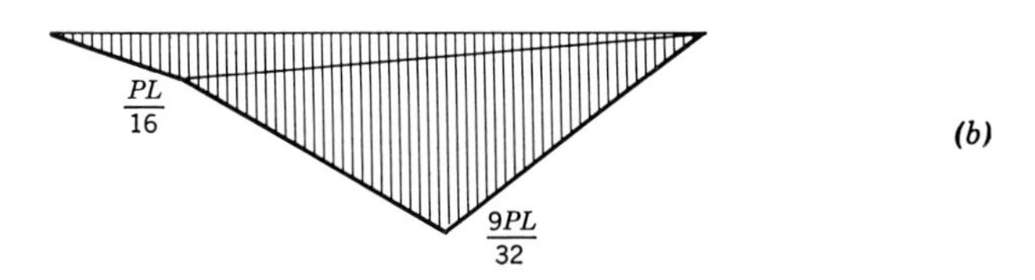

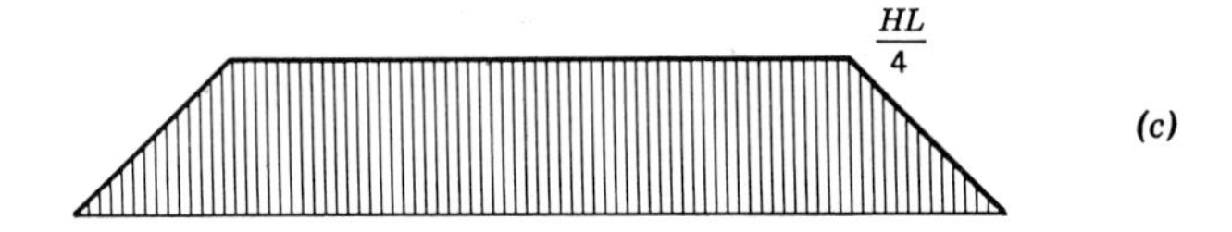

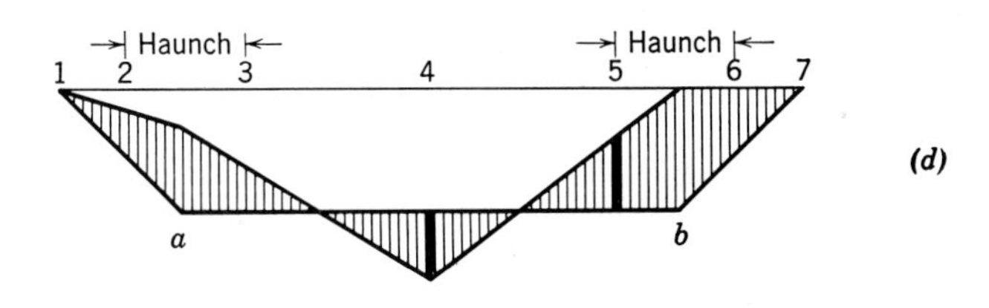

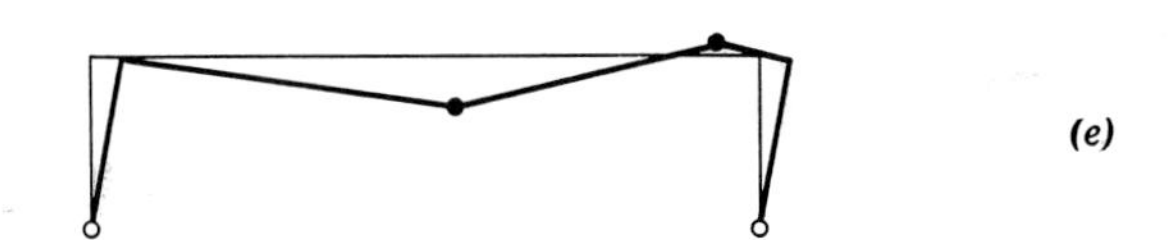

Fig. 5.18. Analysis of portal frame with haunched connections.

where the strength of the member is greater than required according to the resulting moment diagram (shown shaded in sketch *d*). The failure mechanism is shown in sketch (*e*).

To compute the ultimate load P_u one could scale off the value from the moment diagram or it could be computed by taking equilibrium at sections 4 and

5. At section 4,

$$\frac{9}{32}PL = M_p + \frac{HL}{4}$$

At section 5,

$$\frac{HL}{4} = M_p + \left(\frac{L/10}{L/2}\right)\left(\frac{9}{32}PL\right)$$

Eliminating $HL/4$ between these two equations,

$$\tfrac{9}{32}PL = M_p + M_p + (\tfrac{1}{5})(\tfrac{9}{32}PL)$$

from which the ultimate load is computed as

$$P_u = \frac{80}{9}\frac{M_p}{L}$$

Since the ultimate load without a haunch would have been

$$P_u = \frac{64}{9}\frac{M_p}{L}$$

then this frame has been strengthened 25% through the use of a haunch.

The example just completed illustrates an exception to the rule stated in Art. 3.5 that the number of independent mechanisms is equal to the number of possible plastic hinges less the number of redundants. In Fig. 5.18 it is possible for plastic hinges to form at sections 2, 3, 4, 5, and 6. Since the frame is indeterminate to the first degree, then the number of independent mechanisms should be

$$n = N - X = 5 - 1 = 4$$

However, in this structure there are only two independent equations of equilibrium, not four as implied by the above equation.

Actually, the difficulty is resolved when a more general definition for N is used. In all the cases examined up to this point the number of possible plastic hinges has coincided with the number of moments that must be known before the moment diagram may be drawn. However, when haunches are used, this agreement no longer holds. In Fig. 5.18*a*, if the moment is known at section 4 and at one point at both corners, then the moment diagram may be completed. In other words, the correct value of N for this problem is 3, and the number of independent mechanisms is two instead of four. These mechanisms are beam mechanism 3–4–5 and panel mechanism 2–6. The plastic moment condition must of course be satisfied at the other locations at which plastic hinges might form.

In ordinary problems involving single-span pinned-base structures, no problem exists because the mechanism method would not be used.

However, if haunches are used for problems involving fixed bases and for other cases of greater redundancy, then the value of N must be taken as

N = Number of moments that must be known in order that the moment diagram may be drawn.

5.5 INTERIOR BEAM-COLUMN CONNECTIONS

The interior beam-to-column connections are those shown as ② in Fig. 5.1 and in further detail in Fig. 5.19. The function of the "top" and the "interior" connections is to transmit moment from the left to the right beam, the column carrying any unbalanced moment. The "side" connection transmits beam moment to upper and lower columns.

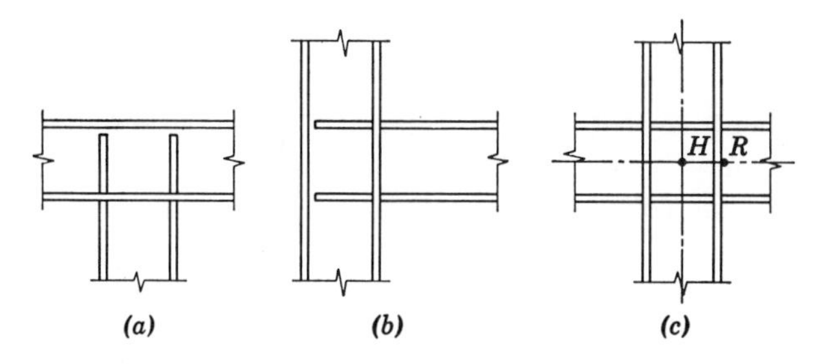

Fig. 5.19. Beam-to-column connections of the (a) "top," (b) "side," and (c) "interior" type.

The two basic types of interior beam-column connections are direct-welded (Fig. 5.19*c*) and top-plate beam-column connections (Fig. 5.20). Both types may be used, although the latter ordinarily may be counted upon for a hinge moment somewhat less than M_p.

The design problem is to provide sufficient stiffening material so that the connection will transmit the desired moment (usually the plastic moment M_p). Therefore, methods must be available for analyzing the joint to predict the resisting moment of unstiffened and stiffened columns. Much information is becoming available from a research project at Lehigh University in which these problems are being studied.[5.14, 5.15] Three possible "solutions" to the moment stiffening problem are shown in Fig. 5.21 and procedures of analysis will now be discussed.

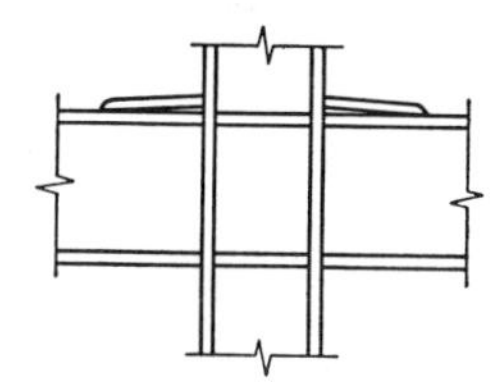

Fig. 5.20. A top-plate interior beam-column connection.

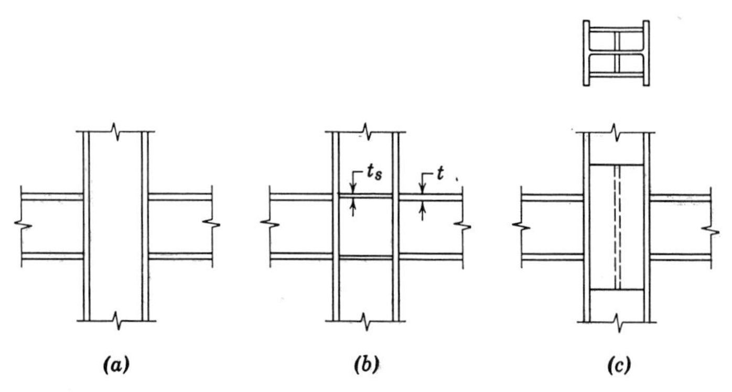

Fig. 5.21. Various methods for stiffening an interior beam-column connection: (a) No stiffener, (b) horizontal flange-type stiffener, and (c) vertical web-type stiffener.

Columns without Stiffeners

The moment capacity of unstiffened beam-to-column connections (Fig. 5.21a) may be computed from a plastic analysis. In the limit, the force that the column web can sustain is equal to the area available

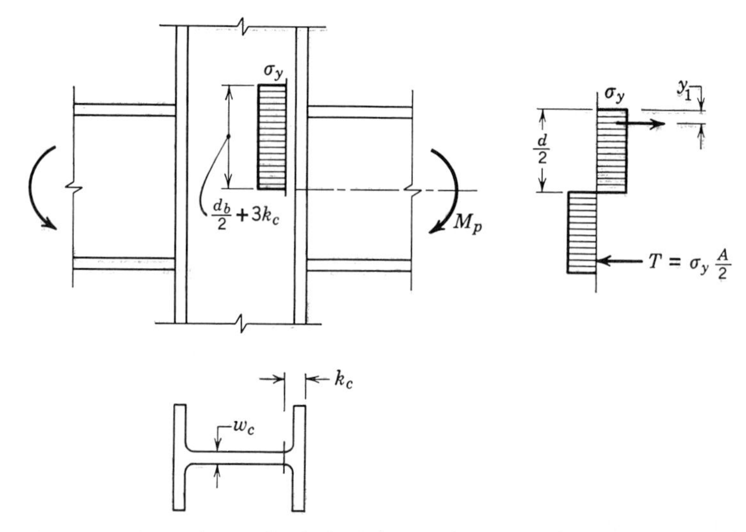

Fig. 5.22. Assumed stress-distribution in beam-column connection with no stiffener.

to carry the reaction times the yield-point stress. Referring to Fig. 5.22, the force that must be transmitted is known,

$$T = \sigma_y \frac{A_{\text{beam}}}{2} \tag{5.36}$$

The reaction width is equal to the column web thickness w_c. Tests of connections show that the length of reaction zone may be taken as one-half the beam depth plus three times the k-distance of the column. Therefore one may write

$$T = (\text{Reaction area}) \times (\sigma_y) \tag{5.37}$$

or

$$\frac{\sigma_y A_b}{2} = \left[w_c \left(\frac{d_b}{2} + 3k_c \right) \right] (\sigma_y) \tag{5.38}$$

From Eq. 5.38 a direct design check may be formulated:

$$w_c \geq \frac{A_b}{d_b + 6k_c} \tag{5.39}$$

which gives the required column web thickness w_c to assure that the plastic moment will be developed in the beam adjoining an unstiffened column. Except for those cases where the columns are relatively heavy in comparison to the beams, the test by Eq. 5.39 will often show inadequate strength of the column. Recourse is then made to flange or web stiffeners of the type shown in Fig. 5.21.

In order to guard against buckling of the web under the action of a compressive thrust equal to the yield value, the ratio of web depth to web thickness should not exceed about 30.

Flange Stiffeners

A plastic analysis of connections with flange stiffeners may be used which results in a direct design procedure for determining the required thickness of stiffener t_s. Assume that a stiffener is required and that

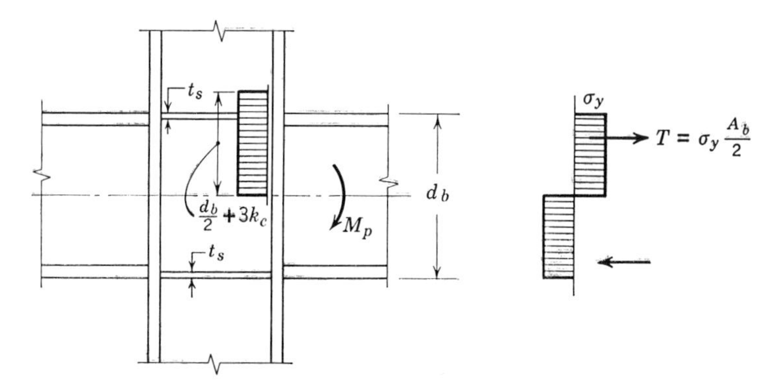

Fig. 5.23. Assumed stress-distribution in beam-column connection with flange-type stiffener.

it will adequately brace the column web against buckling. Then, referring to Fig. 5.23 in which the plastic moment M_p is acting at the end of the beam, the thrust T must be balanced by the strength of the web T_w and of the flange plate T_s or

$$T = T_w + T_s \tag{5.40}$$

With

$$T_w = \text{force resisted by web}$$

$$= \sigma_y w_c \left(\frac{d_b}{2} + 3k_c\right) \tag{5.41}$$

and

$$T_s = \text{force resisted by stiffener plate}$$

$$= \sigma_y t_s b$$

and

$$T = \sigma_y \frac{A_b}{2}$$

a direct solution for required flange stiffener thickness is

$$t_s = \frac{1}{2b}[A_b - w_c(d_b + 6k_c)] \tag{5.42}$$

Web Stiffeners for Resisting Thrust

Web stiffeners may be proportioned on a similar basis to that described for unstiffened connections. For use in Eq. 5.37 the reaction area is made up of the area supplied by the column web and the two inserted auxiliary webs (Fig. 5.21c). In view of the fact that the stiffener plates are close to the edge of the column flanges, it is assumed that the column flange is only half as effective in distributing load to the plates. Thus the reaction area is given by

$$\text{Reaction area} = w_c\left(\frac{d_b}{2} + 3k_c\right) + 2w_s(t_b + 3k_c) \tag{5.43}$$

where w_s is the thickness of the web-type stiffener. Therefore,

$$w_s = \frac{A_b - w_c(d_b + 6k_c)}{4(t_b + 3k_c)} \tag{5.44}$$

The thickness w_s should not be less than that of the columns.

Web Stiffeners for Resisting Shear

A second general type of stiffener that might be needed is that necessary to assist in transmitting shear forces. "Side" connections (Fig. 5.19*b*) or interior connections with large unbalanced moments may require "shear stiffening." In such a case the column web at the joint is called upon to transmit forces much like those of Fig. 5.6. An examination similar to that leading to Eq. 5.7 would therefore be desirable for this case.

Consider the interior connection shown in Fig. 5.24 in which the two beams do not transmit the same moment to the column. In fact the extreme condition is assumed, namely that M is reversed. This will

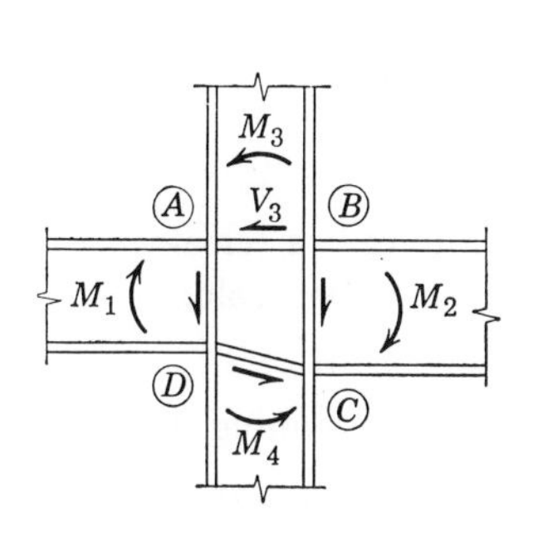

Fig. 5.24. Forces and moments acting on interior beam-column connection.

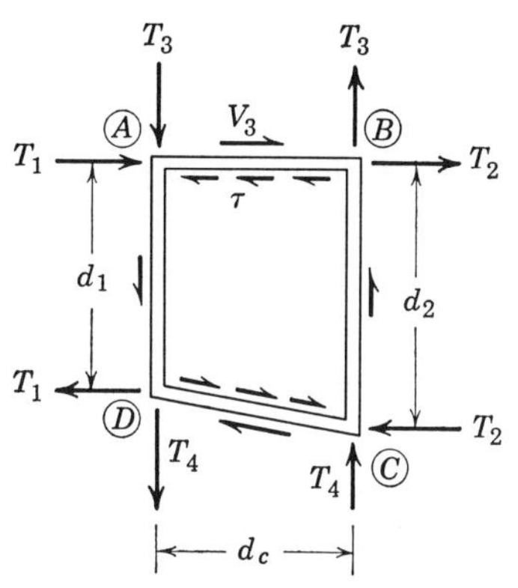

Fig. 5.25. Forces and stresses acting on flanges and web of interior beam-column connection.

produce high shear forces in the web of the connection, and the problem is to determine the web thickness to keep the shear deformation within suitable limits.

The element $ABCD$ is redrawn in Fig. 5.25 with the forces acting upon it, using the same assumptions as for the straight corner connection. The direction of V_3 has been reversed to correspond to the positive direction. It is further assumed that there are no normal forces, that $Z_2 > Z_1$, and that all of the "unbalanced" moment is introduced by the difference in moments of the beams.

Taking equilibrium of the top flange AB of the element,

$$\tau w d_c = T_2 + T_1 + V_3 \tag{5.45}$$

from which

$$w = \frac{T_2 + T_1 + V_3}{\tau d_c} \tag{5.46}$$

Making use of the assumptions stated before, the following relationships may be written:

$$\tau = \frac{\sigma_y}{\sqrt{3}}$$

$$T_2 = \frac{M_2}{d_2}$$

$$T_1 = \frac{M_1}{d_1}$$

Therefore, the required web thickness is given by

$$w = \frac{\sqrt{3}}{\sigma_y d_c}\left[\frac{M_2}{d_2} + \frac{M_1}{d_1} + V_3\right] \tag{5.47}$$

in which σ_y = yield stress level
d_c, d_1, d_2 = depth of respective members
M_2, M_1 = moment applied to joint by the beams (clockwise +)
V_3 = shear applied to joint by the upper column (positive to right)

Eq. 5.47 may be simplified to a certain extent by assuming the following:

(1) V_3 may be expressed in terms of M_3 by assuming that the column height h has a point of inflection at mid-height and that no horizontal loads are applied between the ends. Thus, $V_3 = 2M_3/h$. Further, if $M_3 = M_4$, then since $(M_1 + M_2) = -(M_3 + M_4)$ then

$$V_3 = \frac{-(M_1 + M_2)}{h}$$

(2) The depth d_1 will be replaced by the larger depth d_2. This is slightly unconservative for the case sketched in Fig. 5.25, but, for the more usual case where M_1 is counterclockwise, this assumption is conservative.

If, now, we call ΔM the sum of the moments applied to the connection by the beams, then using the above assumptions, Eq. 5.47 becomes

$$w = \frac{\sqrt{3}\Delta M}{\sigma_y d_c d_2}\left(1 - \frac{d_2}{h}\right) \tag{5.48}$$

If d_2/h is taken as $\frac{1}{15}$, then

$$w = \frac{0.6\ \Delta M}{d_c d_b}, \qquad \left(\text{for } \frac{d_2}{h} \leq \frac{1}{15},\ \sigma_y = 33 \text{ ksi}\right) \tag{5.49}$$

where ΔM = sum of moments applied to connection by the beams (k-ft)
d_c = depth of column in inches
d_b = depth of largest beam in inches

In addition to the connection shown in Fig. 5.24, Eq. 5.47 or 5.49 could also be applied to the "top" and "side" connections of Fig. 5.19. In case members join at other than a right angle, two sides of the connection zone in all likelihood will still be parallel. The quantity $d_c d_b$ could then be taken as the length of the larger of the two parallel sides times the distance separating them.

In the event that the web thickness actually supplied is less than required, then stiffening will be required either with web doublers or with a diagonal stiffener.

Experiments

Figure 5.26 shows the behavior of a number of interior beam-column connections, proportioned by the suggested methods. Figure 5.27 shows one of the connections at the end of the test, indicating the desirable

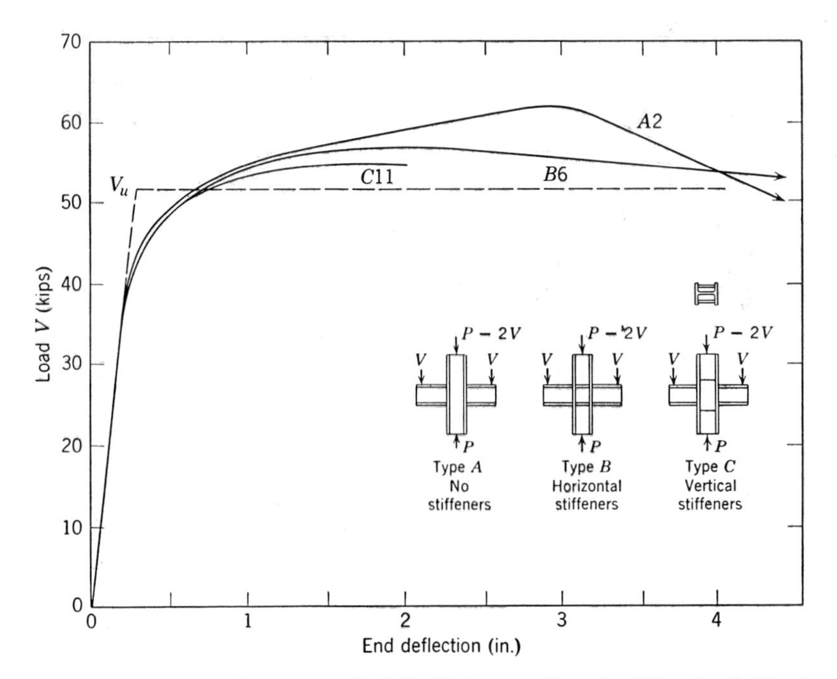

Fig. 5.26. Theoretical and experimental relationship between beam load V and end deflection of beam-column connections. (*Proc. AISC Nat'l Engr. Conf.*, 1956.)

characteristic, namely that the connection proper suffers but little deformation, the real failure occurring in the beam.

The behavior of connection $A2$ shown in Fig. 5.26 joining a 16WF36 beam to an 8WF67 column, will be examined to see if Eq. 5.39 is a reasonable basis for design.

$$\text{Req'd } w_c \geq \frac{A_b}{d_b + 6k_c} = \frac{10.59}{15.85 + 7.86} = 0.45 \text{ in.}$$

Since w_c is actually 0.575 in., the fact that the connection performed satisfactorily without a stiffener is a verification of the procedure.

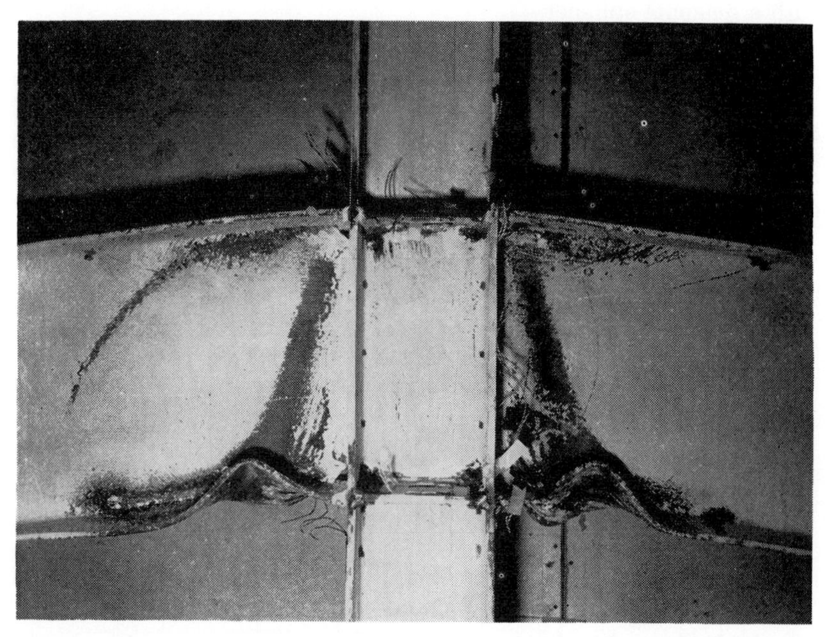

Fig. 5.27. Photograph of beam-column connection after test (Type *B* of Fig. 5.26). (*Proc. AISC Nat'l Engr. Conf.*, 1956.)

Connection $B8$ joins a 16WF36 beam ($A_b = 10.59$ in.^{2}, $b = 6.99$ in.) to a 12WF40 column ($w_c = 0.294$ in., $d_c = 11.94$ in.). Its performance was similar to connection $B6$ (Fig. 5.26). In Eq. 5.42 the required thickness of stiffener is $t_s = 0.28$ in. A stiffener plate $\frac{1}{4}$-in. thick was used in the test, and the connection met all of the design requirements.

Connection $C11$ joining the 16WF36 beam to the 12WF40 column and using the "web stiffener," performed satisfactorily (Fig. 5.26).

According to Eq. 5.44, the required stiffener thickness is given by

$$w_s = \frac{A_b - w_c(d_b + 6k_c)}{4(t_b + 3k_c)}$$

Thus

$$w_s = 0.25 \text{ in.}$$

The actual thickness used was 0.313 in.

Equation 5.49 has had partial experimental confirmation in the test of a two-span gabled frame[5.17] in which 8B13 and 10B17 shapes were used. The required thickness was 0.208 in. as compared with an actual thickness of 0.240 in. for the 10B17 rolled shape. Therefore no additional stiffening was specified. The connection performed satisfactorily although some local web yielding occurred.

PROBLEMS

5.1 Why is rotation capacity important in a connection?

5.2 In addition to the requirement that a stiffener plate in a connection be able to transmit the design force, what other requirement must it meet? How is this accomplished in design?

5.3 A straight corner connection must transmit a plastic moment of 1650 kip-in. Select the needed member and detail the knee connection.

5.4 A 24WF100 shape is found to be just adequate to transmit the needed plastic moment at a corner. Detail the knee (*a*) using a web doubler plate and (*b*) using a diagonal stiffener.

5.5 Detail a knee using a 10B17 shape using all fillet welding.

5.6 Show how the equation for required thickness of web in a straight corner connection is modified when the connection is part of a gabled frame in which the members do not meet at right angles.

5.7 Proportion a connection with tapered haunches to join a 21WF96 girder to a 24WF120 column. The rafter rise is 1:4 and the point of inflection is three times the depth of the section from the point where the prismatic girder and haunch meet. The length of haunch measured in a horizontal direction is 8 ft and the length of haunch measured vertically is 5 ft. Assume that the two rolled shapes just meet the moment requirements.

5.8 Select moment stiffening for a 14WF87 column which must transmit moments of two 18WF96 beams that join on opposite sides with direct welding to the flanges. Compare result with dimensions of beam and explain.

Deflections

6.1 INTRODUCTION

How may one be assured that a structure proportioned by the plastic method will not be bent out of shape to such an extent that it is unserviceable? One cannot be sure without computing the deflections—any more than elastic design can be regarded as satisfactory in this respect without making further computations or by knowing from experience that satisfactory performance may be expected. Just when is a deflection "excessive"? This is a difficult question to answer, even in conventional elastic design; more often than not a precise answer cannot be given. Except for those unusual cases for which some positive deflection limitation exists, the "rules" that are used in elastic design are based on quite arbitrary assumptions. Consequently the engineer's judgment based on his experience plays an important role in the problem of a proper deflection limit.

One of the principal advantages of plastic analysis is the simplicity with which the maximum load-carrying capacity may be determined as compared with procedures of indeterminate elastic frame analysis. If deflections must be calculated, some of this advantage is lost, although fairly simple procedures may be applied in certain cases. Fortunately, this difficulty is not as serious as might be supposed because in most cases a structure designed for ultimate loading by the plastic method will actually deflect no more at working loads (which are nearly always in the elastic range) than a structure designed elastically according to current specifications. A plastically designed continuous beam exhibits *less* deflection than a simply supported beam designed to carry the same loads because of the restraining moments that are present. Figure 6.1, for example, shows three different designs of a beam of 30-ft span to carry a working load of 21 kips. Curve I corresponds to the simple beam design. Curve III represents the plastic design. The deflections at working load for the plastic design are significantly less than those of the simply supported beam, albeit slightly greater than the elastic

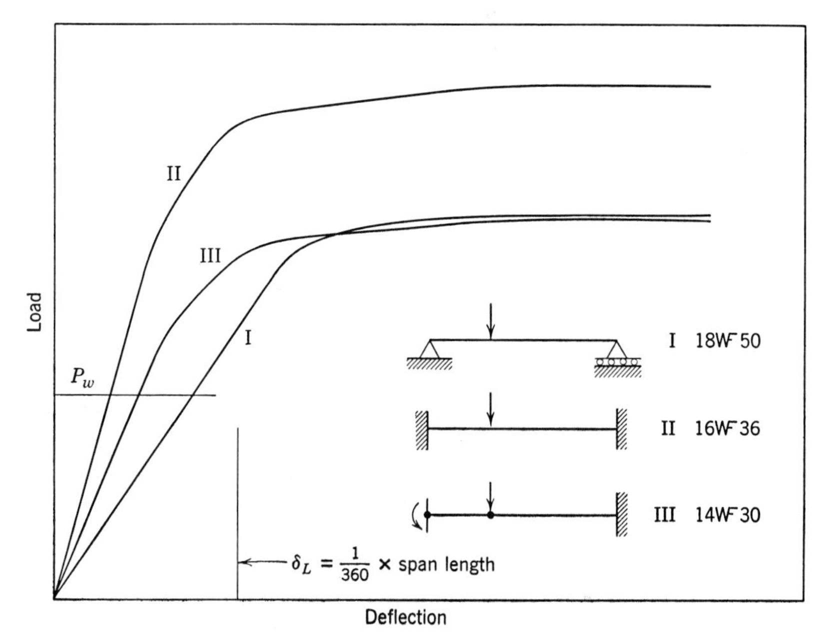

Fig. 6.1. Load–deflection relationship for three different beam designs for supporting the same load.[1.8]

design of the restrained beam (curve II). Incidentally, at working load P_w the plastically designed beam is still elastic.

The primary design requirement is that the structure must carry the assumed load. The deflection requirement is a secondary one: the structure must not deform too much out of shape. Therefore, our needs involving deflection computation may be satisfied with approximations. Such calculations fall into two categories:

(*a*) *Determination of approximate magnitude of deflection at ultimate load.* The load factor of safety does not preclude the rare overload and an estimate of the corresponding deflection would be of value.

(*b*) *Estimate of deflection at working load.* In certain cases, the design requirements may limit deflection at this load.

Fortunately, even though these computations will rarely be required, methods are available that approach in simplicity the methods of analyzing for ultimate load. The problem of deflections has been treated in Refs. 1.6, 6.1, and 6.2; and in Ref. 2.4 various methods have been compared. It has been shown in this previous work that any deflection calculations that may be required may be based on the idealized behavior of a WF shape.

In addition to the assumptions inherent in the simple plastic theory the following articles are based on these concepts:

(1) The M–ϕ relationship is idealized as shown in Fig. 2.6.

(2) As a consequence of assumption 1, each span retains its flexural rigidity EI for the whole length between hinge sections.

(3) Unlimited rotation is possible at hinge sections at a moment-value of $M = M_p$.

Examination of the mechanisms formed in the various structures analyzed in Chapter 3 may well leave the impression that the task of computing deflections is a formidable one. However, an important additional concept makes it possible to compute deformations in a straightforward and orderly manner. Although the deflection of a structure is indeterminate as mechanism motion takes place, this is not true when the mechanism just starts to form. Consider Fig. 2.18, p. 43, for example. As shown by the load-deflection curve of sketch (*d*), until the ultimate load is first reached, the deformations of the structure are still controlled by the elastic portions of it. As shown by the deflected shape at stage ② in sketch (*b*) of Fig. 2.18, continuity still exists at the centerline, and of course this section is the one at which the last plastic hinge forms. Consequently, the following important concept makes it possible to estimate deflections at ultimate load and to obtain an upper limit for the deflection at working load: Although "kinks" form at the other hinge sections, just as the structure attains the computed ultimate load, there is still continuity at that section at which the *last* plastic hinge forms.

The analysis neglects catenary forces (which tend to decrease deflection and increase strength) [6.3] and second-order effects (which tend to increase deflection and decrease strength). Also ignored are any factors that influence the moment–curvature relationship. (In Refs. 1.9, 2.2, and 2.4 may be found discussion of these and other influencing factors.)

This chapter presents a method for computing the approximate magnitude of the deflection at ultimate load and for obtaining an upper limit to the deflection at working load. The closely associated problem of rotation capacity is treated at the end of the chapter.

6.2 DEFLECTION AT ULTIMATE LOAD

Since each span retains its elastic flexural rigidity (EI) for the whole segment between sections at which plastic hinges are located, and since elastic continuity still exists up to the point at which the last plastic hinge forms, the slope-deflection equations may be used to solve for relative deflection of segments of the structure. The moments will have been determined from the plastic analysis. The following form of the slope-deflection equations will be used, the nomenclature being as shown

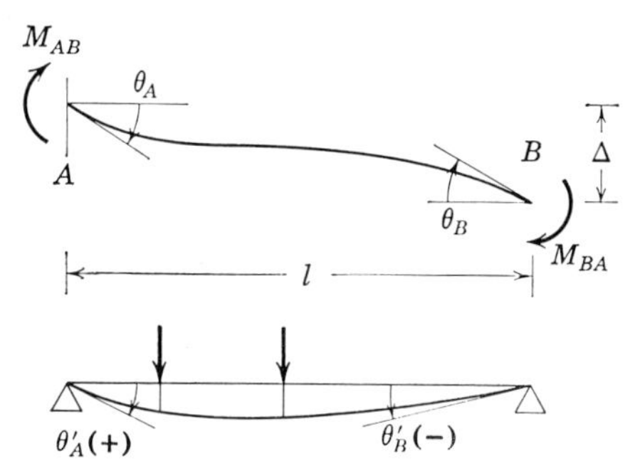

Fig. 6.2. Sign convention and nomenclature for use in the slope–deflection equations.

in Fig. 6.2 with clockwise moment and angle change being positive:

$$\theta_{AB} = \theta_{AB}' + \frac{\Delta}{l} + \frac{l}{3EI}\left(M_{AB} - \frac{M_{BA}}{2}\right) \tag{6.1}$$

The quantity θ_{AB}' is the slope at end A due to similar loading of a simply-supported beam.*

The only remaining question is this: Which hinge is the last to form? An elastic-plastic analysis could be carried out to determine the sequence of formation of hinges—and thus the last hinge. However, a few examples will demonstrate that a simpler method is available: The deflection may be calculated on the assumption that each hinge, in turn, is the last to form; then the correct deflection at ultimate load is the maximum value obtained from the various trials.

In outline, the following summarizes the procedure for computing deflections at ultimate load:

(1) Obtain the ultimate load, the corresponding moment diagram and the mechanism (from the plastic analysis).

* In Ref. 6.3 a number of standard cases have been solved and values of θ_{AB}' have been tabulated.

(2) Compute the deflection of the various frame segments assuming, in turn, that each hinge is the last to form.

(*a*) Draw free-body diagram of segment.

(*b*) Solve slope-deflection equation for the condition of continuity at the assumed last plastic hinge.

(3) Select as the correct deflection the largest value so obtained (corresponds to last plastic hinge).

(4) *A check:* From a deflection calculation based on an arbitrary assumption as to which hinge is the last to form, compute the "kinks" formed due to the incorrect assumption. Remove the "kinks" by mechanism motion and obtain the correct deflection.

The procedure is illustrated by the examples which follow.

EXAMPLE Fixed-Ended, Uniformly Loaded Beam (Fig. 6.3)

This beam with uniform vertical load will be analyzed to determine the center deflection at ultimate load.

(*a*) The ultimate load is computed: see Eq. 2.30.

$$W_u = \frac{16M_p}{L}$$

(*b*) The moment diagram and mechanism are sketched: see Fig. 6.3*a*.
(*c*) Trial computations are made of vertical ℄ deflection:

Trial at section 2: (Section 2 assumed as last hinge to form.)
Free-body diagram: see Fig. 6.3*b*.
Slope-deflection equation for member 2–1 using the condition that $\theta_{21} = 0$:

$$\theta_{21} = \theta_{21}' + \frac{\delta_{V2}}{l} + \frac{l}{3EI}\left(M_{21} - \frac{M_{12}}{2}\right)$$

$$\theta_{21}' = \text{Simple beam end rotation} = -\frac{M_pL}{12EI}$$

$$\delta_{V2} = \text{Vertical deflection with continuity assumed at section 2}$$

$$0 = -\frac{M_pL}{12EI} + \frac{\delta_{V2}}{L/2} + \frac{L/2}{3EI}\left(-M_p + \frac{M_p}{2}\right)$$

$$\boldsymbol{\delta_{V2} = +\frac{M_pL^2}{12EI}} \tag{6.2}$$

Trial at section 1: (Even though it is obvious that the last hinge forms at section 2, a trial calculation assuming continuity at section 1 will show the effect of this incorrect assumption.)
Free body diagram: see Fig. 6.3*c*.

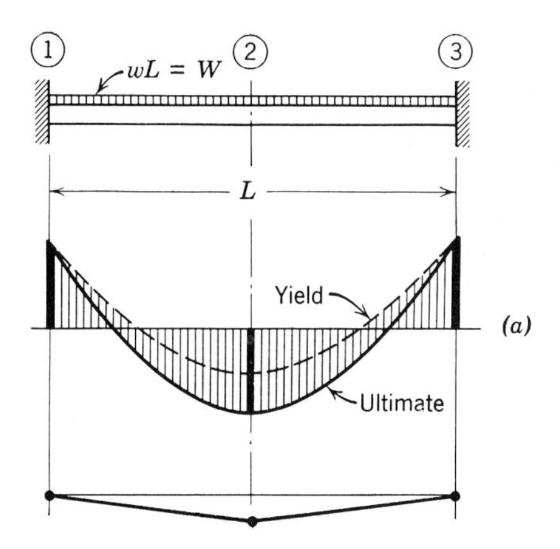

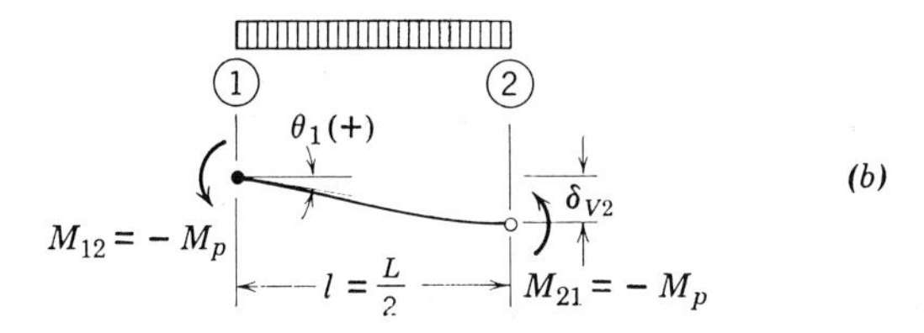

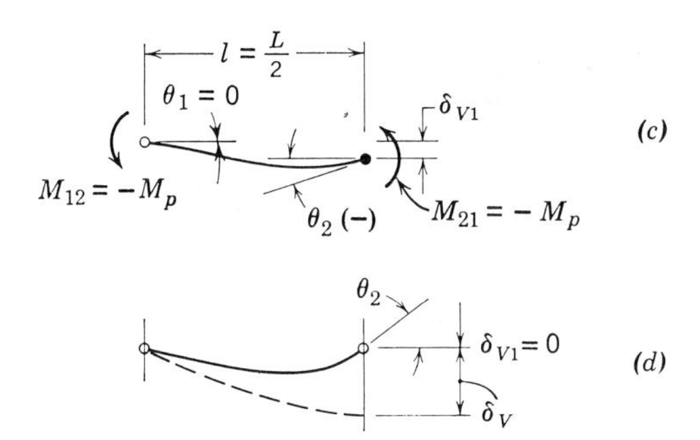

Fig. 6.3. Deflection analysis of fixed-ended, uniformly loaded beam.

Slope-deflection equation for segment 1–2 using the condition that $\theta_1 = 0$:

$$\theta_{12} = \theta_{12}' + \frac{\delta_{V1}}{l} + \frac{l}{3EI}\left(M_{12} - \frac{M_{21}}{2}\right)$$

$$0 = +\frac{M_pL}{12EI} + \frac{\delta_{V1}}{L/2} + \frac{L/2}{3EI}\left(-M_p + \frac{M_p}{2}\right)$$

$$\delta_{V1} = 0$$

Since the deflection obtained for the last trial was less than that obtained on the basis of continuity at section 2, then the first trial was the correct one.

The result obtained in this example confirms what was already known at the outset—the first hinge forms at sections 1 and 3 and the last hinge forms at section 2. The effect of the incorrect assumption as to which plastic hinge was last to form resulted in a lesser deflection because a "kink" or mechanism angle that should have formed at section 1 was removed. Instead, a "negative" slope discontinuity was created at section 2 where continuity should have existed. As a matter of fact, if a rigid body motion were applied, allowing the segments to deflect until the corresponding rotation at section 2 was equal to the negative kink (angle θ_2 in Fig. 6.3*c*), then we would obtain the correct deflection. This may be seen by the following computation:

Slope at section 2 (θ_{21}):

$$\theta_{21} = \theta_2' + \frac{\delta_{V1}}{l} + \frac{l}{3EI}\left(M_{21} - \frac{M_{12}}{2}\right)$$

$$= -\frac{M_pL}{12EI} + 0 + \frac{L/2}{3EI}\left(-M_p + \frac{M_p}{2}\right) = -\frac{M_pL}{6EI}$$

Rigid body rotation (sketch *d*):

$$\delta_V = \delta_{V1} + \delta_V'$$

δ_{V1} = Deflection computed on basis of incorrect assumption $= 0$

δ_V' = Deflection due to rigid body rotation (opposite to θ_2) $= (-\theta_2)(l) = -L/2\,\theta_2$

$$\delta_V = +\frac{M_pL^2}{12EI}$$

which is the same answer as was obtained before.

This same procedure could be used as an alternate method for computing the deflection. An arbitrary assumption as to the last plastic hinge would be made, the "negative kinks" would be computed, and the structure would be moved through a rigid body deformation until all negative kinks had been eliminated.

EXAMPLE Fixed-Ended Beam with Concentrated Load (Fig. 6.4)

This beam with concentrated load off-center will be analyzed to determine the deflection under the load.

(*a*) Compute ultimate load (by equilibrium). See Fig. 6.4*b*.

$$P_u = \frac{9M_p}{L}$$

(*b*) Find moment diagram and mechanism: see sketches (*b*) and (*c*) of Fig. 6.4.
(*c*) Compute trial values of vertical deflection:

Trial at section 1: (Member 1–2, $\theta_1 = 0$.)
Free-body diagram: see sketch (*d*)
Slope-deflection equation:

$$\theta_1 = \theta_1' + \frac{\delta_{V1}}{l} + \frac{l}{3EI}\left(M_{12} - \frac{M_{21}}{2}\right)$$

$$0 = 0 + \frac{\delta_{V1}}{L/3} + \frac{L/3}{3EI}\left(-M_p + \frac{M_p}{2}\right)$$

$$\delta_{V1} = +\frac{M_pL^2}{54EI}$$

Trial at section 2: (Continuity assumed at section 2: $\theta_{21} = \theta_{23}$.)
Free-body diagram: see Fig. 6.4*e*.
Slope-deflection equations for members 2–1 and 2–3:

$$\theta_{21} = \frac{\delta_{V2}}{L/3} + \frac{L/3}{3EI}\left(-M_p + \frac{M_p}{2}\right) = \frac{3\delta_{V2}}{L} - \frac{M_pL}{18EI}$$

$$\theta_{23} = \frac{-\delta_{V2}}{2L/3} + \frac{2L/3}{3EI}\left(M_p - \frac{M_p}{2}\right) = \frac{-3\delta_{V2}}{2L} + \frac{M_pL}{9EI}$$

$$\theta_{21} = \theta_{23}$$

$$\delta_{V2} = \frac{M_pL^2}{27EI}$$

Trial at section 3: (Continuity assumed at section 3.)
Free-body diagram: see Fig. 6.4*f*.
Slope-deflection equations for member 3–2 ($\theta_3 = 0$):

$$\theta_3 = \frac{\delta_{V3}}{l} + \frac{l}{3EI}\left(M_{32} - \frac{M_{23}}{2}\right)$$

$$0 = \frac{-\delta_{V3}}{2L/3} + \frac{2L/3}{3EI}\left(M_p - \frac{M_p}{2}\right)$$

$$\delta_{V3} = +\frac{2}{27}\frac{M_pL^2}{EI}$$ **Maximum deflection (last hinge)**

The maximum deflection occurs when the last hinge is assumed to form at section 3. All other assumptions result in "negative kinks" (which, by comparison with correct mechanism, are impossible) and produce smaller deflections.

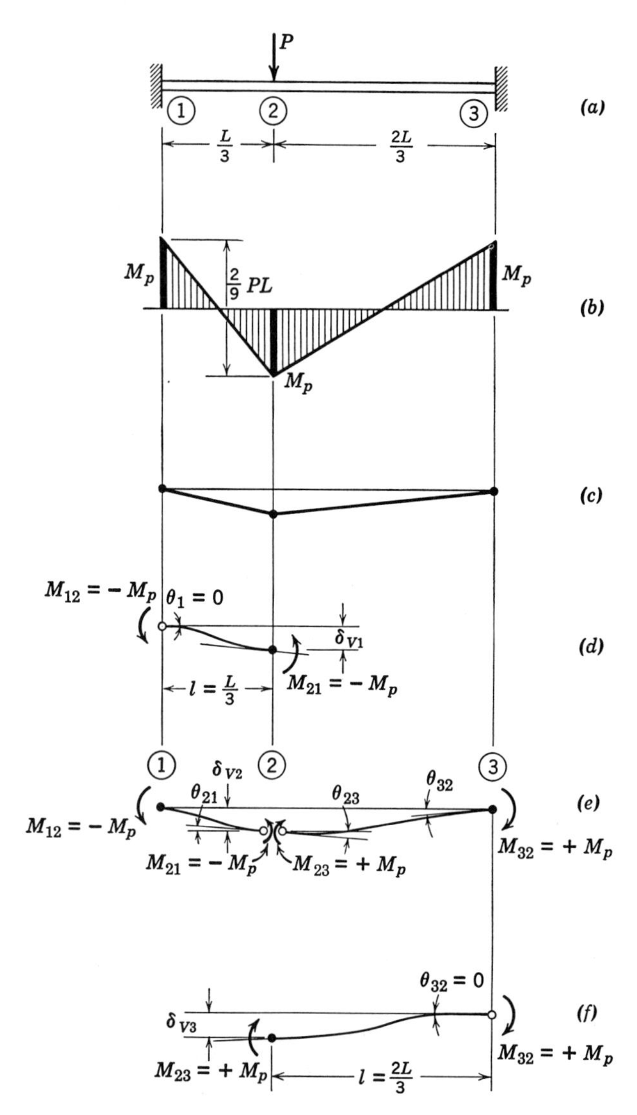

Fig. 6.4. Deflection analysis of fixed-ended beam with concentrated load at the third point.

EXAMPLE Rectangular Fixed-Base Portal Frame (Fig. 6.5)

The problem is to find the vertical deflection under the load applied at the midspan of the girder. One new situation is introduced in this example. Application of the slope-deflection equations to the columns does not give an explicit expression for the vertical deflection. However, by use of the continuity

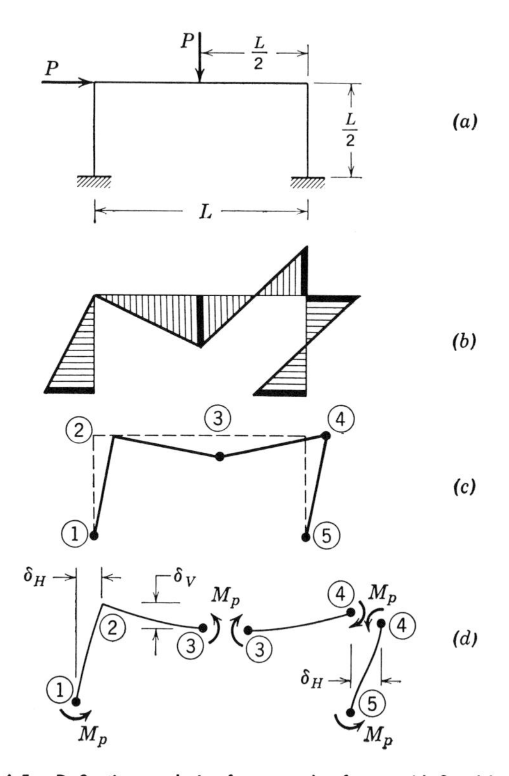

Fig. 6.5. Deflection analysis of rectangular frame with fixed bases.

condition at an elastic joint, a relationship may be found between the horizontal deflection at the column top and the vertical deflection of the beam.

(*a*) Compute the ultimate load by plastic analysis:

$$P_u = \frac{6M_p}{L}$$

(*b*) Obtain the moment diagram and mechanism: see Fig. 6.5*b*, *c*.
(*c*) Free-body diagrams: see Fig. 6.5*d*.
(*d*) Computation of vertical deflection:

Ratio of δ_H and δ_V:

Continuity assumed at section 2 ($\theta_{23} = \theta_{21}$):

$$\theta_A = \theta_A' + \frac{\delta_{V2}}{l} + \frac{l}{3EI}\left(M_{AB} - \frac{M_{BA}}{2}\right)$$

$$\theta_{23} = 0 + \frac{\delta_{V2}}{L/2} + \frac{L/2}{3EI}\left(0 + \frac{M_p}{2}\right)$$

$$\theta_{23} = \frac{2\delta_{V2}}{L} + \frac{M_pL}{12EI}$$

$$\theta_{21} = 0 + \frac{\delta_{H2}}{L/2} + \frac{L/2}{3EI}\left(0 + \frac{M_p}{2}\right)$$

$$\theta_{21} = \frac{2\delta_{H2}}{L} + \frac{M_pL}{12EI}$$

$$\frac{2\delta_{V2}}{L} + \frac{M_pL}{12EI} = \frac{2\delta_{H2}}{L} + \frac{M_pL}{12EI}$$

$$\delta_V = \delta_H$$

Trial at section 1: (Member 1–2, $\theta_1 = 0$.)

$$0 = 0 + \frac{\delta_{H1}}{L/2} + \frac{L/2}{3EI}(-M_p + 0)$$

$$\delta_{H1} = +\frac{M_pL^2}{12EI}$$

$$\boldsymbol{\delta_{V1} = \frac{M_pL^2}{12EI}}$$

Trial at section 3: $\theta_{32} = \theta_{34}$.

$$\theta_{32} = 0 + \frac{\delta_{V3}}{L/2} + \frac{L/2}{3EI}(-M_p + 0) = \frac{2\delta_{V3}}{L} - \frac{M_pL}{6EI}$$

$$\theta_{34} = 0 - \frac{\delta_{V3}}{L/2} + \frac{L/2}{3EI}\left(M_p - \frac{M_p}{2}\right) = \frac{-2\delta_{V3}}{L} + \frac{M_pL}{12EI}$$

$$\theta_{32} = \theta_{34}$$

$$\delta_{V3} = \frac{M_pL^2}{16EI}$$

Trial at section 4: Similar procedure using $\theta_{43} = \theta_{45}$.

$$\delta_{V4} = \frac{M_pL^2}{24EI}$$

Trial at section 5: Similar procedure using $\theta_5 = 0$.

$$\delta_{V5} = \frac{M_pL^2}{24EI}$$

Correct answer:

$$\delta_V = \delta_{\max} = \delta_{V1} = \frac{M_pL^2}{12EI} \qquad \text{(Last hinge at section 1)}$$

Frequently it is possible to shorten the number of trial calculations by making a guess as to which hinge is the last to form. This guess is made by considering the action of the separate loads as they cause deformation at the various hinges.

For example, in the problem just solved for the frame shown in Fig. 6.5, under action of the vertical load P joint 4 would tend to close. The horizontal load P would also tend to close joint 4. Therefore, this is probably one of the first hinges to form. Turning our attention to joint 5, the vertical load would tend to rotate the leeward column clockwise about its base. The side load has the same effect. Consequently joint 5, in all likelihood, is also one of the first hinges to form. The vertical load tends to rotate section 3 in a direction that produces tension on the bottom fiber, but the horizontal load acting alone produces no deformation at section 3. Therefore one would expect that the hinge at section 3 might form after the hinges at sections 4 and 5. At joint 1, the vertical and horizontal loads produce opposite effects: The vertical load tends to rotate the column in a counter-clockwise direction about the base, whereas the horizontal load tends to push it in a clockwise direction. It would be reasonable to guess that joint 1 might be the last hinge to form.

The calculations made in the proceeding example bear out these estimates. "Last hinge" guesses at sections 4 and 5 produced the smallest deflection. A somewhat larger value was obtained for assumed continuity at section 3, and the greatest value was obtained when the last hinge was assumed to form at section 1. The sequence is precisely the same as that "guessed" by a consideration of the action produced by the separate loads.

In Ref. 4.5 charts have been prepared from which the deflection at ultimate load of pinned base gabled portal frames may be obtained for a wide variety of loading conditions and frame geometry.

6.3 DEFLECTION AT WORKING LOAD

For those few instances where deflection calculations must be made, primary interest will probably center around an estimate of deflection at working load. The working load on a structure designed by the plastic method will usually be in the so-called "elastic" range. This implies a need for an elastic analysis; and indeed if a precise value for the deflection is needed, then such an analysis must be made. However, since the magnitude of the deflection is a matter of secondary importance, a rough estimate will usually be suitable; and this article outlines several procedures that may be followed and which circumvent the elastic frame analysis.

An obvious step would be to make use of existing solutions. A number of standard cases of loading and geometry have been solved, formulas for the deflections being tabulated in such places as Ref. 2.1. The procedure in this instance would be as follows:

(1) Divide the computed ultimate load by the load factor of safety F.

(2) Solve for the deflection at this (elastic) working load using formulas that can be obtained from the tables mentioned.

EXAMPLE Fixed-Ended Uniformly Loaded Beam (Fig. 6.6)

To illustrate, using Eq. 2.30 with a load factor of safety of 1.85,

$$W_w = \frac{W_u}{F}$$

$$W_w = \left(\frac{16M_p}{L}\right)\left(\frac{1}{1.85}\right) = 8.65\,\frac{M_p}{L}$$

Since the working load is less than the yield load $\left(W_y = 12\,\dfrac{M_p}{L}\right)$, the beam is elastic. Therefore, from the tables in Ref. 2.1,

$$\delta = \frac{WL^3}{384EI}$$

$$\delta_w = 0.0225\,\frac{M_pL^2}{EI} \qquad (6.3)$$

When end restraint conditions are not known, often they may be estimated and the above technique employed and an approximation obtained that will be sufficiently accurate.

A second procedure is based upon the "last hinge" method of the previous article. The deflection at ultimate load δ_u may be computed by the method of Art. 6.2, and a value that will be *greater* than the true deflection at working load may be obtained by dividing δ_u by the

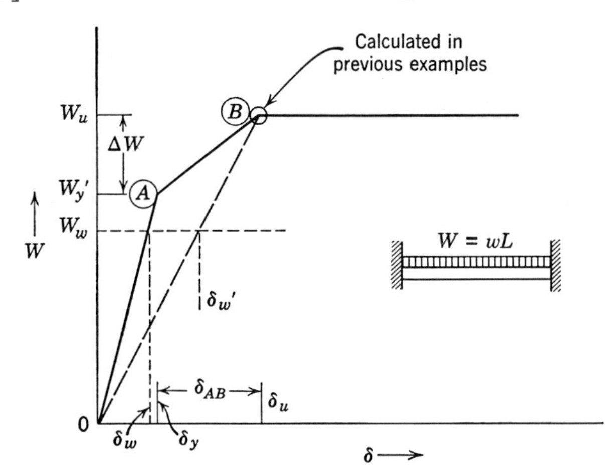

Fig. 6. 6. Idealized load–deflection relationship for fixed-ended beam with uniformly distributed load.

working load factor of safety, or

$$\delta_w' = \frac{\delta_u}{F} \tag{6.4}$$

This is illustrated by the dashed line in Fig. 6.6 for the uniformly loaded, fixed-ended beam. The error may often be greater than 100%, but at least the method gives an upper limit to the magnitude of the deflection at working load and indicates when more refined calculations would be necessary.

Suppose that it was desired to determine the approximate load–deflection relationship as load is gradually applied. This may be done with relative ease, using the same assumptions as in the previous examples. The procedure is to consider the structure just prior to the formation of each hinge. Figure 6.7 shows the load–deflection curve of the beam used in the example on p. 191 (Fig. 6.4). There are 4 distinct phases that may be uniquely determined as follows:

Phase 1 (0–A): (Elastic): Represents slope of deflection curve of structure (a).

Phase 2 (A–B): Represents slope of deflection curve of structure (b).

Phase 3 (B–C): Represents slope of deflection curve of cantilever structure (c).

Phase 4 (C–D): Mechanism.

Thus, each portion of the curve represents the load–deflection curve of a "new" structure containing one less redundant than previously.

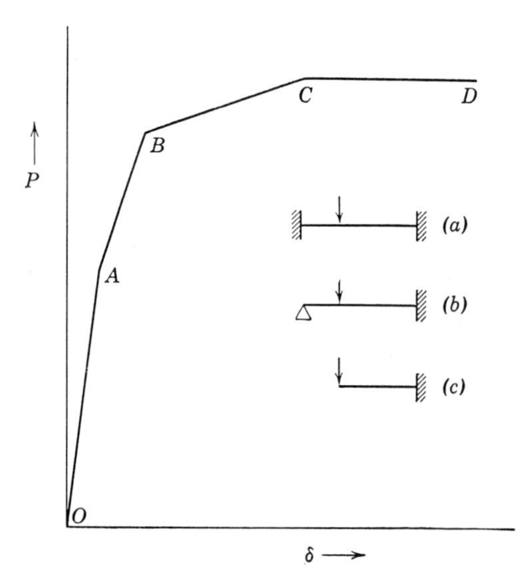

Fig. 6.7. Idealized load–deflection curve for fixed-ended beam with concentrated load off center.[6.1]

The corresponding load–deflection curve may be computed by determining the corresponding deflection increments.

EXAMPLE Load–Deflection Curve of Fixed-Ended Beam (Fig. 6.6)

The method of obtaining the load–deflection relationship will now be illustrated for this beam with uniformly-distributed load. From Eqs. 2.29 and 6.3 the "elastic limit" deflection δ_y' is given by

$$\delta_y' = \frac{M_p L^2}{32EI} \tag{6.5}$$

Above the yield load, the slope of the load–deflection curve is the same as that of a simple beam,

$$\delta_{AB} = \frac{5(\Delta W)L^3}{384EI} \tag{6.6}$$

where the quantity ΔW may be obtained from Eqs. 2.29 and 2.30 as

$$\Delta W = W_u - W_y' = \tfrac{1}{3}(W_y')$$

Comparing Eq. 6.6 with Eq. 6.3 it is seen that the slope of portion AB of Fig. 6.6 is one-fifth of the portion OA.

The total deflection, from Fig. 6.6 is given by

$$\delta_u = \delta_y' + \delta_{AB}$$

Thus

$$\delta_u = \frac{M_pL^2}{32EI} + \frac{1}{3}\left(\frac{12M_p}{L}\right)\frac{5L^3}{384EI}$$

$$\delta_u = \frac{M_pL^2}{12EI}$$

This checks with the value obtained in the example on page 188.

It is of interest to see whether or not the methods given in this article will predict actual load–deflection relationships with a sufficient degree of accuracy. Figures 1.11 and 2.23*d* show load–deflection curves of two indeterminate structures. Although the agreement between the theory based on idealized behavior and the tests is by no means exact, it is considered adequate in view of the fact that the effect of residual stresses, stress-concentrations, and the gradual plastification of the cross section have been neglected in the theory.

6.4 ROTATION CAPACITY

In order that a structure attain the computed ultimate load, it is necessary for redistribution of moment to occur. As pointed out in Art. 2.6 and as emphasized frequently, the necessary transfer of moment is only possible if the plastic moment is maintained at the *first* hinge to form while hinges are developing *elsewhere* in the structure. The term "rotation capacity" characterizes this ability of a structural member to absorb rotations at near-maximum (plastic) moment. The behavior of the connection shown in Fig. 5.10, p. 155, would be described as "adequate," whereas the behavior characterized by curve B in Fig. 5.3, p. 148, would definitely be "inadequate."

It is evident that certain factors such as local or lateral buckling or brittle failure might limit the rotation capacity of a section. One of the primary aims of Chapter 4 in fact, was to make certain that no such secondary design factor would limit the capacity for deformation. Thus, in setting up a procedure for safeguarding against local buckling (Art. 4.4), it was specified that the section must not buckle until the extreme fiber reached the point of strain hardening ϵ_{st}. For uniform moment, the hinge rotation supplied in this case would be about twelve times the rotation that had occurred up to the elastic limit (Art. 1.3).

As might be imagined, the required rotation capacity of a given hinge position depends upon its location in the structure and the loading. Consider a fixed-ended beam, for example. If the load is uniformly distributed, the plastic hinges at the supports will require a considerable amount of rotation—enough to allow the load to increase by one-third. On the other hand, the hinge that will form at the center of the beam requires no rotation capacity since that hinge is the last to form. For a fixed-ended beam with concentrated load at the center, the rotation capacity requirement at both the ends and the center is theoretically zero because all plastic hinges form simultaneously. Therefore it is evident that the rotation capacity requirement will be a function of the loading and the geometry.

A study of rotation capacity for numerous practical structures was carried out in Ref. 4.5, and the theoretical results were compared with the results of tests. It was shown that structural members whose details are consistent with the provisions of Chapters 4 and 7 will, in fact, supply the maximum rotation capacity requirements of practical structures. A partial summary of the results of this study is contained in Fig. 4.19.

Even though a normal design problem will not require the calculation of required rotation at plastic hinges, an example is given on p. 202 to indicate the procedure—a procedure that is based directly upon the methods for computing deflections at ultimate load. Prior to working the problem, however, it is important to distinguish between three angles (or slopes) that may exist. These are:

(*a*) Slope of a member—designated by θ with subscripts as in Fig. 6.2.
(*b*) Hinge angle—designated by the quantity H and constitutes the angle of rotation through which a yielded segment of a beam must sustain its plastic moment value, enabling a structure to reach its ultimate load P_u.
(*c*) Mechanism angle—designated by θ without subscripts and is the virtual angle used to characterize the *additional* deformation of a structure after it reaches the ultimate load.

These three angles (or slopes) are illustrated and distinguished in Fig. 6.8 for a fixed-ended beam. The deflected shape of the beam is shown in three stages: (1) at the yield load W_y, (2) at the ultimate load W_u, and, (3) at a load equal to W_u but with displacement just greater than δ_u. Finally the mechanism is shown. At stage 1 just prior to formation of the first hinge at the end of the beam, elastic continuity exists throughout, and slopes $\theta_{23} = \theta_{32} = 0$. Upon further loading, rotation occurs at the hinges at sections 1 and 3. The magnitude of this angle (through which M must be maintained equal to M_p) is H_3 as shown in the sketch for stage 2. The slope at section 3 is no longer

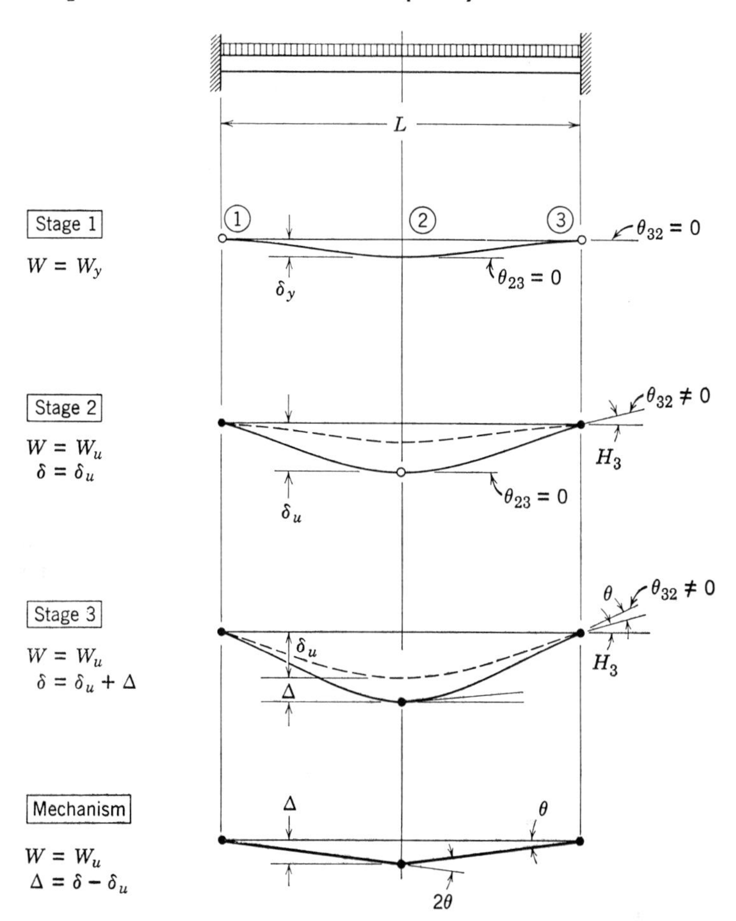

Fig. 6.8. Deformed shape of fixed-ended, uniformly loaded beam as load is increased to elastic limit (stage 1), to ultimate load (stage 2), and into the mechanism condition (stage 3).

zero, but since continuity still exists at section 2, the slope θ_{23} is still equal to zero. Stage 3 represents mechanism motion, and the corresponding additional virtual angles at the plastic hinges are the values θ. θ in the "mechanism" sketch is of course the same as θ in the sketch for stage 3.

The quantity in which our interest centers in this article is the hinge angle H representing the required rotation capacity for hinges at the supports of this particular structure. Referring to the sketch for stage 2 in Fig. 6.8, it is seen that H_3 is uniquely associated with the deflection δ_u. Therefore one can compute the magnitude of H from Eq. 6.1, replacing the subscripts AB by 32. Thus,

$$\theta_{32} = H_3 = \theta_{32}' + \frac{\delta_u}{L/2} + \frac{L/2}{3EI}\left(M_{32} - \frac{M_{23}}{2}\right)$$

$$H_3 = -\frac{M_pL}{12EI} + \frac{-M_pL^2/12EI}{L/2} + \frac{L/2}{3EI}\left(+M_p - \frac{M_p}{2}\right)$$

$$H_3 = -\frac{M_pL}{6EI} \tag{6.7}$$

where the quantity δ_u is obtained from Eq. 6.2. Equation 6.7 may be nondimensionalized by dividing both sides by ϕ_pL, giving

$$\frac{H_3}{\phi_pL} = -\frac{1}{6} \tag{6.8}$$

The nondimensional function, H/ϕ_pL, is the form that was used in the tabulation presented in Fig. 4.19. The same form was also used in Art. 4.5 to relate the critical lateral buckling length to the extent of rotation required at a plastic hinge.

EXAMPLE Hinge Angle for Fixed-Ended Beam (Fig. 6.4)

The above procedure will now be used to compute the hinge angle at section 1 of the example shown on p. 191. The maximum rotation requirement for this structure will thus be obtained, since the first hinge forms at section 1. The hinge angle H_1 will be equal to the change in slope at that section (Fig. 6.9). Therefore, making use of Eq. 6.1,

$$H_1 = \theta_{12} = \theta_{12}' + \frac{\delta_V}{l} + \frac{l}{3EI}\left(M_{12} - \frac{M_{21}}{2}\right)$$

From the example of Fig. 6.4 the quantity δ_V is known:

$$\delta_V = \frac{2M_pL^2}{27EI}$$

Therefore,

$$H_1 = \left(\frac{2M_pL^2}{27EI}\right)\frac{1}{L/3} + \frac{L/3}{3EI}\left(-M_p + \frac{M_p}{2}\right) = \frac{M_pL}{6EI}$$

or

$$\frac{H_1}{\phi_pL} = \frac{1}{6} \tag{6.9}$$

At the beginning of this article it was noted that some of the design guides were based on the desirable characteristic that the section be capable of rotating to the point of strain-hardening. The answer just found cannot be compared with this criterion in other than an approximate manner, because that answer was obtained on the assumption that all of the rotation occurred at a point. Actually the yield zone is distributed along the beam with strains varying all the way from the elastic limit to the point of strain-hardening or beyond. Nevertheless, an approximate comparison may be made by computing the average

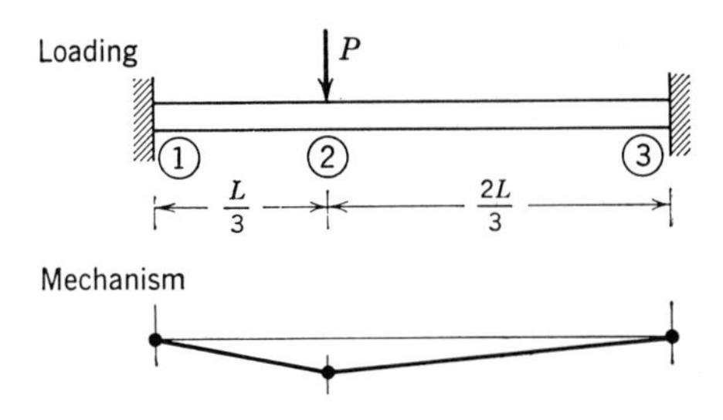

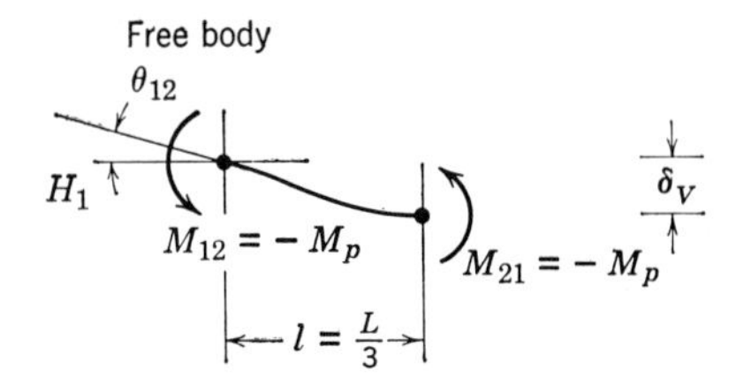

Fig. 6.9. Computation of hinge rotation for a fixed-ended beam.

unit rotation, ϕ_A, on the basis that the total rotation is divided by the "hinge length," defined in Art. 2.5 as the length along the beam in which the moment is greater than the yield moment. Referring to the previous example,

$$\phi_A = \frac{H_1}{\Delta L}$$

or

$$H_1 = \phi_A \Delta L \tag{6.10}$$

ΔL may be determined from Fig. 2.17b and Fig. 6.9 as

$$\Delta L = \left(\frac{L}{6}\right)\left(\frac{1}{8}\right) = \frac{L}{48}$$

Substituting into Eq. 6.9,

$$\frac{\phi_A \Delta L}{\phi_p L} = \frac{1}{6} \tag{6.11}$$

$$\frac{\phi_A}{\phi_p} = 8$$

Since ϕ_p represents the rotation up to the elastic limit (Fig. 2.6), the answer just obtained indicates that the yield zone must absorb a subsequent average unit rotation that is 8 times the value at the elastic limit—a magnitude that is well within the specified value of 12.

PROBLEMS

6.1 Derive the slope–deflection equations.

6.2 Compute the deflection at ultimate load for the beams shown in Fig. 1.1*a* and 1.1*b*. Compute the deflections at working load.

6.3 Compute the deflection under the load of magnitude $2P$ of Fig. 3.6 when ultimate loading has been reached. Estimate the deflection at working load.

6.4 A beam of length L is fixed at one end and simply-supported at the other. Concentrated loads P are applied at the third-points. Compute the deflection at ultimate load.

6.5 Compute the required hinge angle H for Problem 6.3.

6.6 A fixed-ended beam is loaded at one of the third-points by a concentrated load P. Draw the complete load–deflection relationship.

6.7 A rectangular frame with two equal spans of length L has columns of length L. The loading consists of a horizontal load P acting to the right at the top of the left column, a concentrated vertical load $2P$ acting at the center of the left girder, and a uniformly distributed load $wL = 4P$ acting on the right girder. The columns have a moment capacity M_p, and moment of inertia I_1; the beams have a moment capacity of $2M_p$ and moment of inertia I_2. Find the maximum vertical deflection.

7 Design Guides

It is the purpose of this chapter to focus attention on problems encountered in design. Up to this point the emphasis has been on methods of analysis—both as regards analyzing the structure for maximum strength and as regards details. Now, the emphasis will be on design, and the information developed thus far will be used in the solution of such problems.

Since most of the theoretical and experimental data have already been given, the information in this chapter is presented primarily in tabular form for ready reference. The first tabulations (Tables 7.1, 7.2, and 7.3) are summaries or outlines of provisions and procedures. The second group (Tables 7.4 to 7.8) are "design guides" for the proportioning of members and connections and for details.

Of first interest are the "General Provisions." What basic conditions must be satisfied before a plastic design procedure may be set up? In this category would be included questions regarding types of construction, materials, fabrication methods, the yield stress level and corresponding plastic moment, loads and forces, and the load factor.

A summary of the steps in design follows, and of interest here is the choice of relative strength of the different members. Since the design procedures rest directly on the methods of analysis, a summary of procedures of analysis follows.

Finally, design guides are presented that are of use in accounting for axial force and shear force, for proportioning beams and girders, columns and beam-columns, connections, and for details.

Actually few new principles are involved in this chapter. Although there is some explanation of the provisions, procedures, and guides in the text of this chapter, reference is given throughout to the appropriate articles that appear earlier in this book or to other sources for additional information.

7.1 GENERAL PROVISIONS

Table 7.1 is an outline of the general provisions.

Provision 1, types of construction. Plastic analysis should be applied in the design of those structures and structural components which have been studied by analysis and/or tests sufficiently to give confidence that the calculated maximum strengths will be realized.

Subject to the other provisions of this article, plastic design may be applied to continuous structures suggested in Provision 1 using cross-sectional forms meeting the requirements of strength and reserve ductility. Both military and civil structures are included. Continuity may be achieved by welding, or in some cases by riveting or bolting. The application to tier buildings is limited at present in that the lateral forces should be carried by panel bracing.

Provision 2 (material) and Provision 3 (fabrication). These two provisions, in part, constitute safeguards against brittle failure. In specifying any other material than that suggested in Provision 2, it would be most important to verify it insofar as the other requirements suggested in the book are concerned. If punched holes are required for erection, they may frequently be located in regions that would not be subjected to large tensile forces.

Provision 4, yield stress level. The magnitude of the yield stress level (σ_y = 33,000 psi) corresponds to the minimum yield point permitted in a mill-type acceptance test. Since the value is very close to the average yield stress level obtained in a series of tests conducted on WF shapes to determine the basic static yield stress of ASTM A7 steel,[2.3] it is thus reasonable to use a similar philosophy when specifying σ_y for some other material for use in plastic design. Alternatively, a representative series of "stub column" tests may be performed on such "new" material and the results can be correlated with mill tests as was done in work reported in Ref. 2.3.

Although 33,000 psi represents the *minimum* yield point permitted in the mill, it represents about the *average* basic yield stress level of material rolled to ASTM A7 specifications. Thus the factor of safety or "load factor" actually covers the possibility of variation below this average value.

Provision 5, plastic moment and plastic modulus. In addition to the basic equation, formulas have been given for computing the plastic modulus of a pair of cover plates and for computing the net value of the plastic modulus when holes are provided in the flange or web. In the latter case, when Eq. 7.2 is used, it is not necessary to compute the shift in neutral axis because the resulting error is less than about 2%. Formulas for computing Z for H- or I-shaped members and other approximations to the plastic modulus are presented in Art. 2.5. In some cases it may be found advantageous to use two channels welded together along the toes for increased torsional resistance. Equation 2.23 would be appropriate in that case.

There is considerable variation in the shape factor ($1.10 < f < 1.23$) and this matter is discussed under Provision 7.

Provision 6, loads and forces. It is assumed that the loading is static and proportional, even for the ordinary fluctuations of load found in buildings. For unusual cases, the references of Art. 4.8 may be consulted.

Ordinarily the loading conditions that must be investigated are total live load plus dead load, live load plus dead load plus wind acting from the left, and live load plus dead load plus wind acting from the right for unsymmetrical conditions. In continuous beams the usual "checkerboard" loading need not be investigated, since the load-carrying capacity of any one span is not a function of the load on adjacent spans. This may be illustrated by referring to Fig. 3.5. If the load P is applied to span 1–3 alone, then the required plastic moment M_p is equal to $PL/6$. No matter what load is applied to span 3–5, as long as the moment capacity of that span is at least equal, then the strength of span 1–3 is unaffected by the adjacent loads.

Provision 7, load factor. In order to determine the loads to be used in the design of a structure by the plastic method, the expected or design loads must be multiplied by a factor of safety. This gives the ultimate load on the basis of which the section is selected.

As suggested in Art. 1.7 the philosophy by which this load factor is selected is as follows: If the present conventional elastic design of a simply supported beam is satisfactory, then an indeterminate structure should be designed with a load factor equal to that of a simple beam. There would certainly be no point in requiring any greater margin of safety simply because the structure is redundant. Now, the load factor of a simple beam is equal to the ratio of the ultimate load P_u divided by the working load P_w; thus $F = P_u/P_w$. In a simple beam the bending moment varies linearly with the load, or

$$F = \frac{P_u}{P_w} = \frac{M_p}{M_w}$$

TABLE 7.1. OUTLINE OF GENERAL PROVISIONS

Provision Number, Title, Outline of Provision, and Formulas	Definition of Terms, Units, Sketches, and Explanatory Notes	References
① Types of construction (1) Continuous beams. (2) Industrial building frames of limited number of stories (single- or multi-span, one or two stories). (3) Tier buildings with wall bracing. (4) Structures intended to absorb dynamic loads (bomb burst, collision, earthquake).		Art. 1.1
② Materials The material would be structural grade steel having the characteristics of ASTM A7, with modifications in specifications when needed to insure weldability and freedom from cracking at lowest service temperature anticipated.	(1) AWS has promulgated specification for weldability. (2) Other steels, such as high-strength steels, with characteristic diagram may be used if the other requirements are met.	ASTM A7 ASTM A373 Art. 4.7 Art. 1.3
③ Fabrication The fabrication process must not inhibit ductility. To this end, sheared edges and punched holes in zones of plastic tension are not permitted. Punched and reamed holes for connecting devices will be suitable if the reaming removes the cold-worked material.	The design of details should not inhibit ductility. Triaxial tension is to be minimized.	Art. 4.7 Ref. 4.16 Ref. 7.2
④ Yield stress level For ASTM A7 steel, $\sigma_y = 33{,}000$ psi $\tau_y = 19{,}000$ psi	$\tau_y = \dfrac{\sigma_y}{\sqrt{3}}$ = shear yield stress σ σ_y = Yield stress level ϵ	Ref. 2.3 Art. 2.5

Provision	Definitions and notes	Reference
(5) Plastic moment and plastic modulus $$M_p = \sigma_y Z \qquad (2.11)$$ $$Z\ (\text{in.}^3) = 0.364 M_p\ (\text{kip-ft}) \qquad (2.11a)$$ WF's: See Appendix 2 for values of Z. Cover plates: $$\Delta Z = A_p d_p = A_p(d + t_p) \qquad (7.1)$$ Net Z for member with holes in flange or web, $$Z_{\text{net}} = Z - \frac{t(d - t)\ (\text{diam.})}{2} - wy\ (\text{diam.}) \qquad (7.2)$$	M_p = Plastic moment. Z = Plastic modulus. $= f \cdot S$. 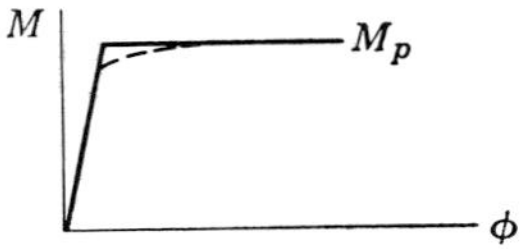 f = Shape factor. For WF's, $1.10 < f < 1.23$. t,d,w = Flange thickness, section depth, web thickness. A_p, d_p = Area and separation distance of plates. y = Distance from centroid to hole. For welded I's, channels, etc., see Art. 2.5, Tables 7.5, and 7.8.	Art. 2.5
(6) Loads and forces Design (working) loads and forces are to be those presently specified for the particular type of construction. Members are selected on basis of the most critical ultimate load condition $$P_u = FP_w \qquad (7.3)$$ Loading is assumed to be static and proportional.	F = Load factor. See Provision 7. Loading conditions: (1) Live Load + Dead Load. (2) L.L. + D.L. + Wind from left. (3) L.L. + D.L. + Wind from right. (Unsymmetrical structure.) NOTE: "Checkerboard" loading need not be investigated in continuous beams.	
(7) Load factor For building construction: D.L. + L.L., $F = 1.85$ D.L. + L.L. + Wind, $F = 1.40$ (7.5) For other forms of construction, $$F = \frac{\sigma_y}{\sigma_w} f \qquad (7.6)$$	σ_w = Working stress permitted according to conventional specification.	Ref. 1.9, p. 11.1 Ref. 6.1

Substituting the known values for M_p and M_w,

$$F = \frac{M_p}{M_w} = \frac{\sigma_y Z}{\sigma_w S} = \frac{33}{20} f = 1.65f \tag{7.4}$$

Thus the magnitude of the load factor is dependent upon the shape factor.

The variation of the shape factor for WF shapes used as beams, for WF shapes used as columns, and for I shapes is shown in Fig. 7.1. For

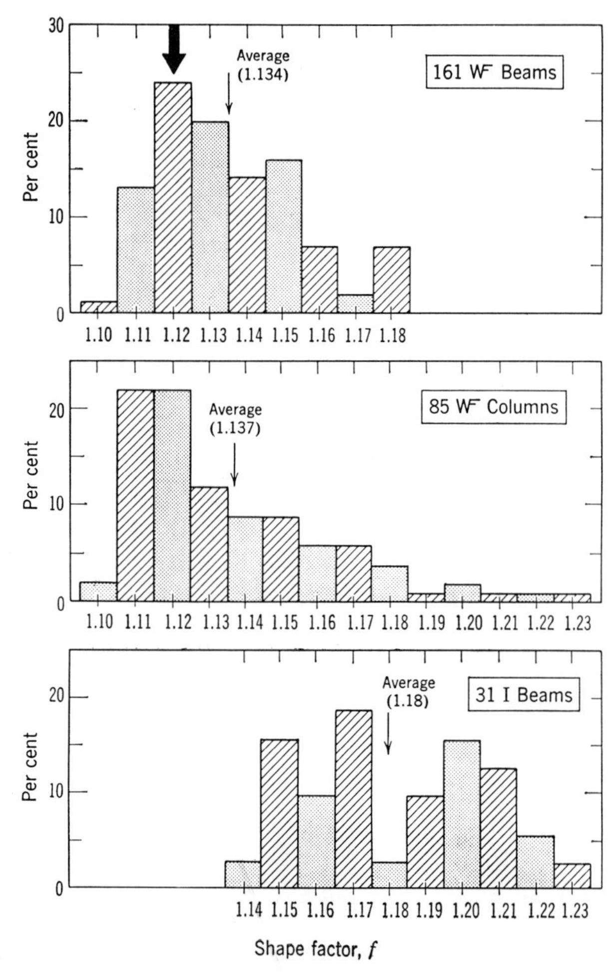

Fig. 7.1. Variation of shape factor for rolled WF and I shapes.

WF shapes normally used as beams (WF or other H-shaped members listed in the "Section Economy" table in the AISC Manual *Steel Construction*)[2.1] the shape factor varies from 1.10 to 1.18 with an average value of 1.134 and a mode of 1.12. For WF shapes normally used as columns (members that appear in the "column" tables of Ref. 2.1), the shape factor varies from 1.10 to 1.23 with an average value of 1.137 and a mode of 1.115. The shape factor distribution of American Standard I beams is shown in the lower portion of Fig. 7.1. The minimum is 1.14 and the maximum is 1.23, the average being 1.18.

The accompanying table shows the possible values of the load factor, depending on the choice of the shape factor.

Shape Factor	Factor of Safety *	Load Factor
1.10—Minimum value	1.65	1.81
1.12—Mode for WF beams	1.65	**1.85**
1.14—Average for WF beams and columns	1.65	1.88
1.18—I beams	1.65	1.95
1.23—Maximum value	1.65	2.03

* Yield stress divided by working stress.

The two most reasonable values for the load factor are 1.85 and 1.88. The former is selected because it represents the shape factor that will recur most frequently in beams. The number 1.88 implies an accuracy in the knowledge of the general problem of safety that is not justified.

In the case of loading in combination with such forces as those due to wind and earthquake, elastic design specifications normally permit a one-third increase in stresses. Consistent with this allowance, the value of F for combined dead, live, and wind loading would be $\frac{3}{4} \times 1.85 = 1.40$.

In summary, then, the load factors are:

$$\left.\begin{array}{ll} \text{Dead load plus live load,} & F = 1.85 \\ \text{Dead load plus live load, plus wind or earthquake forces.} & F = 1.40 \end{array}\right\} \tag{7.5}$$

Although safety is achieved in design by multiplying the working loads by a term that is called a load factor, it is important to recognize that, besides variation in load, there are many other possible sources of error in design.[6.1] Some of these are:

(1) Overrun in computed dead loads.
(2) Future increases in live loads.
(3) Loss of section due to corrosion.
(4) Approximations in analysis.
(5) Underrun in dimensions.
(6) Underrun in physical properties.
(7) Inadequate design theory.
(8) Errors in distribution of load.
(9) Errors in fabrication and erection.
(10) Presence of residual stresses and stress-concentrations.
(11) Time effects.

Depending upon the type of structure and the use intended, the variation due to one error may be larger than another. Although one might arrive at a new "factor of safety" by determining the magnitude of possible errors and making a statistical estimate of the total factor, any change should also be reflected in the allowable stresses permitted in conventional elastic design. Therefore, all of the possible variations have been combined into a "load factor" which assures that a plastically designed structure will have the same degree of safety as is inherent in a simply-supported beam designed by elastic methods.

The term "load factor" also serves to emphasize that the final test of the suitability of a structure rests upon its ability to support the load.

7.2 DESIGN

In Table 7.2 are outlined certain procedures of design. Some aspects of the procedures are discussed in the following paragraphs.

Procedure 1, preliminary design. On what basis is the first choice of relative plastic moment values made? In the various examples used to illustrate methods of *analysis*, the problem was to find the ultimate load for a given structure with known plastic moment values of its members. In *design*, the problem is reversed. Given a certain set of loads, the problem is to select members of suitable strength. Since "uniform section throughout" may not be the most economical solution, some guide may be needed for selecting the ratio or ratios of plastic moment strength of the various members.

Of course, this problem exists in elastic design, so it is not a matter that is unique to design on the basis of ultimate load. However, a few simple techniques will occur to the designer which, coupled with his experience, will enable him to make a preliminary economical choice of relative moment strength without too many trials. Some general principles are as follows:

(1) In the event the critical mechanism involves partial failure with a portion of the structure remaining redundant, the material in the redundant portion of the frame is not being used to full capacity. This suggests that a more efficient choice of moment ratios may be made such that the critical mechanism is a "composite mechanism" involving plastic hinges in more members.

(2) Adjacent spans of continuous beams will often be most economically proportioned when the independent beam mechanism forms in each span simultaneously. This is illustrated in Plate II, p. 256. Numerous examples of the design of continuous beams are given in Ref. 7.3.

(3) The absolute minimum beam section for vertical load is obtained if the joints provide complete plastic restraint (i.e., restraining members supply a restraining plastic moment equal to that of the beam). Similarly, the minimum column sections are obtained under the action of sway forces when full plastic restraint is provided at the ends. This suggests that, if the important loads are the vertical loads, the design might well be commenced on the basis that all joints are restrained as described, the ratio of beam sections be determined on this basis and that the columns be proportioned to provide the needed joint moment balance and resistance to side load.[3.4] Plate XI, p. 314, is an illustration. Alternatively, if the important loads were side loads, the design could start instead with the columns.

(4) Finally, it should be kept in mind that maximum over-all economy is not necessarily associated with a choice of lightest section for each span. It is necessary to consider fabrication conditions which may dictate uniform section where, theoretically, sections of different weight might be used.

Procedure 2, general design procedure. Although there will be variations as to specific procedure and detail, the following steps will be a part of practically every design:

(1) Determine possible loading conditions on the structure.
(2) Compute the ultimate load(s).
(3) Estimate the plastic moment ratios of frame members.
(4) Analyze each loading condition for maximum M_p.
(5) Compute reactions.
(6) Select the sections.
(7) Check the result according to "Design Guides."
(8) Estimate the deflection (if required).

These steps are discussed briefly.

The *first* step is to determine the possible loading conditions using the suggestions of Table 7.1, Provision 6. At this stage it is decided whether to treat distributed loads as such or to consider them as concentrated. (See Art. 3.7.)

The *second* step, to compute the ultimate load, represents a departure from present practice in that the working loads are multiplied by the load factor to assure an adequate margin of safety. This has been discussed in Art. 7.1, Provision 7.

The *third* step is to make an estimate of the plastic moment ratio of the frame members. This has been discussed in the previous section and has been outlined in Procedure 1.

In the *fourth* step each loading condition is analyzed for the maximum required M_p. Either the statical method outlined in Table 7.3, Procedure 1, or the mechanism method (Table 7.3, Procedure 2) may be used. Alternatively, the simplified methods outlined in Procedure 6 below may be used for "standard" geometrical and loading conditions for which charts and graphs have been developed. The only difference in this step and in the analysis procedures of Chapter 3 is that the *lowest failure load* was sought in the latter, whereas now we are looking for the *maximum required plastic moment* as a basis for selecting the section.

The *fifth* step is to compute the reactions for each loading condition. Whereas in most cases it is appropriate to use the reactions for the critical loading condition when checking the secondary design factors, there are some exceptions. An example would be the case in which two different loading conditions result in practically equivalent M_p values but with quite different reactions.

The *sixth* step is to select the section. The equation $M_p = \sigma_y Z$ is solved for Z and the section is selected from the economy table in Appendix 2. Note that when $\sigma_y = 33.0$ ksi, $Z = 0.364\ M_p$, where Z is expressed in in.^{3} and M_p in kip-ft. The table of Appendix 2 also could be prepared in terms of the corresponding plastic moment M_p. It would have the advantage of eliminating one step in the design procedure. The reason the Z value has been used is that it makes Appendix 2 applicable for a material with a yield stress other than 33.0 ksi.

The *seventh* step (and a most important one) is to check the design to see that it satisfies the "guides" contained in Tables 7.4 through 7.8.

An *eighth* step is occasionally required if the designer is concerned about the flexibility of the structure. Procedures 11 and 12 of Table 7.3 are concerned with estimates of deflections.

Procedure 3, continuous beams of uniform section. For continuous beams the statical method will usually be the simplest analytical technique.

TABLE 7.2 OUTLINE OF DESIGN PROCEDURES

Procedure Number, Title, Outline of Procedure, and Formulas	Definition of Terms, Units, Sketches, and Explanatory Notes	References
① Preliminary design (First choice of plastic moment ratios) (1) Determine absolute plastic moment values for separate loading conditions. (Assume all joints fixed against rotation, but frame free to sway.) (*a*) Beams: Solve beam mechanism equation. (*b*) Columns: Solve panel mechanism equation. Actual section will be greater than or at least equal to these values. (2) Select plastic moment ratios using the following guides: (*a*) Beams: Use values determined in step 1 above. (*b*) Columns: At corner connections, M_p (column) $= M_p$ (beam). (*c*) Joints: Establish equilibrium. (3) Analyze for maximum required plastic moment values (see Procedure 2). (4) Examine frame for further economies as may be apparent from consideration of relative beam and sway moments.		Plate XI, p. 314
② General design procedure (1) Determine possible loading conditions (Table 7.1, Provision 6). (2) Compute ultimate loads by multiplying working loads by F (Provision 7). (3) Estimate plastic moment ratio of frame members (Procedure 1, above). (4) Analyze each loading condition for maximum M_p (Procedure 6, or Table 7.3, Procedures 1 or 2). (5) Compute reactions for each loading condition. (6) Select section (Provision 5, $Z = 0.364M_p$). (7) Check design to see that it satisfies applicable Design Guides (Tables 7.4–7.8). (8) Check deflection if necessary (Table 7.3, Procedures 11, 12).		Chapters 8, 9

TABLE 7.2. OUTLINE OF DESIGN PROCEDURES (*Continued*)

③ Continuous beams—uniform section Use the *statical* method of analysis (Procedure 1 of Table 7.3). On the ultimate load moment diagram draw fixing lines for the end span and interior span with the largest determinate bending moments. Then select greater of required M_p's.	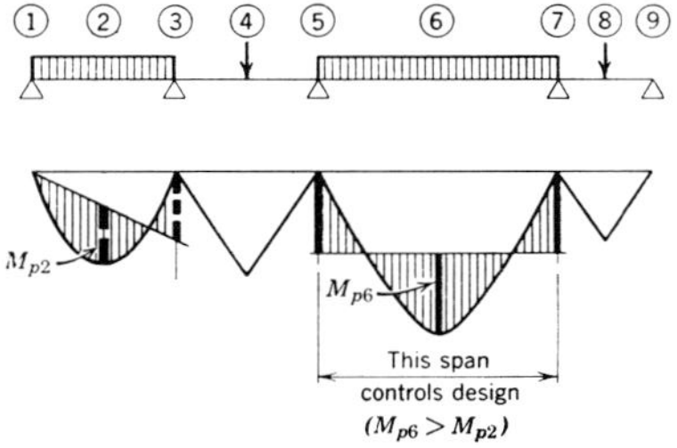 ① ② ③ ④ ⑤ ⑥ ⑦ ⑧ ⑨ M_{p2} M_{p6} This span controls design ($M_{p6} > M_{p2}$) NOTE: See Procedure 1, Table 7.3, for moment convention. See Fig. 8.1 for standard cases.	
④ Continuous beams—nonuniform section For maximum economy of material, proportion members such that mechanism forms in each span. *Method 1:* Select section to suit maximum required M_p as in Procedure 3 above and select corresponding sections for remaining spans. Locate splices to suit. See Illustration *A*. *Method 2:* Select section to suit a smaller plastic moment requirement and reinforce with cover plates where $M > M_p$. See Illustration *B*. *Method 3:* Use members of variable cross-section. See Procedure 3, Table 7.5.	Illustration *A* 1 2 3 4 5 6 7 8 *a* *d* M_{p2} *A* *b* M_{p4} *c* *B* M_{p7} Illustration *B* *a* *i* *e* *j* *f* *k* *b* *l* *g* *m* *h* *n* *c* M_{p5} *d* M_{p7} Cover plates Cover plates	Ref. 7.3 Plate II, p. 256 Plate III, p. 260 Plate IV, p. 263

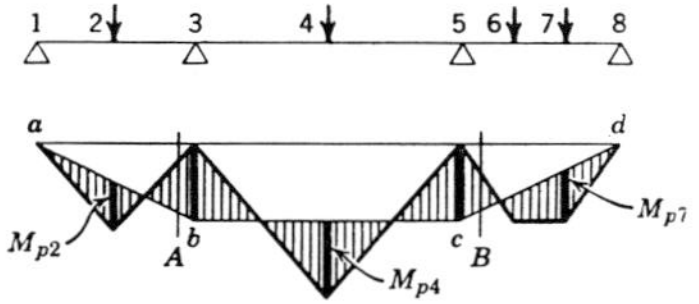

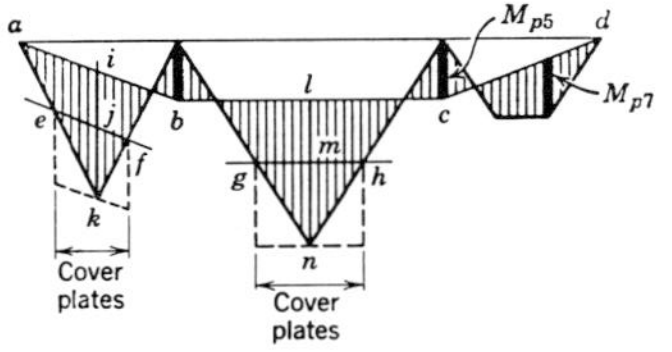

Procedure Number, Title, Outline of Procedure, and Formulas	Definition of Terms, Units, Sketches, and Explanatory Notes	References
(5) Tier buildings Proportion beams and girders by plastic methods. Proportion columns according to conventional methods.	(1) Diagonal bracing in walls to resist shear. (2) See also Procedure 3 of Table 7.6.	Art. 9.6
(6) Simplified methods For "standard" dimensions and loading conditions, design aids in the form of charts, formulas, and graphs may be used. From such aids (see Arts. 8.5, 9.3, 9.5) determine required member sizes. Check Design Guides to see that applicable requirements are met.		Ref. 7.1 Ref. 7.4 Art. 8.5 Arts. 9.3, 9.5

Usually the critical span requiring the largest plastic moment ratio can be determined by inspection, in which case only one equilibrium equation need be solved. Having sketched the moment diagrams to scale, the fixing lines are next drawn for the end span that carries the largest determinate bending moment and similarly for the interior span. The greater of the resulting M_p values indicates the critical span, and a section is selected accordingly. This has been illustrated in Table 7.2 in the sketch. Span 1–3 has a greater bending moment than span 7–9 and therefore a fixing line is drawn for the former span. In the interior, the fixing line is drawn for span 5–7 because it supports the largest determinate moment. Since $M_{p6} > M_{p2}$, span 5–7 controls the design of this uniform continuous beam and the section is selected on that basis.

Procedure 4, continuous beams of nonuniform section. For economy of steel, the members of a continuous beam could be proportioned in such a way that a mechanism formed in each span. There are several methods of accomplishing this. Using prismatic members, sections can be selected that will just form a mechanism for each span (using the statical method), and splices may be located near the point of inflection. In illustration A of Procedure 4, for example, the maximum required plastic moment is controlled by span 3–5. By drawing the redundant line a–b–c–d the required plastic moment values at sections 2 and 7 are determined, and the required sections can then be selected. The splice in span 1–3 would be located anywhere between the point of inflection and section A where $M = M_{p2}$. Similarly for the splice in span 5–8. Of course, if the splices were located at supports 3 and 5 there would be insufficient restraining moments for the central span. Plate II, p. 258, illustrates the procedure.

Another method is to select a section that will suit the smallest plastic moment requirement and then to reinforce the member with cover plates. Thus as sketched in illustration B of Procedure 4, after the determinate moment diagrams are drawn, the fixing line a–b–c–d is drawn in such a way that $M_{p7} = M_{p5}$, with line b–c parallel to line a–d. In order to determine the required length of cover plates in spans 1–3 and 3–5, the distances i–j and l–m are laid off equal to M_{p5} representing the moment capacity of the beam. The cover plates make up the difference and would extend over distances e–f and g–h, respectively. The required moment capacity of the cover plates is then given by j–k and m–n, and the corresponding required areas of the plates would be obtained from Eq. 7.1 of Provision 5. Plate III, p. 260, is illustrative.

In some instances, members of variable cross section may be used. This procedure has been described in connection with Plate IV, p. 263.

Procedure 5, tier buildings. In multi-story buildings plastic design could be applied to the proportioning of beams and girders so long as horizontal forces are resisted by diagonal bracing in wall panels. The subject is discussed further in Art. 9.6.

Procedure 6, simplified methods. As in the case of elastic design, it has been possible to solve certain "standard" cases and present the results in the form of charts, formulas, or graphs, thus eliminating the necessity for making a complete plastic analysis. The development of such design aids has been discussed in Arts. 8.5, 9.3, and 9.5, and several examples have been given to illustrate the methods.

The majority of the examples in this book have been solved, however, by the direct prccedure in order to illustrate the principles of plastic analysis and design.

7.3 ANALYSIS

Table 7.3 is a partial outline of Chapter 3. As before, the material is arranged in the form of a procedure for each topic—a procedure that is intended for reference purposes after the principles have been understood.

Procedure 1, statical method. The choice of the appropriate method of analysis is governed mainly by the personal preference of the designer. Usually it will be found convenient to use the statical method for continuous beams and for frames with only a few redundancies.

The convention for drawing moment diagrams, shown in illustration A, was described in Art. 3.4.

In step 5 of this procedure the appropriate equilibrium equations of Procedure 8, p. 223, could be used, but it will probably be simpler to use the moment diagram directly to formulate the equilibrium equation. Thus for the problem shown in illustration A, at section 4,

$$M_p = b\text{–}c = a\text{–}c - a\text{–}b$$

from which

$$M_p = \frac{P_2 L}{4} - \frac{M_p}{2}$$

and this equation may be solved for M_p.

In drawing the moment diagram due to loading by a horizontal redundant force at the column base of pinned-base frames with gabled or sloping roofs, it is convenient to locate point 0, the center of a radius along which segment 2–3 must lie. (See illustration B.) From similar

TABLE 7.3. OUTLINE OF ANALYSIS PROCEDURES

Procedure Number, Title, Outline of Procedure, and Formulas	Definition of Terms, Units, Sketches, and Explanatory Notes	References
① Statical method By the following procedure find an equilibrium moment diagram in which $M \leq M_p$ such that a mechanism is formed: (1) Select redundants. * (2) Draw moment diagram for determinate structure. * (3) Draw moment diagram for structure loaded by redundants. (4) Sketch composite moment diagram in such a way that a mechanism is formed (sketch mechanism). (5) Compute value of M_p by solving equilibrium equations. (6) Check to see that $M \leq M_p$.	Method based on lower bound theorem. By requiring that a mechanism form it also satisfies upper bound theorem. *Moment diagram convention:* Moment diagram drawn on the side of the member that is in tension. Illustration *A* 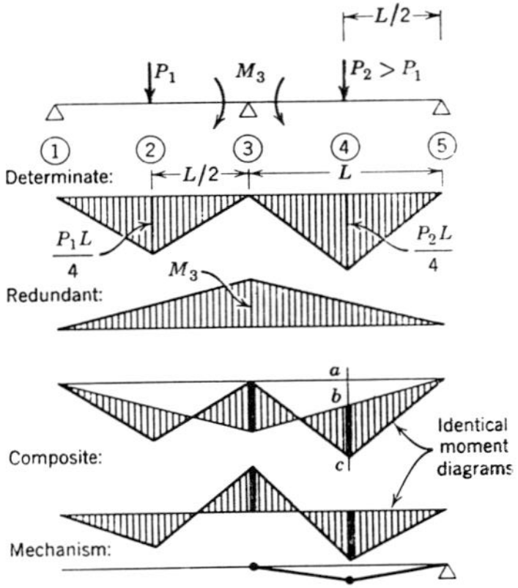	Art. 3.4
$x = \frac{a}{b} L_c$ (7.8)	Illustration *B* 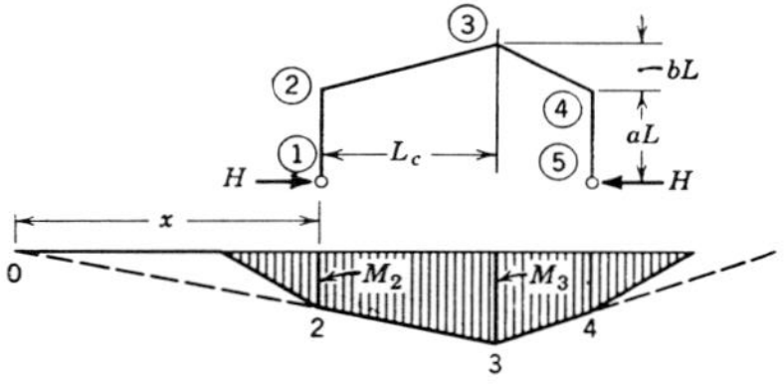	
* These two steps are usually omitted. In routine design one may go directly from step 1 to step 4.		

Procedure	Remarks	Reference
② Mechanism method By the following procedure find a mechanism (independent or composite) such that $M \leq M_p$: (1) Determine number and location of possible plastic hinges (Procedure 6). (2) Determine number of redundants (Procedure 3). (3) Find number of independent mechanisms (Procedure 5). (4) Select possible independent and composite mechanisms. (5) Solve virtual work equations for maximum M_p. (6) Carry out moment check to see that $M \leq M_p$ (Procedure 7).	1. This method is based on upper bound theorem. If $M \leq M_p$ it also satisfies lower bound theorem. 2. See Art. 3.5 for types of mechanisms. 3. Tabular solution will often be convenient. See Fig. 3.19b, for example.	Art. 3.5
③ Indeterminacy To determine the number of redundants, cut sufficient supports and structural members such that all loads are carried by simple beam or cantilever action. X = Forces + moments required to restore continuity (7.9)	Examples given in Fig. 3.10 X = Number of redundants.	Art. 3.5, p. 75
④ Partial redundancy For an assumed mechanism, the number of remaining redundancies at failure is given by $I = X - (M - 1)$ (3.23)	I = Remaining redundancies. X = Redundancies in original structure. M = Number of plastic hinges necessary to develop a mechanism.	Art. 3.8
⑤ Number of independent mechanisms $n = N - X$ (3.17)	n = Number of independent mechanisms. N = Number of possible plastic hinges (see Procedure 6). X = Redundancies.	Art. 3.5, p. 72

TABLE 7.3. OUTLINE OF ANALYSIS PROCEDURES (*Continued*)

<table>
<tr><th>Procedure Number, Title, Outline of Procedure, and Formulas</th><th>Definition of Terms, Units, Sketches, and Explanatory Notes</th><th>References</th></tr>
<tr><td>(6) Location and number of possible plastic hinges
Hinges may form at:
(1) Points of concentrated load.
(2) End of each member meeting at connection.
(3) Point of zero shear in span under distributed load.</td><td>NOTE: For haunches and tapered members see Art. 5.4, p. 174.
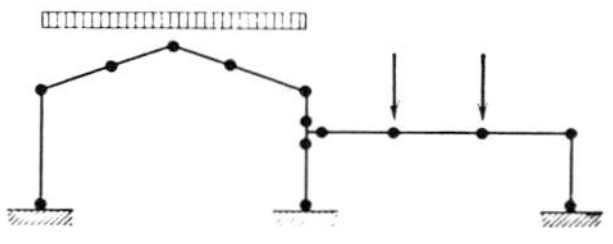
</td><td>Art. 2.5</td></tr>
<tr><td>(7) Moment check
When it is believed that the correct solution has been found, draw moment diagram to make certain that the "plastic moment" condition is satisfied (that is, $M \leq M_p$).
For structure that is determinate at failure: Compute reactions and find unknown moments by static equilibrium.
For partially redundant structure (Procedure 4) use one of the following methods:
Trial and error method: Assume values for "I" unknown moments (as determined from Procedure 4) and compute remaining values from equilibrium equations.
Moment-balancing method: Use a pseudo moment-distribution process as outlined in Art. 3.8 and illustrated on p. 92 (Fig. 3.19).
Semi-graphical method: Draw the "known" portion of the moment diagram and find possible values for remaining moments by shifting the various "simple-span" and column sway moment diagrams, always keeping within + and − values of M_p. See Art. 3.8, p. 93 (Fig. 3.20).</td><td>Sign convention for moments: Clockwise end moments are positive. Positive interior moments produce tension on lower side.
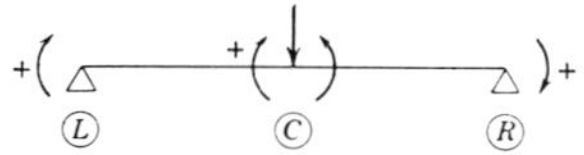

Carry-over factors for "moment-balancing"
<table><tr><th>ΔM_L</th><th>ΔM_C</th><th>ΔM_R</th></tr><tr><td>1</td><td>½</td><td>0</td></tr><tr><td>1</td><td>0</td><td>1</td></tr><tr><td>0</td><td>− ½</td><td>1</td></tr></table>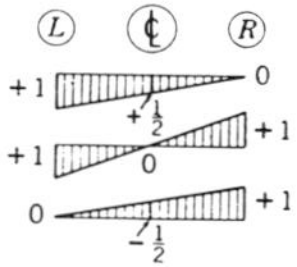

For partially redundant structure attempt redesign that will involve complete failure of frame.</td><td>Art. 3.8</td></tr>
</table>

(8) Equilibrium moment equations

Beams (Typical):

$$M_C = \frac{M_L}{2} - \frac{M_R}{2} + \frac{P_u L}{4} \qquad (7.10)$$

Joint:

$$M_a + M_b + M_c = 0 \qquad (7.11)$$

Panel (Sway):

$$M_1 + M_2 + M_3 + M_4 + M_5 + M_6 + Ph = 0 \qquad (7.12)$$

Sign convention for moments: Same as above.

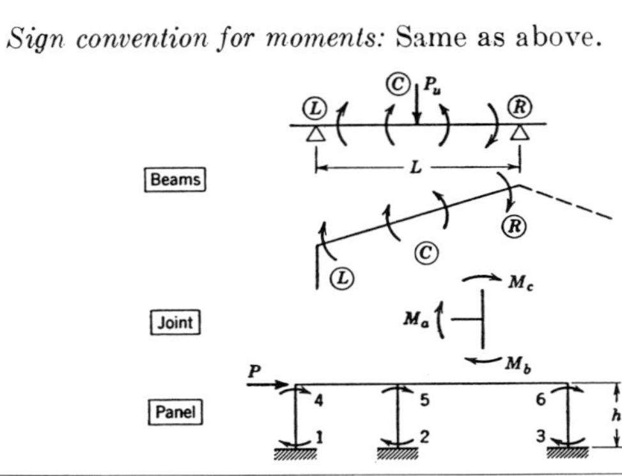

(9) Distributed load

A conservative result is obtained if distributed load is replaced by equivalent concentrated loads. (See typical example.)

If correct mechanism involves hinge in beam, further economy is gained by working out mechanism by differentiation or trial and error.

Another procedure is to assume that hinge forms at mid-span and then revise design after correct mechanism has been found.

Where load is brought to main frame by purlins, distributed load may be converted at the outset to actual concentrated loads applied at the assumed purlin spacing.

In mechanism method,

$$W_E = w \text{ lb/ft} \times \text{area swept} \qquad (7.20)$$

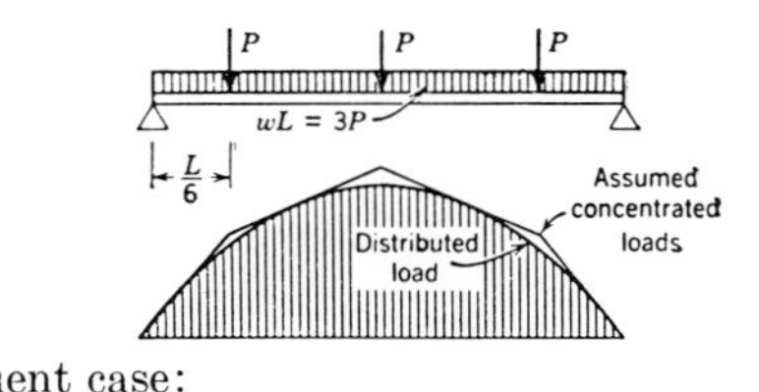

A frequent case:

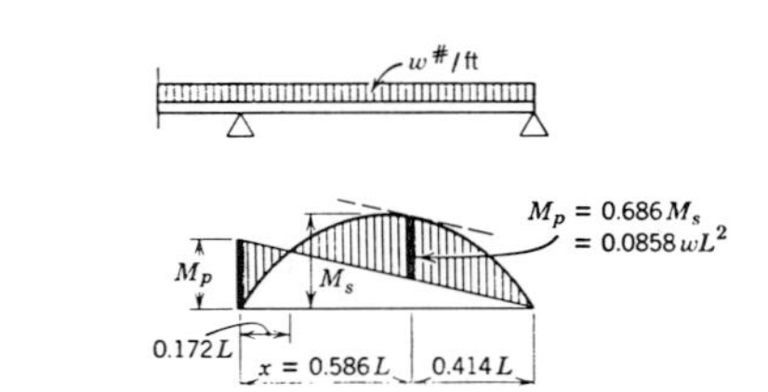

Art. 3.7

TABLE 7.3. OUTLINE OF ANALYSIS PROCEDURES (*Continued*)

Procedure Number, Title, Outline of Procedure, and Formulas	Definition of Terms, Units, Sketches, and Explanatory Notes	References
(10) Variable repeated loading These procedures are intended for cases normally considered as "static" loading. For such cases the problem of repeated loading may be disregarded because the actual stabilizing load ("shakedown" load) will not be exceeded.	NOTE: Where the full magnitude of vertical load on a member is expected to vary, the ultimate load may be modified according to analysis of deflection stability.	Art. 4.8
(11) Deflection at ultimate load If computation of deflection at ultimate load is required, the following steps may be used: (1) Obtain P_u, moment diagram, and mechanism (Procedures 1, 2, and 7). (2) Compute deflection of frame segments assuming, in turn, that each hinge forms last. (3) Correct deflection is largest value. (4) Check by "kink-removal" process.	Slope–deflection equation: $\theta_A = \theta_A' + \frac{\Delta}{l} + \frac{l}{3EI}\left(M_{AB} - \frac{M_{BA}}{2}\right)$ (6.1) (A) (B) M_{AB} θ_A θ_B Δ M_{BA} l	Art. 6.2
(12) Deflection at working load If computation of beam deflection at working load is required, refer to handbook tables. See example on p. 196, Fig. 6.6. Upper limit of deflection of frame at working load is obtained by dividing the deflection at ultimate load (Procedure 11) by F.	NOTE: Deflection of plastically designed structure is usually less than that of a similar structure designed on a "simple beam" basis.	Art. 6.3

triangles,

$$\frac{x}{M_2} = \frac{L_c}{M_3 - M_2} \tag{7.7}$$

from which

$$x = \frac{a}{b} L_c \tag{7.8}$$

In Ref. 7.1 a similar procedure is described for fixed-base frames.

Procedure 4, partial redundancy. Although there are some exceptions, usually the best possible design will not be obtained if the remaining structure is redundant. The aim is to make the structure fail in a manner that uses all of the material most effectively.

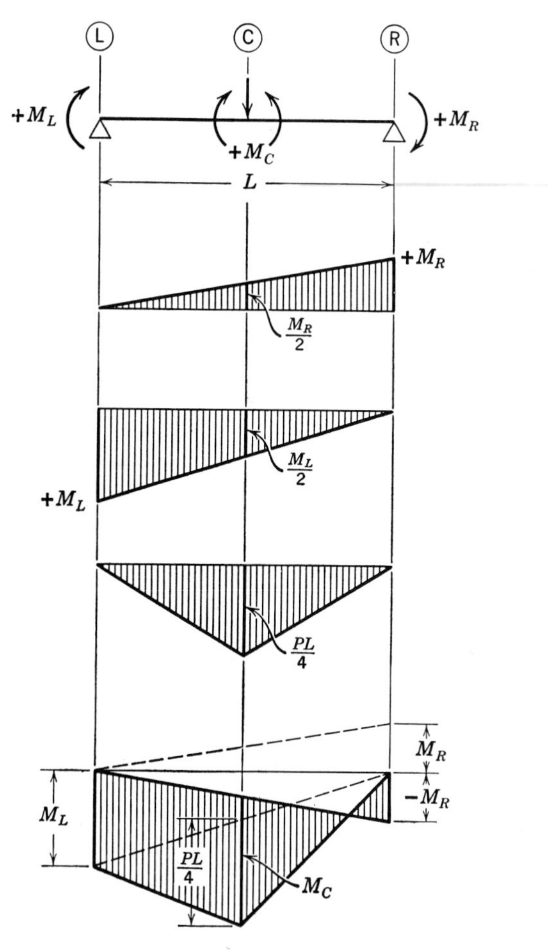

Fig. 7.2. Basis for development of typical moment equilibrium equations.

Procedure 8, equilibrium moment equations. Equations 7.10, 7.11, and 7.12 may be derived from the sketches shown in this procedure. Referring also to Fig. 7.2, if a beam is loaded with a center concentrated load and if the sign convention described in Procedure 7 is adopted, then a summation of moments at the centerline yields

$$M_c = \frac{PL}{4} + \frac{M_L}{2} - \frac{M_R}{2} \tag{7.10}$$

This equation would be modified, of course, for a different loading condition.

Equation 7.11 indicates that the sum of all moments applied to a joint must be equal to zero.

Equation 7.12 is a sway equilibrium expression. The horizontal force in member 1–4 is equal to $(M_1 + M_4)/h$, and similarly for the other columns. Equilibrium of horizontal shear, therefore, requires that Eq. 7.12 be satisfied.

Procedure 9, distributed load. Shown in the lower sketch of Procedure 9 is a frequently encountered case—the end span of a uniformly loaded continuous beam. The solution is obtained by making use of the principles presented in Art. 3.7. The position of the plastic hinge is first found by locating the point of zero shear. Knowing this distance, the magnitude of M_p in terms of the simple span moment M_s may be found. Thus, referring to Fig. 7.3 and taking moments about section 3,

$$M_p = \frac{wy^2}{2} \tag{7.13}$$

Taking a summation of vertical forces for segment 2–3,

$$R = wy \tag{7.14}$$

Taking moments about section 1,

$$RL + M_p - \frac{wL^2}{2} = 0 \tag{7.15}$$

Substituting Eq. 7.13 and 7.14 into Eq. 7.15,

$$wyL + \frac{wy^2}{2} - \frac{wL^2}{2} = 0$$

which reduces to

$$y^2 + 2yL - L^2 = 0 \tag{7.16}$$

Solving this quadratic,

$$y = L(\sqrt{2} - 1) = 0.414L \tag{7.17}$$

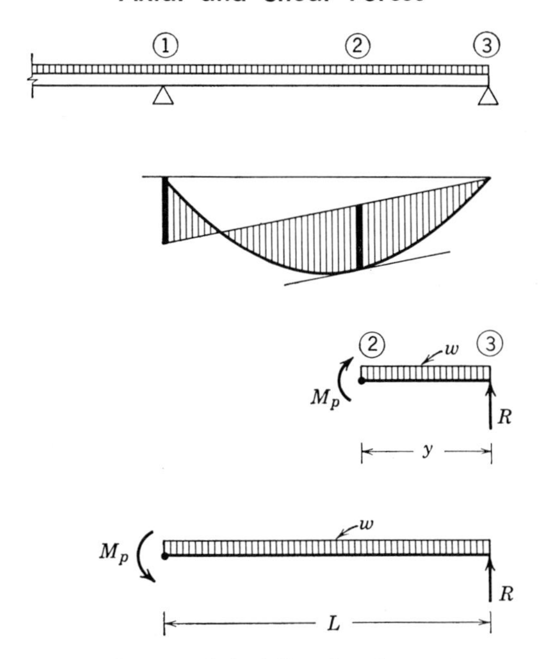

Fig. 7.3. Determination of position of plastic hinge in end span of uniformly loaded beam.

To determine the value of M_p, Eq. 7.17 is substituted into Eq. 7.13 and

$$M_p = w\frac{(0.1716)L^2}{2} = 0.0858\ wL^2 \tag{7.18}$$

In terms of the simple span moment $\left(M_s = \frac{wL^2}{8}\right)$,

$$M_p = 0.686M_s \tag{7.19}$$

7.4 AXIAL AND SHEAR FORCES

Table 7.4 presents the guides for accounting for the influence of axial force and shear force in the design of a structure. Some of these guides are discussed in the following paragraphs.

Design guide 2, shear force. In many elastic design specifications there is a different factor of safety with respect to shear yielding than with respect to flexural yielding. The factor of safety against flexural yielding given by Eq. 7.6 is 1.65 for building construction. On the other hand, using the AISC specifications with shear stress at working load (τ_w equal to 13,000 psi), then the factor of safety against shear

TABLE 7.4. DESIGN GUIDES: AXIAL AND SHEAR FORCE

Design Guide Number, Title, Form, and Manner of Loading of Member	Statement of Process, Formulas, Definition of Terms, Units, and Explanatory Notes	References
(1) Axial force P, M_{pc}	For WF shapes bent about strong axis, the required plastic modulus necessary to transmit applied moment is given by: $\boldsymbol{Z} = \boldsymbol{Z}_t \quad \left(0 < \frac{P}{P_y} < 0.15\right)$ (4.11) $\boldsymbol{Z} = \boldsymbol{Z}_t\left(\frac{\boldsymbol{P}}{\boldsymbol{P}_y} + \mathbf{0.85}\right) \quad \left(0.15 < \frac{P}{P_y} < 1.0\right)$ Z = Required plastic modulus. Z_t = Trial value of Z, neglecting axial force. P = Axial load in member. $P_y = \sigma_y A$, member selected neglecting axial force. For rectangle and other H-shapes see Fig. 4.4. For WF shapes bent about weak axis, use curve for rectangle, Fig. 4.4. *Influence of axial force* on M_p of WF shapes: 	Art. 4.2

② Shear force

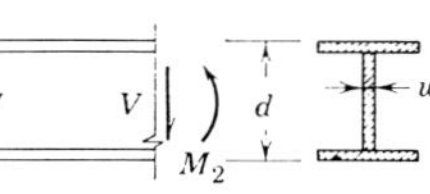

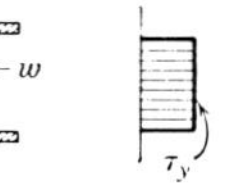

$$V_{max} = \mathbf{18.0}\ \boldsymbol{wd} \quad (4.20)$$

Art. 4.3

w = Web thickness (in.).
d = Section depth (in.).
V_{max} = Max. allowable shear force (kips).

Alternate procedures in case shear is "critical":

(1) Increase section to provide adequate web area.
(2) Provide diagonal web reinforcement.
(3) Provide web doubler plate.
(4) Modify allowable ultimate shear stress.

yielding on an "average stress" basis is

$$F_s = \frac{\tau_y}{\tau_w} = \frac{\sigma_y/\sqrt{3}}{13{,}000} = \frac{19{,}050}{13{,}000} = 1.47 \tag{7.21}$$

Therefore a simply-supported beam, loaded in such a way that the average shear stress at the conventional working load is 13,000 psi, will have an average shear stress greater than the yield value when the flexural yield point is reached, or

$$\tau_{\max} = \tau_w \frac{\sigma_y}{\sigma_w} = (13.0)\frac{33.0}{20.0} = 21{,}500 \text{ psi} \tag{7.22}$$

The beam can support this average stress because of strain-hardening.

One could go a step further and inquire as to the average shear stress on the web of a similar typical simple beam which is loaded until a plastic hinge forms. This shear stress is given by

$$\tau_{\max} = \tau_w \frac{M_p}{M_w} = (13{,}000)(1.85) = 24{,}100 \text{ psi} \tag{7.23}$$

In formulating a plastic design procedure for determining the maximum allowable shear force in an indeterminate structure at ultimate load, it appears that Eq. 4.20, derived on the basis that the shear yield stress should not be exceeded, is too conservative because we are limiting the resultant average shear stress at failure to a value that is considerably less than what is already permitted in present practice. If one based the design on Eq. 7.22, then by making the same assumptions as those which led to Eq. 4.20 the maximum allowable shear force would be given by

$$V_{\max(2)} = \frac{21{,}500}{1.07} wd = 20{,}000wd \tag{7.24}$$

This would assure that the maximum shear force at failure would not be any greater than that which inherently would be allowed in a simple beam when the maximum flexural stress reached the yield point.

As a matter of fact, since the choice of load factor (1.85) was based on the concept that the required margin of safety in plastic design should be the same as in an elastically designed simple beam, then a plastically designed simple beam would be penalized when compared with its elastic counterpart unless the maximum allowable shear force at failure were selected on the basis of Eq. 7.23. In this case,

$$V_{\max(3)} = \frac{24{,}100}{1.07} wd = 22{,}500wd \tag{7.25}$$

Since there are relatively few designs that require revision because the shear force is too high, and since the only difference in the behavior of the structure is that the deflections increase nonlinearly (see Art. 4.3), then it is left to the discretion of the designer to use Eq. 7.25 in lieu of changing a design that otherwise would be satisfactory.

7.5 BEAMS AND GIRDERS

Table 7.5 presents the design guides for beams and girders.

Design guide 2, built-up members. Further research will extend the applicability of plastic design to members whose webs and flanges are of more slender proportion than those of rolled beams. Ship girders and frames with spans greater than those possible with rolled beams are examples.

The built-up members shown in sketches B and C represent a built-up welded box beam and a member fabricated by welding channels toe to toe. Although somewhat less efficient than the I-shape, they will nonetheless exhibit considerably better lateral buckling characteristics.

Design guide 3, beams of nonuniform cross section. Although it will often be more economical to use uniform prismatic members, this guide shows three schemes for decreasing the weight of steel required in a plastic design (at the expense of additional fabrication cost). The design is based on the equation

$$M_s = M_{p1} + M_{p2}$$

Part of the moment check is to make sure that the plastic moment condition is not violated at any point along the beam.

7.6 COLUMNS

Design guides for columns are presented in Table 7.6.

Design guide 1, simple columns. Equations 4.34 are based on the results of recent research on representative rolled WF shapes. If several loading conditions must be studied on a structure, there may not be much application of this procedure. In certain cases the design may appear to require only a strut; but in the selection of such a member other loading conditions must also be considered, although they may not control the selection of other member sizes.

Procedure 2, framed columns for industrial buildings. The procedure for three loading cases is described in Table 7.6. It calls for the use of one

TABLE 7.5. DESIGN GUIDES: BEAMS AND GIRDERS

<table>
<tr><th>Design Guide Number, Title, Form, and Manner of Loading of Beam</th><th>Statement of Process, Formulas, Definition of Terms, Units, and Explanatory Notes</th><th>References</th></tr>
<tr><td>(1) Rolled beams 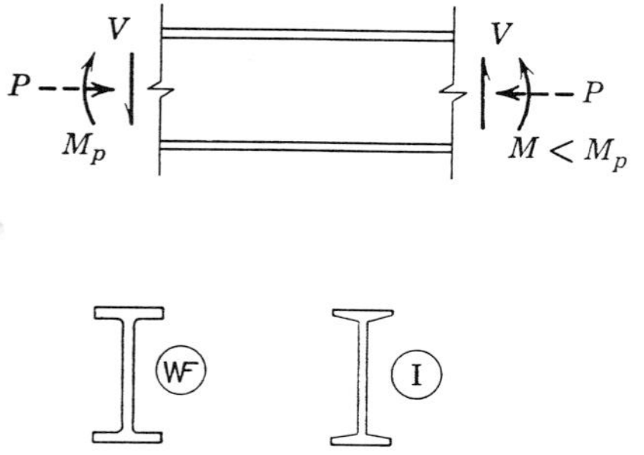

V, P, M_p — V, P, $M < M_p$
WF   I</td><td>Required plastic modulus:

$$\underset{(\text{in.}^3)}{Z} = \frac{M_p}{\sigma_y} = \underset{(\text{k-ft})}{\mathbf{0.364}\ M_p} \qquad (2.11a)$$

Z for WF and I shapes is given in Appendix 2.
WF beams bent about weak axis: $Z = 1.50S$.
For other forms of cross section, see Design Guide 2, Table 7.5, and Guide 2, Table 7.8.
Standard Loading cases: Fig. 8.1.
Other considerations for beams:
Axial Force — Guide 1, Table 7.4
Shear Force — Guide 2, Table 7.4
Bracing — Guide 3, Table 7.8
Cross-Section Proportions — Guide 1, Table 7.8</td><td></td></tr>
<tr><td>(2) Built-up members 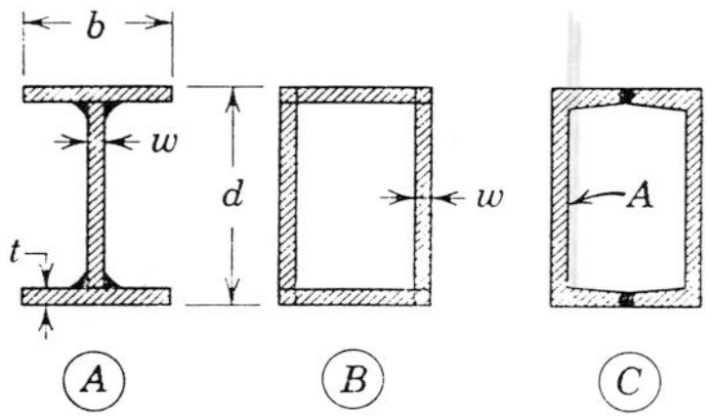

b, w, t, d, w, A
(A)   (B)   (C)</td><td>Members fabricated by welding, riveting, or bolting may be used so long as they comply with the other applicable design guides in this chapter.

(A) $$Z \cong bt(d - t) + \frac{w}{4}(d - 2t)^2 \qquad (2.24)$$
(B) $$Z \cong bt(d - t) + \frac{w}{2}(d - 2t)^2 \qquad (7.26)$$
(C) $$Z \cong 2\left[(A - wd)\left(\frac{d - t}{2}\right) + \frac{wd^2}{4}\right] \qquad (2.23)$$

Other shapes: See Design Guide 2, Table 7.8.</td><td></td></tr>
</table>

(3) Beams of nonuniform cross section 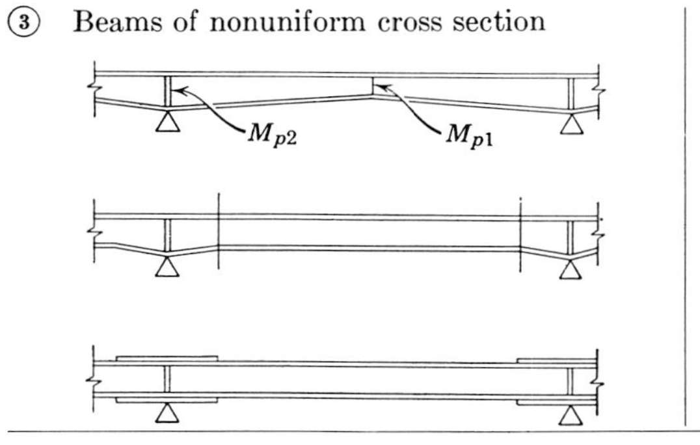	The design is carried out on basis of M_{p1} and M_{p2}. Check moment along member to make sure that plastic moment condition is not violated. For H-shaped cross sections, Z given by Eq. 2.24. For other shapes see Design Guide 2, Table 7.8. For cover-plated members: $$Z = Z_{WF} + A_p(d + t_p) \qquad (3.27)$$	Art. 3.9 Plate III, p. 260 Plate IV, p. 263

TABLE 7.6. DESIGN GUIDES: COLUMNS

Design Guide Number, Title, Form, and Manner of Loading of Column	Statement of Process, Formulas, Definition of Terms, Units, Explanatory Notes	References
(1) Simple columns P, x, y x-x = Strong axis y-y = Weak axis	After frame members have been selected, any columns subjected only to axial force (with flexible connection) may be checked according to: $\frac{P}{P_y} = 1 - \frac{(L/r_x)^2}{35{,}600} \quad \left(\frac{L}{r_x} \leq 120\right)$ $\frac{P}{P_y} = 1 - \frac{L/r_y}{330} \quad \left(\frac{L}{r_y} \leq 120\right)$ (4.34) $\frac{P}{P_y} = \frac{8850}{(L/r)^2} \quad \left(\frac{L}{r_y} > 120\right)$ L = Unbraced length. r = Radius of gyration. $E = 29.6 \times 10^3$ ksi, $\sigma_y = 33.0$ ksi, $\sigma_{rc} = 13.0$ ksi.	Art. 4.6
(2) Framed columns for industrial buildings P, M_o (A) (B) (C)	For $P/P_y \leq 0.15$, cases B and C $M_o = M_p \quad \left(0 < \frac{P}{P_y} \leq 0.15\right)$ (7.27) Otherwise, for adequately braced columns, select section on basis that: $Z = Z_t \Big/ \left(\frac{M_o}{M_p}\right)$ (7.28) Z = Required plastic modulus. Z_t = Trial value of Z neglecting axial force. $\left(\frac{M_o}{M_p}\right)$ = Value determined from Fig. 4.22 (case A) or Fig. 4.23 (case B) for "trial" values of L/r_x and P/P_y. For Case C use Fig. 4.23 with $L = \frac{1}{2}$ column height. Limitations: $\frac{P}{P_y} \leq 0.60 \qquad \frac{L}{r_x} \leq 120$	Art. 4.6

③ Framed columns for multi-story buildings

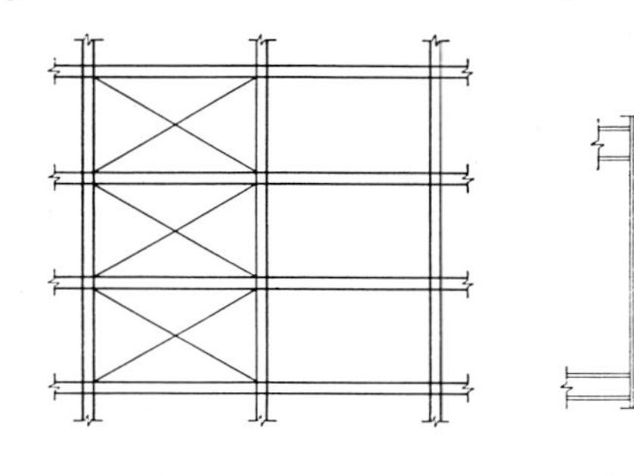

Where wind-bracing in multi-story structures consists of diagonals in wall panels, columns may be proportioned by present methods.

To determine design loading on columns, divide ultimate load and bending moments by load factor F.

④ Checking the "weak axis"

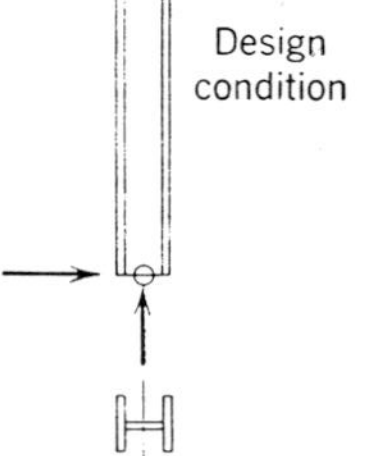

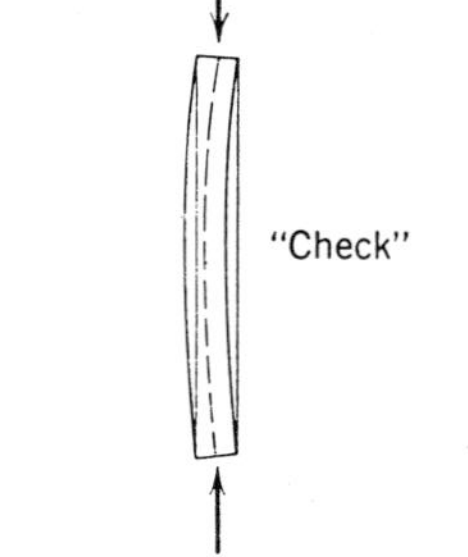

For columns whose design is controlled by combined bending and axial load, check for failure of column in plane normal to principle plane of bending, assuming columns pinned at the ends:

$$\frac{L}{r_y} < \sqrt{\frac{8850}{P/P_y}} \qquad \left(\frac{P}{P_y} < 0.625\right) \qquad (4.35)$$

L = Unbraced length $\qquad L/r_y > 120$

of two graphs to determine the allowable end bending moment. Approximations to the information presented in Figs. 4.22 and 4.23 have been developed in Ref. 4.12 and the information presented in tabular form. Reference 7.1 also presents the solution to the column problem in tabular form.

7.7 CONNECTIONS

Design guides for connections are presented in Table 7.7, supplemented where needed by the information given in the following paragraphs.

Design guide 14, welds. One could arrive at an allowable stress for fillet welds by one of two approaches. The stress at ultimate load could be limited to the yield stress in shear, or it could be 1.65 times the shear stress that is allowed in a conventional design.

If the maximum shear is limited to the yield value, then the maximum allowable shear stress would be $\sigma_y/\sqrt{3} = 19{,}050$ psi. Since the present allowable stress on the throat of a fillet weld is 13,600 psi, such a procedure would require welds that appear to be unnecessarily large.

Using the second approach, the maximum shear stress would be 1.65 times the shear stress that is allowed upon a weld in conventional design. Then

$$\tau_{\max} = (13{,}600)(1.65) = 22{,}400 \text{ psi}$$

Similarly, the force allowed per inch of fillet weld per 16th of leg is given by

$$\text{Force} = (600)(1.65) = 1000 \text{ lb/in./16th of leg}$$

Of course, for combined forces, present practice would be used in which the size of weld is determined from a resolution of the force per inch acting in the several directions.

The allowable tensile value of butt joints is the tensile yield value, 33,000 psi.

Numerous examples of the plastic design of fillet welds may be found in Ref. 5.4.

Design guide 15, rivets. The allowable stresses that can be permitted in tension and in shear on rivets at ultimate load may be selected by the same philosophy used for welds. In conventional practice for building construction an allowable stress of 15,000 psi in shear is permitted at working load. At the elastic limit in flexure the average stress on a group of rivets loaded in shear is

$$\tau_{\max} = 15{,}000 \times \tfrac{33}{20} = 24{,}800 \text{ psi}$$

Therefore, in plastic design at ultimate load, the allowable shear stress on a rivet could be taken as 25,000 psi.

The tension value of a rivet would be taken as 33,000 psi.

Design guide 16, bolts. Often it will be appropriate to use high-strength bolts for the field connections. When erected in accordance with current recommended practice [7.8,7.9] then the allowable load/bolt at ultimate load may be taken as the guaranteed minimum proof load of the bolt (ASTM A325). These values for the commonly used bolt sizes are given in Table 7.7.

With regard to resistance to shearing forces, high-strength bolts used with unpainted contact surfaces can be counted upon to transmit, without slip, the shear force due to service loading on the basis of one bolt substituted for one rivet of the same nominal diameter. After slipping into bearing (with some joint rotation but no loss in moment capacity) the bolts will not fail in shear at 1.65 times the working value of the rivets they replace. Work currently underway at Lehigh University, sponsored by the Research Council on Riveted and Bolted Structural Joints and the Pennsylvania Department of Highways is aimed at taking further advantage of the theoretically greater strength of bolts when compared with rivets.

TABLE 7.7. DESIGN GUIDES: CONNECTIONS

Design Guide Number, Title, Form, and Manner of Loading of Connection	Formulas, Statement of Process, Definition of Terms, Units, Explanatory Notes	References
(1) Straight corner connections 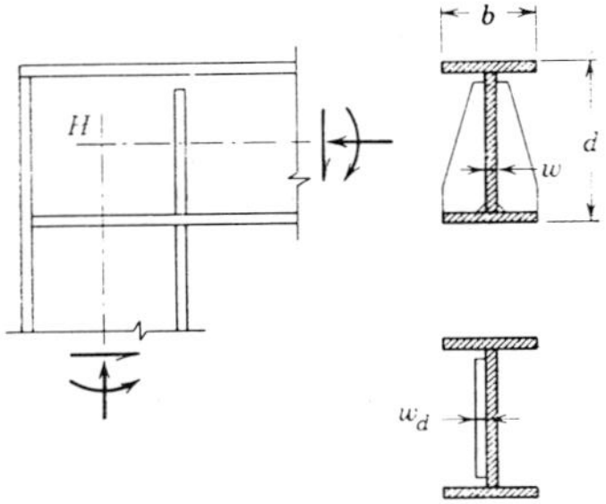	To develop full M_p at point H $w \geq \sqrt{3}\,\frac{S}{d^2}$ (5.7) S = Section modulus, I/c *Connection details:* Fig. 5.7, 5.8, Art. 5.3. *Welding:* See Design Guide *14*. *Inadequate connections:* (*a*) Use web plates (doublers). $w_d = \frac{\sqrt{3}\,S}{d^2} - w$ (5.8) (*b*) See Guide 2.	Art. 5.3, p. 150
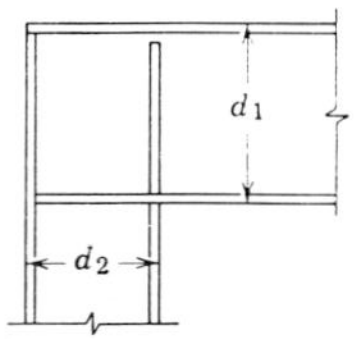	$w \geq \frac{\sqrt{3}\,S_{min}}{d_1 d_2}$ (5.7*a*) S_{min} = Section modulus of weaker member. Alternate form for Eqs. 5.7 and 5.7*a*: $w \geq \frac{0.6M}{d_1 d_2}$ (5.49*a*)	
(2) Diagonal plate stiffeners 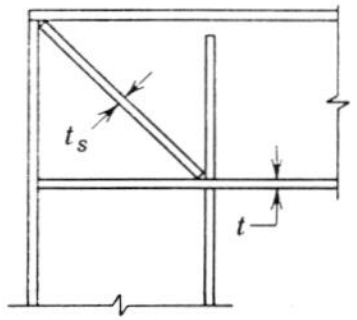	$t_s = \frac{\sqrt{2}}{b}\left(\frac{S}{d} - \frac{wd}{\sqrt{3}}\right)$ (5.12) $t_s \geq b/17$ An approximation that is conservative is to use $t_s = t$.	Art. 5.3, p. 153

③ Tapered haunches—proportions	(1) Select general layout (note architectural features). Brace at sections R and H. (2) Check strength at section 1—develop M_p at R. $Z_1 = bt(d_1 - t) + \frac{w}{4}(d_1 - 2t)^2$ (5.16) $Z_1 \geq M_1/\sigma_y$ (3) For $Z \cong M_1/\sigma_y$ check haunch stability according to the following possible alternates: (*a*) For $s/b \leq 4.0$, $t_t = t$ $t_c = t/\cos\beta$ (5.15) (*b*) For $s/b > 4.0$, use Eq. 5.15 and supply intermediate bracing, or alternatively, (*c*) For $4.0 < s/b < 17.0$ $t_t = \left[1 + 0.1\left(\frac{s}{b} - 4\right)\right] t$ (5.30) $t_c = t_t/\cos\beta$ (*d*) Deepen haunch at section 1 such that $Z_1 \gg \frac{M_1}{\sigma_y}$ $\beta > 24°$ (4) For $Z_1 \gg M_1/\sigma_y$, and $s/b < 17$ $t_t = t$ $t_c = t/\cos\beta$ (5.15) (5) Provide web, w_h, equal to w_{WF}. (6) Select bracing and stiffeners as per Guide 4 and Table 7.8, Guide 3.	Art. 5.4, p. 157 Plate VIII, p. 295

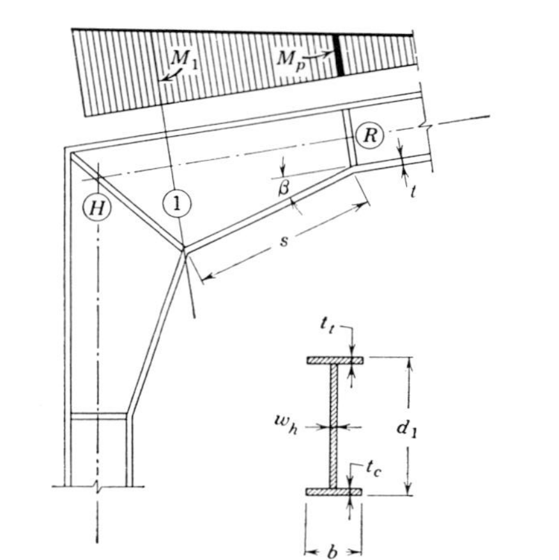

NOTE: Width of haunch flange = width of rolled shape.

TABLE 7.7. DESIGN GUIDES: CONNECTIONS (*Continued*)

Design Guide Number, Title, Form, and Manner of Loading of Connection	Formulas, Statement of Process, Definition of Terms, Units, Explanatory Notes	References
(4) Tapered haunches—stiffeners	$t_s \geq \sqrt{2}\, t_t - 0.82\, wd_1/b$ (5.25) $t_s \geq \sqrt{2}\,(1 - \tan\beta)t$ (5.27) $t_s \geq b/17$ $t_{tr} \geq t_c \sin\beta$ (5.29) $t_{tr} \geq b/17$ Approximation: $t_s = t_{tr} = t$	
(5) Curved knees	(1) Select general layout; brace at sections C, H, G. (2) Check haunch radius $R \leq 4b/\alpha$ (5.31) For $4 < \frac{R\alpha}{b} < 17$, increase tension and compression flange thickness according to $\Delta t = 0.1\left(\frac{R\alpha}{b} - 4\right)$ (5.32) or provide additional bracing. (3) Select flange thickness to provide plastic moment at C (or G). $t_t = t_c = \frac{d_m}{2} - \sqrt{\frac{d_m^2 b}{4(b-w)} - \frac{M_m/\sigma_y}{b-w}}$ (5.33) $d_m = d + 0.02R$ M_m = moment at $\alpha_m = 0.2$ rad. An approximation to Eq. 5.33 is: $t_t = t_c = \frac{4}{3}t$ (4) Check cross-bending $t_c \geq b^2/2R$ (5.35) (5) Provide haunch web thickness and flange width equal to rolled section. (6) Select stiffeners according to Eqs. 5.25 and 4.21.	Art. 5.4, p. 168

⑥ Interior (beam-column) connections without stiffeners 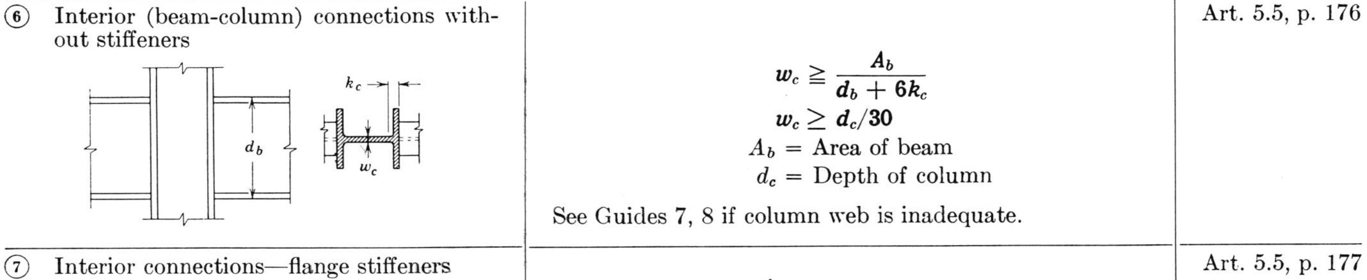	$w_c \geqq \dfrac{A_b}{d_b + 6k_c}$ $w_c \geq d_c/30$ A_b = Area of beam d_c = Depth of column See Guides 7, 8 if column web is inadequate.	Art. 5.5, p. 176
⑦ Interior connections—flange stiffeners 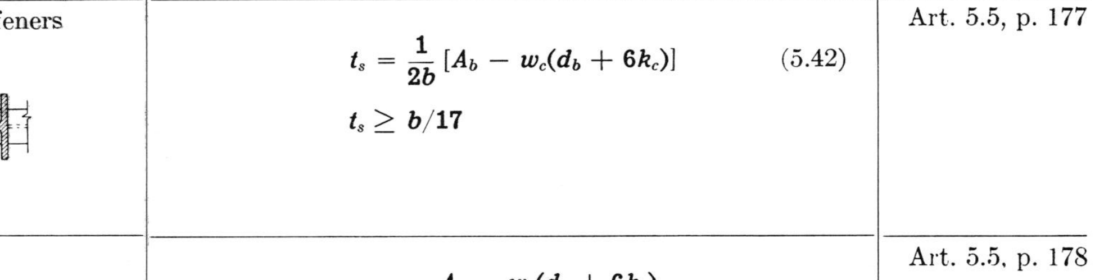	$t_s = \dfrac{1}{2b}[A_b - w_c(d_b + 6k_c)]$ (5.42) $t_s \geq b/17$	Art. 5.5, p. 177
⑧ Web stiffeners to resist thrust 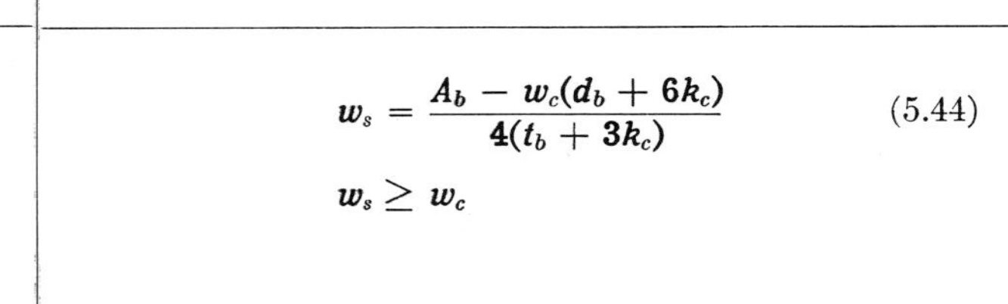	$w_s = \dfrac{A_b - w_c(d_b + 6k_c)}{4(t_b + 3k_c)}$ (5.44) $w_s \geq w_c$	Art. 5.5, p. 178
⑨ Top plate connections 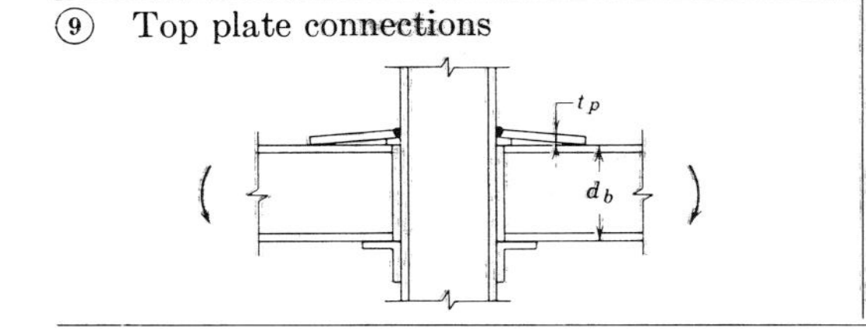	$A_p = t_p b_p = \dfrac{M_{pr}}{\sigma_y d_b}$ (7.29) A_p = Area of top plate. b_p = Width of top plate. M_{pr} = Required plastic moment (reduced).	Ref. 7.5

TABLE 7.7. DESIGN GUIDES: CONNECTIONS (*Continued*)

Design Guide Number, Title, Form, and Manner of Loading of Connection	Formulas, Statement of Process, Definition of Terms, Units. Explanatory Notes	References
(10) Web stiffeners for shear 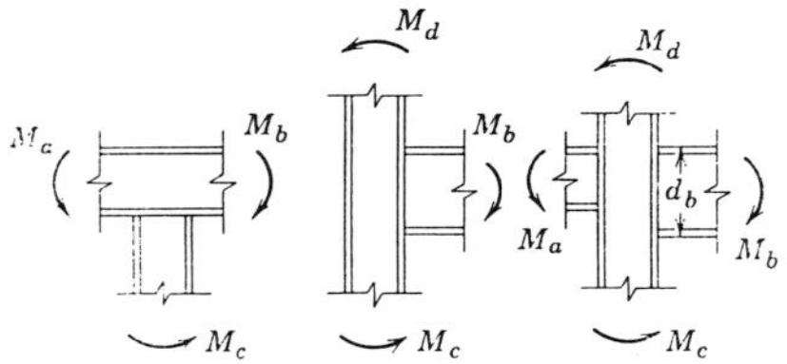 $\Delta M = M_b + M_a$ (Clockwise +)	$w = \dfrac{0.6\,\Delta M}{d_c d_b}$ (5.49) w = Required web thickness (in.). ΔM = Sum of moments applied to connection by beams (kip-ft). d_c = Depth of column (in.). d_b = Depth of largest beam (in.). NOTE: Eq. 5.49 uses $\sigma_y = 33.0$ ksi and assumes ratio of beam depth to story height $\leq 1/15$.	Plate XV, p. 371
(11) Beam-to-girder connections 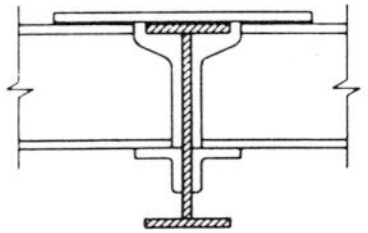	For beam-to-girder connection shown, use top plate equation (7.29)—see Guide 9. For equal moments in beams joining to girders, connections may be proportioned for full continuity and development of plastic hinges. Girder must be adequately braced transversely in order to provide stability for unequal moments.	
(12) Miscellaneous building connections 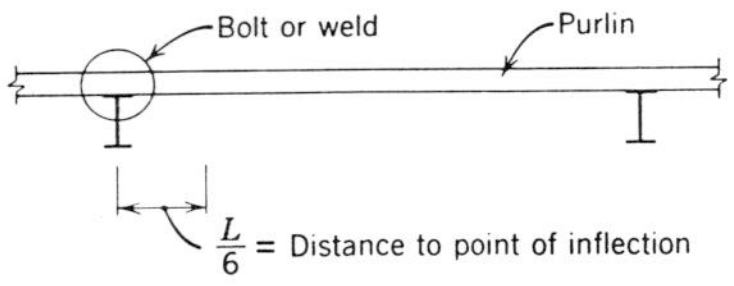 Bolt or weld Purlin $\frac{L}{6}$ = Distance to point of inflection	Purlins and girts designed for plastic action may either be proportioned with full strength connections or may be spliced for shear at points of inflection.	

(13) Splices

Column splices should be similar to conventional practice, adequate to resist applied moment.
In beams and girders, splices at points of maximum moment should be adequate to develop full plastic strength. Shear splices are adequate at point of inflection. Elsewhere splices may be proportioned for partial continuity.
Connections may be welded, bolted, or riveted.

Ref. 7.6

(14) Welds

τ_{max} = 22,400 psi

σ_{max} = 33,000 psi

At ultimate load:

Fillet weld value = 1000 lb/in./16th (7.30)

Tension on butt welds = 33,000 psi

Continuous welds should be used in regions of plastic hinges.
Follow applicable procedures of A.W.S. Code.
For examples of the plastic design of welds, see Ref. 5.4.

Ref. 7.7

(15) Rivets

At ultimate load:

Rivet shear = 25,000 psi

Rivet tension = 33,000 psi

(16) Bolts

For ASTM A325 high-strength bolts, at ultimate load, allowable shear and tensile loads/bolt as shown in table:

Bolt Size	Tension, lb, (Guar. min. proof load)	Shear, lb, (Area × 25,000 psi)
3/4	28,400	11,100
7/8	36,000	15,000
1	47,300	19,600
1 1/8	56,500	24,800

Ref. 7.8

TABLE 7.8. DESIGN GUIDES: DETAILS

Design Guide Number, Title, Form and Type of Detail	Formulas, Statement of Process, Definition of Terms, Units, Explanatory Notes	References
(1) Cross-section proportions	To assure that compressive strains may reach ϵ_{st} $\frac{\boldsymbol{b}}{\boldsymbol{t}} \leq \mathbf{17}$ **(Beams, columns)** $\frac{\boldsymbol{d}}{\boldsymbol{w}} \leq \mathbf{43}$ **(Columns in direct compression)** $\frac{\boldsymbol{d}}{\boldsymbol{w}} \leq \mathbf{55}$ $\left(\textbf{Beams where } \frac{\boldsymbol{P}}{\boldsymbol{P_y}} \cong \mathbf{0.15}\right)$ (4.21) For $\frac{d}{w} > 55$ or for $0.15 < \frac{P}{P_y} < 0.27$: $\frac{\boldsymbol{d}}{\boldsymbol{w}} = \mathbf{70} - \mathbf{100}\frac{\boldsymbol{P}}{\boldsymbol{P_y}} \quad \left(0 < \frac{P}{P_y} < 0.27\right)$ (4.22) Stiffening an otherwise unsatisfactory shape may be done by one of schemes A–E, depending on whether the problem is flange or web buckling. Compare their expense with that of alternate rolled shape. Width–thickness ratio of outstanding flange ≤ 8.5.	Art. 4. 4

② Cross-sectional form	*Symmetry:* Desirable about both principal axes. Necessary about axis in plane of bending, else lateral–torsional failure is accentuated.	Art. 2.5
	Proportions: Table 7.8, Guide 1.	Art. 3.9
	Built-up members: Table 7.5, Guide 2.	
	Nonuniform cross section: Table 7.5, Guide 3.	
	Holes in flange or web: Table 7.1, Provision 5.	
	Unsymmetrical members:	

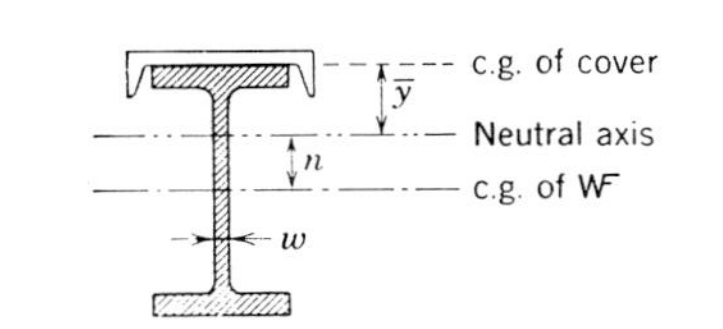

$$Z = \frac{A}{2} a \qquad (2.27)$$

A = Total area.
a = Distance between centroids of areas above and be-below neutral axis.

For built-up WF's with neutral axis in web, ($A_{\text{cover}} \leq wd$).

$$n = A_c/2w \qquad (7.31)$$

$$Z = Z_{WF} + wn^2 + A_c\bar{y}_c \qquad (7.32)$$

A_c = Area of cover material.
n = Shift of neutral axis.
$\bar{y}_c$ = Distance from neutral axis to center of gravity of cover (see Ref. 2.1).

For neutral axis in flange,

$$Z \cong btd + \frac{wd^2}{2} \qquad (7.33)$$

NOTE: Position of neutral axis is such that $A_{\text{tension}} = A_{\text{compression}}$

TABLE 7.8. DESIGN GUIDES: DETAILS (*Continued*)

Design Guide Number, Title, Form and Type of Detail	Formulas, Statement of Process, Definition of Terms, Units, Explanatory Notes	References
(3) Bracing Brace (plastic) Conventional practice	(1) *Function:* Lateral support must be provided at close enough intervals to prevent premature lateral buckling. (2) *Location:* Brace at plastic hinges and check "elastic" portions according to conventional methods. (3) *Position:* Bracing must prevent twisting prior to achieving necessary rotation. At changes in section: Brace tension and compression sides In prismatic beams: Brace compression flange only unless rotation is large (4) *Spacing* (Required at all hinges except at the last to form): Elastic segment: Use conventional practice (see, for example, L_u values of Ref. 2.1). *At plastic hinges:* (1st trial) $\frac{L}{r_y} = 30 \quad \left(1.0 > \frac{M}{M_p} > 0.6\right)$ (4.28) $\frac{L}{r_y} = 48 - 30\frac{M}{M_p} \quad \left(0.6 > \frac{M}{M_p} > -1.0\right)$ L = Distance between bracing points. r_y = Radius of gyration in "weak" direction. M/M_p = Moment ratio in length, L, between bracing point	Art. 4.5

③ Bracing (*Continued*)

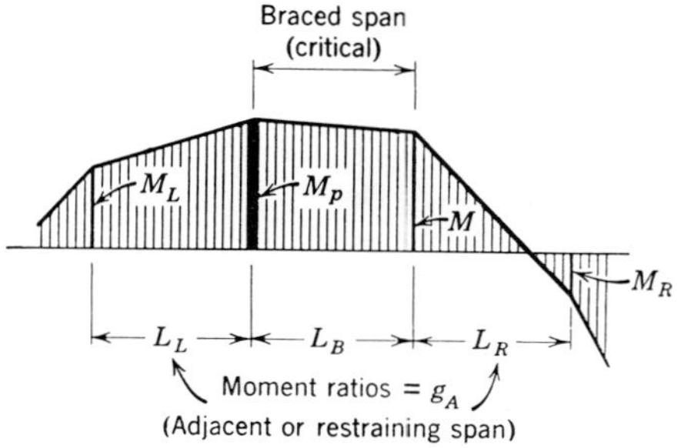

Braced span (critical)

M_L M_p M M_R

L_L L_B L_R

Moment ratios = g_A

(Adjacent or restraining span)

At plastic hinges: (Refinement to increase allowable L/r_y.) See procedure outlined in Appendix 1, using:

$$C_f = \mathbf{1.0} + \mathbf{0.2}\left[1 - \left(\frac{L}{L_l}\right)^2\right]\left[\mathbf{0.9} + \frac{1 - \left(\frac{L}{L_s}\right)^2}{1 + \left(\frac{L}{L_l}\right)^2}\right] \tag{A.1}$$

$$\frac{L_A}{r_y} = \frac{134}{\sqrt{M/M_p}} + 60(1 - g_A) - \underbrace{1100\left(\frac{M}{M_p} - 0.9\right)}_{\text{Delete if } \frac{M}{M_p} < 0.9} \tag{A.3}$$

where Eq. A.3 applies to an adjacent elastic segment.

C_f = Fixity correction factor. Eq. 4.28 would be used in Eq. A.1 for an adjacent plastic segment.

g_A = Moment ratio in adjacent segment.

L_l, L_s = Critical length (with $C_f = 1.0$) of adjacent segment which has longer (and shorter) critical length.

Design approximation (in lieu of restraint correction).

$$\frac{L}{r_y} = 35 \qquad \left(1.0 > \frac{M}{M_p} > 0.625\right)$$

$$\frac{L}{r_y} = 60 - 40\frac{M}{M_p} \qquad \left(0.625 > \frac{M}{M_p} > -1.0\right) \tag{4.29}$$

Appendix 1

TABLE 7.8. DESIGN GUIDES: DETAILS (*Continued*)

Design Guide Number, Title, Form and Type of Detail	Formulas, Statement of Process, Definition of Terms, Units, Explanatory Notes	References
③ Bracing (*Continued*)	(5) *Strength and stiffness:* $$A_b \geq \mathbf{0.04}A$$ Usually the force required to prevent failure is small. Maximum allowable slenderness ratio requirements normally govern. NOTE: Bracing members must themselves be held from lateral motion with respect to the members they brace.	
	(6) *Enclosed structure:* Adequate lateral support is provided when structure is enclosed by walls or slabs normal to plane of frame. Partial enclosure, when coupled with plate stiffeners (see sketch), is adequate.	
	(7) *Corner connections:* Positive lateral support should be provided at all changes in section. See solid circles in sketches. For haunched connections see also Table 7.7, Guide 3.	Art. 5.4
	(8) *Columns:* Equations in Table 7.6, Guide 2, assume wall system provides adequate torsional resistance. In absence of support from wall, bracing provided by procedures of item 4 of this guide is assumed adequate for columns. Check for bracing in weak direction, using Table 7.6, Guide 4.	
	(9) *Beams and girders:* Step 4 above. For uniformly loaded beams of equal span, elastic continuous beam diagrams show that last hinge is always between supports (Ref. 2.1).	

7.8 DETAILS

In Table 7.8 are presented the design guides that pertain to details.

Design guide 2, cross-sectional form. Equations 7.31 and 7.32 given in Table 7.8 are convenient for computing Z of an unsymmetrical built-up member formed by adding material to one flange. They are applicable only when the neutral axis remains in the web, but this covers a large majority of cases. Equation 7.31 is derived from the condition that the compression area must equal the tension area, or

$$A_w + nw = A_w - nw + A_{\text{cover}}$$

from which

$$n = A_{\text{cover}}/2w \tag{7.31}$$

When the neutral axis lies in the flange of the basic WF shape to which cover material is added, then a close approximation to Z is given by

$$Z \cong btd + \frac{wd^2}{2} \tag{7.33}$$

This equation is obtained by assuming that the neutral axis is at the outside face of the WF shape and neglects the statical moment of the cover material.

8 Continuous Beam Design

8.1 INTRODUCTION

This chapter will treat continuous beam problems for the purpose of illustrating the principles of plastic design. In addition to obtaining the required member size following the general procedures suggested in Art. 7.3, each design will be examined in the light of the "secondary design considerations" (Chapters 4, 5, 6, and 7). In the process of analyzing each structure the problem will be worked first by a direct and complete plastic analysis. The experienced designer will, of course, want to use all possible techniques to shorten the design time, and at the end of this chapter will be mentioned some of the available "short-cuts" for speeding up the design process. Reference 7.1 will also be of assistance in this regard. Just as in conventional elastic design where the engineer has available various formulas, tables and charts with which to analyze standard cases, so also it has been possible to arrange convenient design aids for the rapid selection of member sizes.

To assist in arriving at the required section, a table of Z-values has been included in Appendix 2. When the required M_p value has been determined, Z is computed, and the section is selected from this table. An alternate procedure that would save a step in the calculations would have been to arrange the sections according to M_p values instead of Z values. This would have limited the use of the table, however, to a single value of the yield stress level, σ_y. Still another method would have been to use the presently available tables of section modulus S. This would have involved making a guess as to the proper value of the shape factor f, a guess that would be corrected if necessary in the final step.

The load factor of safety has been discussed in Art. 7.1. A value of 1.85 is used for dead load plus live load and a value of 1.40 for these loads plus wind or earthquake forces.

As a convenience for later reference, the examples are all worked in "plates," the discussion of the steps being included in the text. The reader will find that the steps that are considered to be obvious from the plates and which follow directly from the appropriate "provisions," "procedures," or "design guides" of Chapter 7 are not discussed in the text. Therefore the best continuity for reviewing the design examples is probably achieved by following the steps in the plates. Where further explanation has seemed advisable, the discussion in the text has been keyed into the appropriate place in the "plates" by the use of "step numbers."

8.2 CONTINUOUS BEAMS OF UNIFORM SECTION THROUGHOUT

PLATE I Five Equal Spans with Uniformly Distributed Load

This problem is solved in Plate I and illustrates the design of a beam of uniform section throughout.

Step 1: As was demonstrated in Art. 7.2, the most critical loading case involves loading on the entire span. "Checkerboard" loading need not be investigated.

Step 2: Following Procedure 3, Table 7.2, and selecting as redundants the moments at sections 3 and 5, the moment diagram resulting from the determinate loading is drawn using the convention that moments are drawn on the side of the beam that would be in tension. The fixing line a–b is then drawn for the end span in such a way that the moment at section 2 is equal to the moment at section 3. The resulting required plastic moment is greater than that which would have been obtained had fixing lines been drawn for an interior span. Therefore the end span controls the design. Since the end span is critical, the moments and reactions for the three central spans cannot be determined by statics alone. However, this is not serious because the semi-graphical construction demonstrates that the plastic moment value of the beam is not exceeded. Therefore the selection of the 21WF68 will provide adequate bending strength.

The "net" or resultant moment diagram is shown by the shaded portion in sketch (b). The broken line a–b–c–d becomes the new base line and, as before, the shaded portion is on the tension side of the beam. As shown in Procedure 1, Table 7.3, the resulting moments are exactly the same as if the moment diagram had been redrawn, using line a–b–c–d as a horizontal base line.

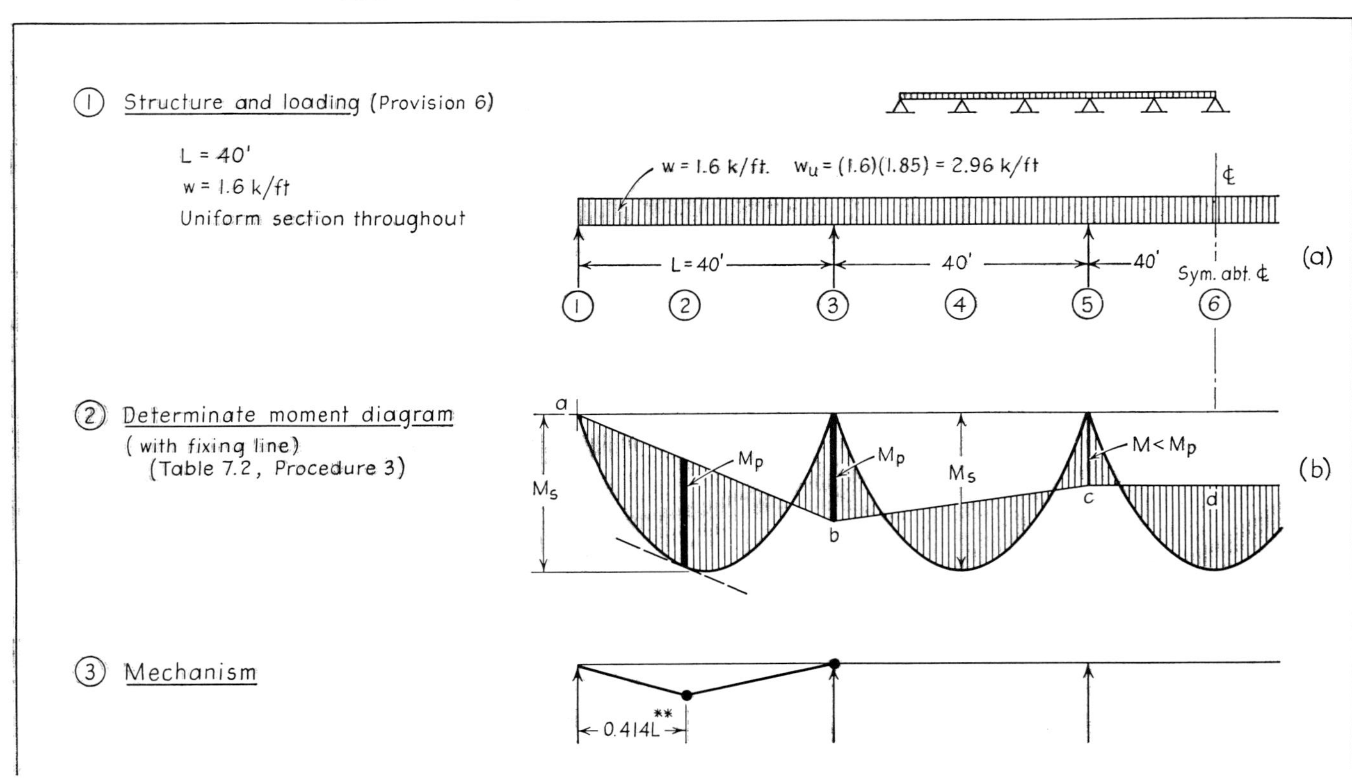
PLATE I. FIVE EQUAL SPANS WITH UNIFORMLY DISTRIBUTED LOAD
1 Structure and loading (Provision 6)
L = 40'
w = 1.6 k/ft
Uniform section throughout
w = 1.6 k/ft. w_u = (1.6)(1.85) = 2.96 k/ft
℄
L = 40'
40'
40'
Sym. abt. ℄
1 2 3 4 5 6
(a)
2 Determinate moment diagram
(with fixing line)
(Table 7.2, Procedure 3)
a
M_s
M_p
M_p
M_s
$M < M_p$
b
c
d
(b)
3 Mechanism
**
0.414L

(4) Required plastic moment (Table 7.3, Procedure 9)

$$M_5 = \frac{w_u L^2}{8}$$

$$M_p = 0.686\, M_5 = 0.0858\, w_u L^2$$
$$= 406^{k'}$$

(5) Reactions (At ultimate load)

$$R_1 (0.414L) - w_u \frac{(0.414L)^2}{2} = M_p$$

$R_1 = 48.9^k$

$$R_3 = V_{32} + V_{34} = (-48.9 + 118.4) + 59.2$$

$R_3 = 128.7^k$ $\quad R_5 = w_u L \quad$ $R_5 = 118.4^k$

(6) Section (Provision 5)

$$Z = \frac{M_p}{\sigma_y} = 0.364\, M_p = \frac{(406)(12)^*}{33.0} = 148.0 \text{ in.}^3$$

Use 21 WF 68
$Z = 159.8 \text{ in}^3$

$w = 0.430$ in.
$d = 21.13$ in.
$I = 1478.3$ in.^{4}

(7) Maximum shear (Table 7.4, Guide 2)

$$V_{max} = V_{32} = 69.5^k$$

$$V_{allow.} = 18.0\, wd = 163^k > 69.5^k \quad \text{ok}$$

* For later calculations use $0.364 = \frac{12}{33}$
** See Procedure 9 of Table 7.3

PLATE I (Continued)

(8) Cross-section proportions
(Table 7.8, Guide I)

$b/t = 12.07$
$d/w = 49.1$ ok

(9) Bracing requirements
(Table 7.8, Guide 3)

NOTE: Beam supports concrete slab, bottom flange exposed.

Provide welded vertical plates at section 3 and 5 over supports.

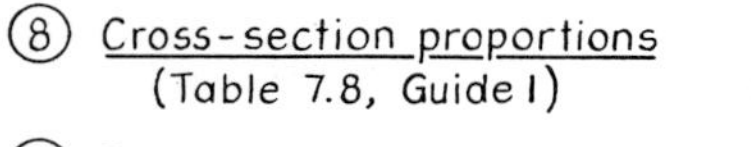

(c)

(10) Splices

←47.5'→←40'→←40'→←25'→←47.5'→

Provide shear splices at points indicated. Alternatively splice at convenient location for moment indicated in moment diagram.

(11) Deflection at working load
(Table 7.3, Procedure 12)

$$\delta \cong \frac{wL^4}{185EI} = \frac{(1.6)(40)^4(1728)}{(185)(30)(10^3)(1478.3)} = 0.86''$$

(d)

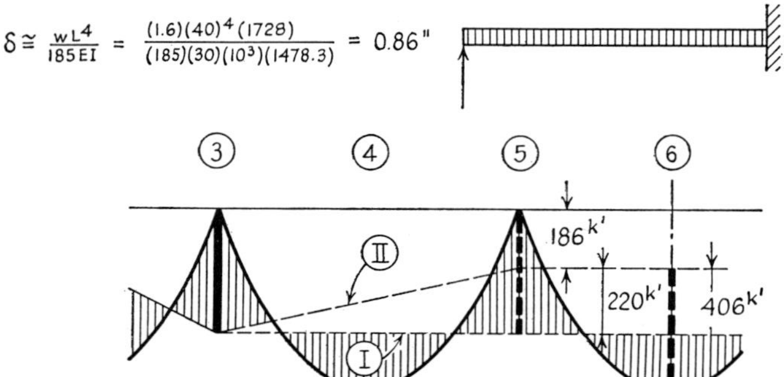

(e)

Steps 3 and 4: As derived in Chapter 7 (Procedure 9, Table 7.3) the required plastic moment for an end span is equal to $0.0858wL^2$, and the plastic hinge is located a distance equal to $0.414L$ from the end of the beam.

Step 5: A *precise* determination of the reactions at ultimate load would require an elastic analysis. The reactions are computed in this problem, however, on the assumption that the load on the interior span 3–5 is divided evenly between the two supports 3 and 5. Since the total load on span 3–5 is 118.4 kips, then the shear V_{34} becomes 59.2 kips. Actually the shear in span 3–5 cannot vary too much and must fall somewhere between values that would correspond to the two limiting conditions indicated by cases I and II in the portion of the moment diagram re-plotted in Plate I, sketch (*e*). The moment diagram shown for case I corresponds to equal moments M_p at sections 3 and 5 and also corresponds to the assumption just made that V_{34} is equal to 59.2 kips. The moment diagram for case II is shown dashed in sketch (*e*) and represents the other limiting case—namely, the attainment of the plastic moment 406 kips at section 6. V_{34} for case II is computed as

$$V_{34} = \frac{w_uL}{2} + \frac{(M_3 - M_5)}{L} = 59.2 + \frac{220}{40} = 64.7 \text{ kips}$$

Thus V_{35} may vary between the assumed value of 59.2 kips (case I) and 64.7 kips (case II), and the reaction at support 3 must lie between 128.7 kips and 134.2 kips.

Steps 7 and 8: The maximum shear (69.5-kips to the left of support 3) is well within the permitted value of 163 kips for this shape. The cross-sectional proportions are also found to be adequate.

Step 9: With regard to bracing, the structure is assumed to be enclosed. Thus the top flange is continuously braced. Vertical plates are supplied at sections 3 and 5 to brace the web over the support. At section 2 similar vertical plates would be needed only if the hinge at section 2 were the first one to form. Since the first hinge will form at the point of maximum elastic moment, continuous beam diagrams such as those given in Ref. 2.1 will show whether or not such bracing is needed. For equal spans and uniform loading, the maximum moment is not located at an intermediate point in a beam (such as section 2) and therefore such stiffening will not be required.

Step 10: Splices for shear will be adequate at the indicated sections in the splice sketch. Moment splices also could be made if desired. Shear splices would be located at a distance equal to 8 ft from the supports at the indicated points. The locations are selected from the moment diagram. Any variation in moment due to shift of the actual point of inflection cannot be of serious consequence to load-carrying capacity. If moment splices were used, they would be located to facilitate field erection.

Step 11: Whether or not the deflection calculation is made depends on the design conditions. The greatest deflection will be found in the end spans and will probably not be far from the value 0.86 in., which is computed on the basis that the beam will deflect about the same as a beam fixed at one end and simply supported at the other.

PLATE II. THREE-SPAN BEAM WITH DISSIMILAR SECTIONS

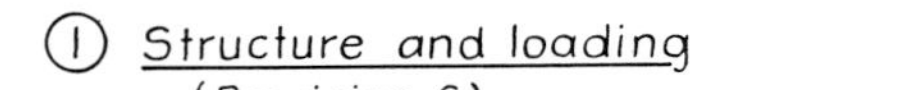

① Structure and loading

(Provision 6)

3-span continuous beam.
Dissimilar sections will be specified to suit the moment diagram.

$L = 20'$ $\quad L_1 = 1.25L$

$L_2 = 1.5L$

$L_3 = 1.0L$

$w_u = (1.0)(1.85) = 1.85$ k/ft. $\quad w_{1u} = w_u$

$w_{2u} = 1.5w_u$

Design for single loading condition.

② Moment diagram

(Table 7.2, Procedure 4)

$$M_{S1} = \frac{w_u L_1^2}{8} = \frac{(1.85)(25)^2}{8} = 144.6^{k'}$$

$$M_{S2} = \qquad = 312.0^{k'}$$

$$M_{S3} = \qquad = 139.0^{k'}$$

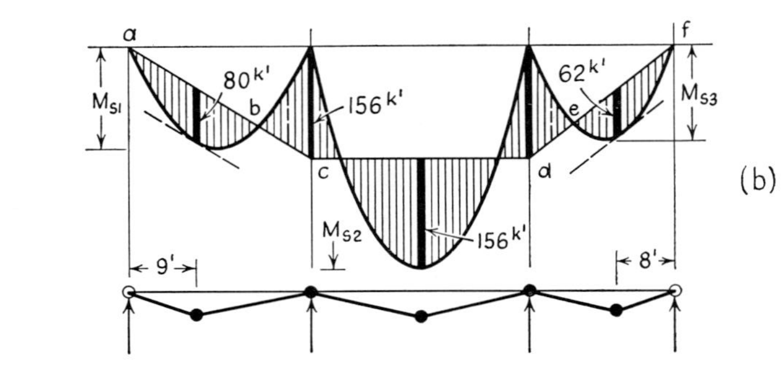

③ Mechanism (See sketch b)

④ Plastic moment values

(Table 7.3, Procedures 1 & 9)

$M_{p4} = \frac{1.5 w_u L_2^2}{16} = \frac{M_{s2}}{2} = \underline{156.0^{k'}}$ $\qquad Z_4 = \frac{M_p}{\sigma_y} = (156)(0.364) = \underline{56.8} \text{ in.}^3$ $\qquad$ **Use 16 WF 36, $Z = 63.9$ in.^{3}**

M_{p2} (determined by scale) $= \underline{80^{k'}}$ $\qquad Z_2 = \underline{29.1} \text{ in.}^3$ $\qquad$ **Use 12 B 22, $Z = 29.35$**

M_{p6} (determined by scale) $= \underline{62^{k'}}$ $\qquad Z_6 = \underline{22.6} \text{ in.}^3$ $\qquad$ **Use 12 B 19, $Z = 24.78$**

⑤ Reactions (Ultimate load)

$R_1(9) - \frac{w_u(9)^2}{2} = M_{p2}, \quad R_1 = \frac{80}{9} + \frac{(1.85)9}{2}$ $\qquad \underline{R_1 = 17.2^k}$

$R_3 = V_{32} + V_{34} = (w_u L_1 - R_1) + \left(\frac{1.5 w_u L_2}{2}\right)$ $\qquad \underline{R_3 = 70.8^k}$

$R_7(8) - (1.5)\frac{w_u(8)^2}{2} = M_{p6}$ $\qquad \underline{R_7 = 18.8^k}$

$R_5 = V_{56} + V_{54} = (1.5 w_u L_3 - R_7) + \frac{(1.5 w_u L_2)}{2}$ $\qquad \underline{R_5 = 78.3^k}$

NOTE: Total $R = 185.1^k$ ck

PLATE II. (*Continued*)

(7) Shear force
(Table 7.4, Guide 2)

$V_{max} = V_{34} = 41.7^k$

V_{allow} (16 WF 36) = (18.0)(0.299)(15.85) = $85.2^k > 41.7^k$ ok

12 B 19: $V_{max} = V_{76} = 18.8^k$

$V_{allow.}$ = (18.0)(0.260)(12.31) = $57.6^k > 18.8^k$ ok

(8) Cross-section proportions
(Table 7.8, Guide 1)

Section	b/t	d/w
16 WF 36	16.34	53.0
12 B 22	9.50	47.3
12 B 19	11.50	50.7

All ok $b/t < 17.0$, $d/w < 55.0$

(9) Splices

Provide shear splices at points indicated.

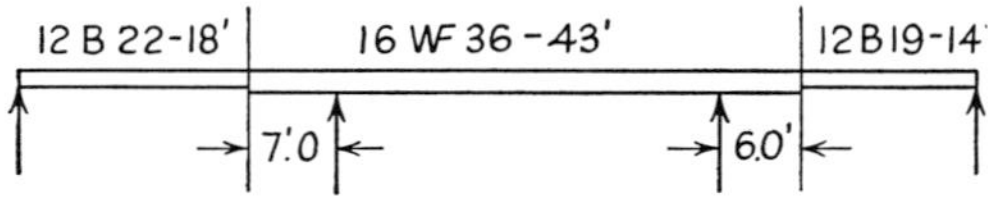

12 B 22-18′ | 16 WF 36 - 43′ | 12 B 19-14′

7.0′ | 6.0′

(c)

(10) Bracing requirements
(Table 7.8, Guide 3)

Top flange continuously supported by concrete slab as in Plate I
Provide welded vertical plates at sections 3 and 5 as in Plate I
Similiar plates are advisable at sections 2 and 6.

8.3 CONTINUOUS BEAMS WITH DIFFERENT CROSS SECTIONS

PLATE II Three-Span Beam with Dissimilar Sections

In the problem of a three-span continuous beam developed in Plate II, dissimilar sections are specified to meet the needs of the different spans.

Step 2: The simple-span moment diagrams are first laid out to scale to facilitate the semi-graphical solution. Lines *a–c* and *d–f* are drawn and the moment diagram due to the "redundant loading" is thus *a–c–d–f*. The shaded portion represents the difference between the determinate and redundant diagrams. The center span is the critical one and requires a 16WF36 shape.

Since it is only planned to splice for shear, the 16WF36 member will extend into the side spans to the points of inflection at *b* and *e*. The required moment capacity of these two spans will be determined by the moments at sections 2 and 6. The magnitude of these moments (80 and 62 kip-ft) are selected graphically as are the distances to hinge points 2 and 6.

Step 7: All sections are satisfactory with regard to shear force. The maximum shear in each of the outer spans is equal to the outer reaction. Since the 12B22 beam is larger and yet is required to carry a smaller shear force, it need not be checked.

Step 9: Splices are located at points of inflection and need be designed for shear only. Alternatively, if full moment splices were desirable, then the length of the heavier 16WF36 beam could be decreased 4 ft on the left and $2\frac{1}{2}$ ft on the right. The position of the splices are indicated by the dotted ordinates in the moment diagram. It is doubtful if the additional fabrication cost warrants the saving in weight of main material unless the latter is of paramount importance.

Step 10: No additional bracing at section 4 was specified because in this configuration, this hinge will not form prior to that at 3 and 5. Plate stiffeners are needed at 3 and 5 and have been suggested for sections 2 and 6 in lieu of making a calculation to determine if they are absolutely necessary.

PLATE III Uniform Beam with Cover Plates

The problem is the same as Plate II except that a uniform section is used throughout, reinforced where necessary with cover plates on top and bottom flanges.

Step 2: The left-hand span 1–3 requires a plastic moment that is only slightly greater than for the right-hand end span 5–7. Therefore a uniform section is chosen that will just meet the moment requirements of the left span (12WF27) and this shape will also be adequate for the right span.

To carry the moment at section 4, cover plates are required with a moment capacity of 114 kip-ft ($M_s - 2M_p$). Two plates $5\frac{1}{2}'' \times \frac{5}{8}''$ will be adequate. They should extend somewhat beyond the point at which $M = M_p$. This distance is selected as about 6 in. and the plates should therefore be 19′-0″ long.

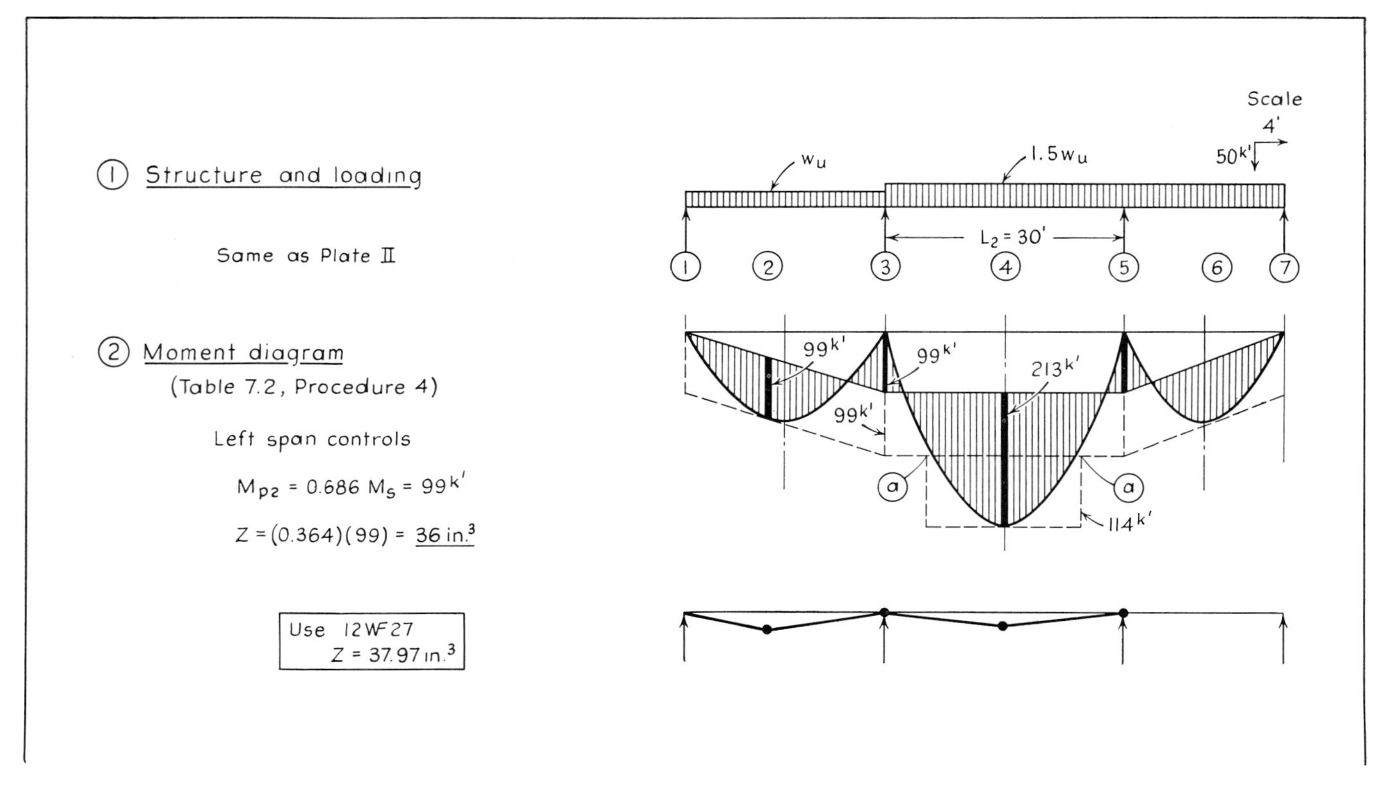
PLATE III. UNIFORM BEAM WITH COVER PLATES
① Structure and loading
Same as Plate II
② Moment diagram
(Table 7.2, Procedure 4)
Left span controls
M_p2 = 0.686 M_s = 99k'
Z = (0.364)(99) = 36 in.³
Use 12WF27
Z = 37.97 in.³
Scale
4'
50k'
w_u
1.5w_u
L_2 = 30'
1
2
3
4
5
6
7
99k'
99k'
99k'
213k'
a
a
114k'

Reinforcing ℘s:

$\Delta M_p = M_s - M_p = 312^{k'} - 198^{k'} = 114^{k'}$

$A_{℘} = \frac{\Delta M_p}{\sigma_y d} = \frac{(114)(12)}{(33)(12.0)} = 3.45^{\square''}$

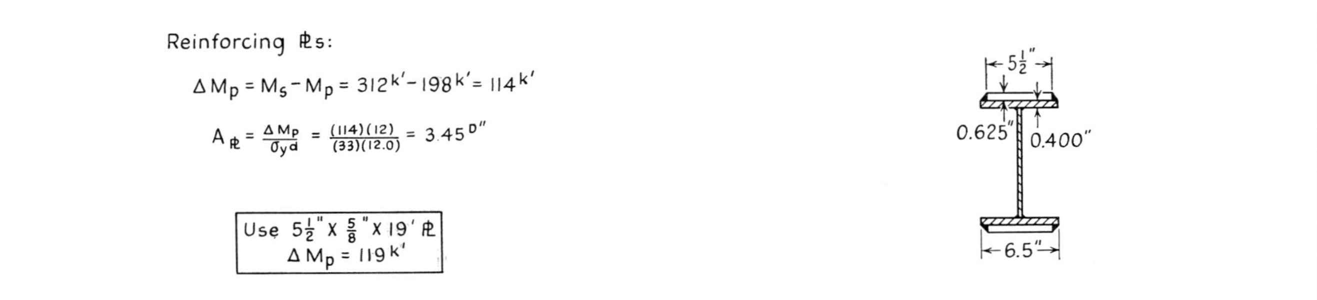

Use $5\frac{1}{2}'' \times \frac{5}{8}'' \times 19'$ ℘
$\Delta M_p = 119^{k'}$

③ Mechanism — 2 beam mechanisms

④ Reactions — Compute by statics. See previous examples

⑤ Shear force (Table 7.4, Guide 2)

$V_{max} = V_{34} = 41.7^{k}$ *

$V_{allow.}\,(12\ WF\ 27) = (18.0)(0.240)(11.95) = 51.6^{k} > 41.7^{k}$ ok

⑥ Cross-section (Table 7.8, Guide 1)

$b/t = 16.25, \quad d/w = 49.8$ ok

⑦ Splices — Splice for shear at inflection point

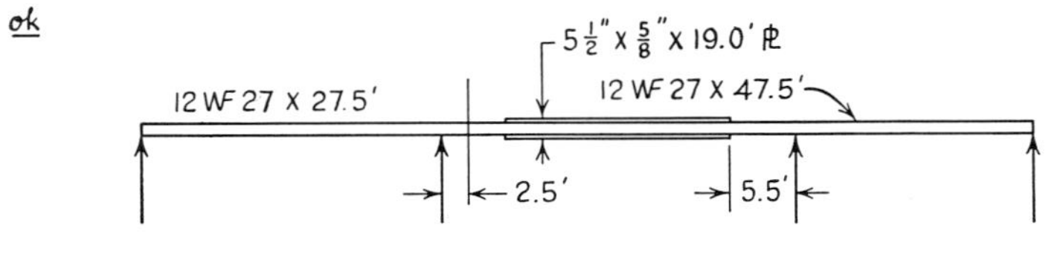

* See Plate II, $V = 1.5 w_u \frac{L_2}{2}$

Step 3: Two local (beam) mechanisms result. The reactions were not computed in this example. The same procedures would be used as in Plate II, p. 257.

Step 5: The shear force begins to approach (but does not reach) a critical value in this problem. Had it exceeded 51.6 kips, then local stiffening of the web would have been required in the region in which $V > V_{max}$.

Step 7: The position of the splice(s) is controlled in this problem by shipping requirements. A single splice for shear is shown at the point of inflection in segment 3–4. The cover plates are to be fillet-welded to the beam flanges.

Comparing the weights of three designs (uniform section, Plate II, p. 256, and Plate III, p. 260) the following is obtained:

Design	Shapes	Weight
Uniform section	16WF36	2700 lb
Dissimilar section (Plate II)	16WF36 12B22 12B19	2222 lb
Uniform with cover plates (Plate III)	12WF27 $5\frac{1}{2}'' \times \frac{5}{8}''$ Cover plates	2247 lb

Therefore the lightest design is the one in which dissimilar sections are used (Plate II). However, local conditions would dictate whether or not the extra splices in this design would be a more economical choice than the fillet welding of the cover plates of Plate III.

8.4 MEMBERS OF VARIABLE CROSS SECTION

PLATE IV Tapered Beam

A fixed-ended, uniformly loaded beam will now be designed in which the member is tapered from the ends to midspan.* The resulting design will be compared with the weight of a member of uniform section throughout. The two designs are shown in Plate IV, the span being 40 ft and the total distributed ultimate load being 375 kips.

For the design of the uniform beam according to the equation,

$$M_p = \frac{W_u L}{16}$$

the required plastic modulus Z is 340 in.^{3}

* The analysis of structures with tapered members is discussed in Art. 3.9.

PLATE IV. TAPERED BEAMS

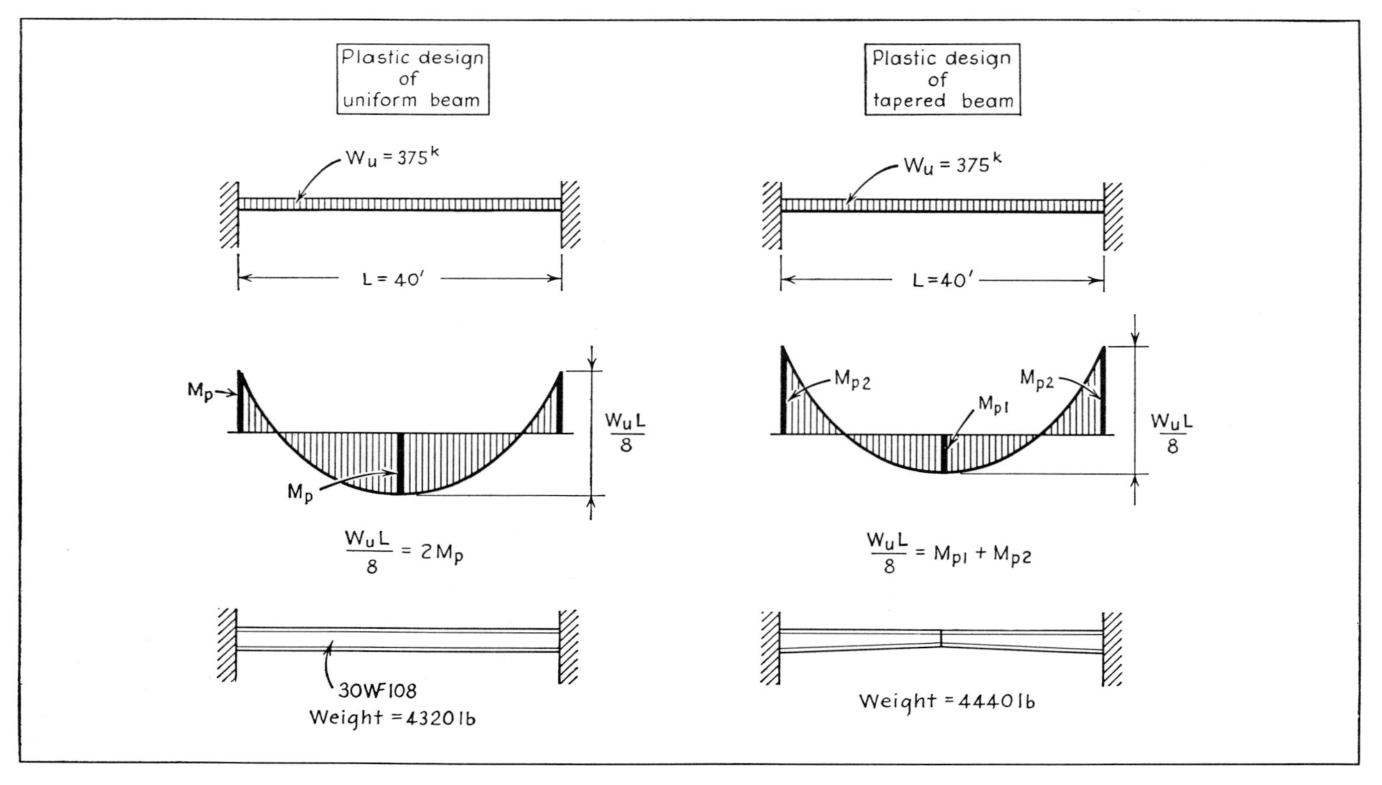

In designing the tapered member, equilibrium of moment requires that

$$\frac{W_u L}{8} = M_{p1} + M_{p2} \tag{3.26}$$

where M_{p1} and M_{p2} are plastic moments at the center and the ends, respectively. It is assumed that the member has a uniform taper, is smaller at the center, and is symmetrical about the center line as shown. The ratio of M_{p2} to M_{p1} can be selected as any reasonable value, and in this case it is assumed that $M_{p2} = 3M_{p1}$. Since the plastic modulus Z bears a linear relationship to M_p (Eq. 2.11), then $Z_2 = 3Z_1$. Equation 3.26 may now be solved for Z_1:

$$Z_1 = \frac{W_u L}{32\sigma_y} = \frac{(375)(40)(12)}{(32)(33)} = 170 \text{ in.}^3$$

and Z_2 equals 510 in.^{3}

For a built-up member, the value of Z may be determined from Eq. 2.24. If the flange dimensions and the web thickness are selected to be about those values that would be common for such members ($b = 11.0''$, $t = 0.75''$, $w = 0.625''$), then the only unknown in Eq. 2.24 is the depth d, which may be obtained by solving the quadratic:

$$\frac{w}{4} d^2 + td(b - w) - Z - t^2(b - w) = 0$$

For the values of Z_1 and Z_2 determined above, the required depth at the center is 16.8 in. and at the ends is 38 in.

Comparing the weight of the two beams, the uniform member, according to the plastic design, weighs 4320 lb. The weight of the plastically designed tapered beam is 4440 lb. Thus the two designs turn out to have about the same weight with a slight advantage (3%) for the uniform member. The uniform member has a further advantage in view of the additional fabrication cost for the tapered member. Incidentally, conventional elastic design of the uniform beam would have required a section modulus of 429 in.^{3} A 33WF141 section would have been used, the total weight being 5640 lb or an increase of 31% over the similar plastic design.

In addition to showing the economy of material of plastic over elastic design, this example demonstrates that plastic design achieves its economy without the necessity of tapering the members.

8.5 SIMPLIFIED PROCEDURES

One of the advantages of plastic design is that the engineer is able to complete the analysis in less time than required by present elastic procedures. It is possible, however, to shorten the design time even further, by taking advantage of the same technique that is used in conventional design and one that is frequently used whenever a procedure

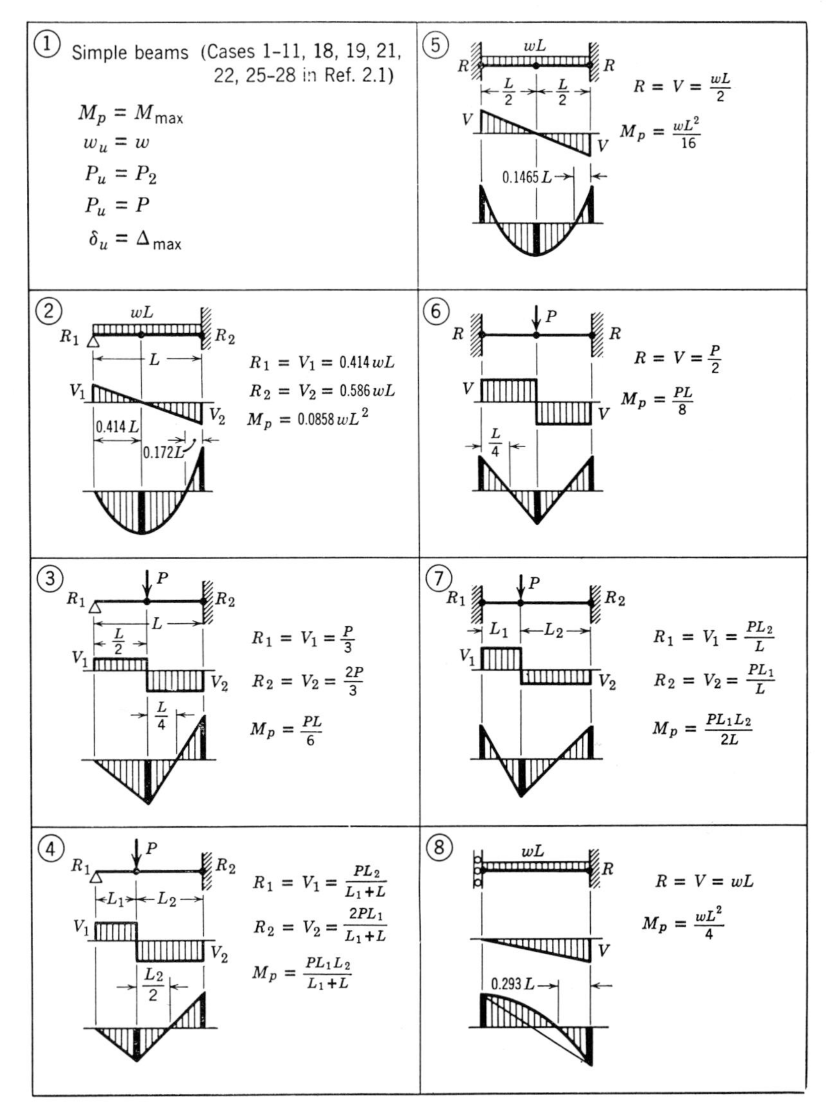

Fig. 8.1. Beam diagrams and formulas.

becomes time-consuming. The solution of frequently encountered standard cases may be given in formula or in chart form. In this section one such technique is described; others will be presented later. The presentation is by no means a complete one and the interested reader is referred to Ref. 7.1. Two words of caution are appropriate at this point:

(1) Since superposition does not hold in plastic analysis, generally it is not possible to combine two separate solutions as is done so commonly in elastic design. Any "formula" or "chart" can only assist in the solution of the particular loading and geometry for which it is developed.

(2) Even though the formulas and charts are correct in themselves, it is a good rule to check the plastic moment condition by drawing the moment diagram. In this way one is assured of the correct answer.

In Fig. 8.1 are beam diagrams and formulas for certain loading conditions on indeterminate beams. The table is patterned after similar tables contained in Ref. 2.1. In addition to the reactions and M_p values for these standard cases, the position of plastic hinges and points of inflection are indicated. As before, the moment diagrams are plotted on the tension side of the member.

Steel Frame Design

9.1 INTRODUCTION

The presentation of design examples is continued in this chapter and steel frames of various configurations are considered. The general objective is the same as described in Art. 8.1. Required member sizes will be determined, followed by an examination of the secondary design considerations. A direct plastic analysis will be made for each problem and in certain cases, the answers will be checked by "short cuts" using charts or formulas that have been developed.

The design of single-span frames with flat and gabled roofs will be covered in Art. 9.2. The effect of fixing column bases will be discussed. The use of haunches is illustrated. In Art. 9.3 simplified procedures are presented for the more rapid solution of single-span frame problems. Multi-span frames will be treated in Art. 9.4, and the corresponding formulas and charts for the quick solution of these problems follow in Art. 9.5. Multi-story frames are briefly treated in Art. 9.6.

9.2 SINGLE-SPAN FRAMES

PLATE V Pinned-Base Rectangular Frame

A single-span, pinned-base frame with flat roof is to be designed to withstand vertical and horizontal loads. All applicable "rules" will be checked. The solution is shown in Plate V.

PLATE V. PINNED-BASE RECTANGULAR FRAME

① Structure and loading:

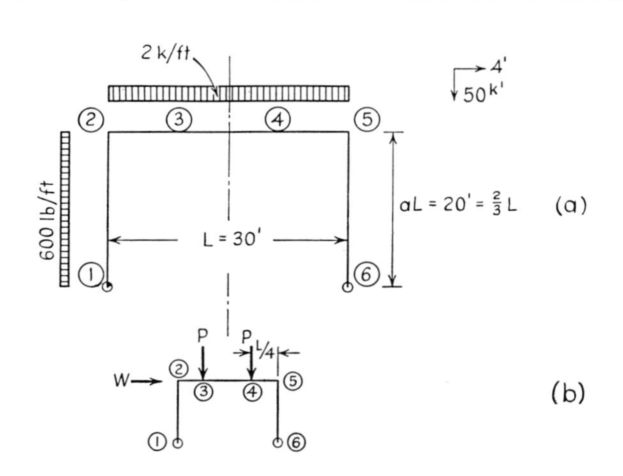

② Loading conditions: (Table 7.1, Provision 6)

Case I: (DL + LL), F = 1.85

$w_u = 2.0 \times 1.85 = 3.70$ k/ft

Case II: (DL + LL + wind), F = 1.40

$w_v = 2.0 \times 1.40 = 2.80$ k/ft $= w_u$

$w_h = 0.60 \times 1.40 = 0.84$ k/ft $= 0.3\, w_u$

Replace uniform vertical load by concentrated loads at quarter-point (sk. b) $P = \frac{w_u L}{2}$

Replace horizontal load with concentrated load with equal overturning moment

$$W = \frac{w_h (aL)\left(\frac{aL}{2}\right)}{aL} = 0.1\, w_u L = 8.4^k$$

③ Plastic moment ratios: Uniform section throughout (Table 7.2, Procedure 1)

④ Case I analysis
(Table 7.3, Procedure I)

Moment diagram:
(Redudant = H_6)

$M_S = \frac{PL}{4} = \frac{w_u L^2}{8} = \frac{(3.70)(30)^2}{8}$

$M_S = 417^{k'}$

$M_p = \frac{M_s}{2} = \underline{209^{k'}}$

Mechanism: sketch (d)

Reactions:

$H_6 = H_1 = \frac{M_p}{aL} = \frac{209^{k'}}{20'} = 10.5^k$

$V_1 = V_6 = P = (3.70)\left(\frac{30}{2}\right) = 55.5^k$

NOTE: Above values to be compared with values for Case II and maximum values used.

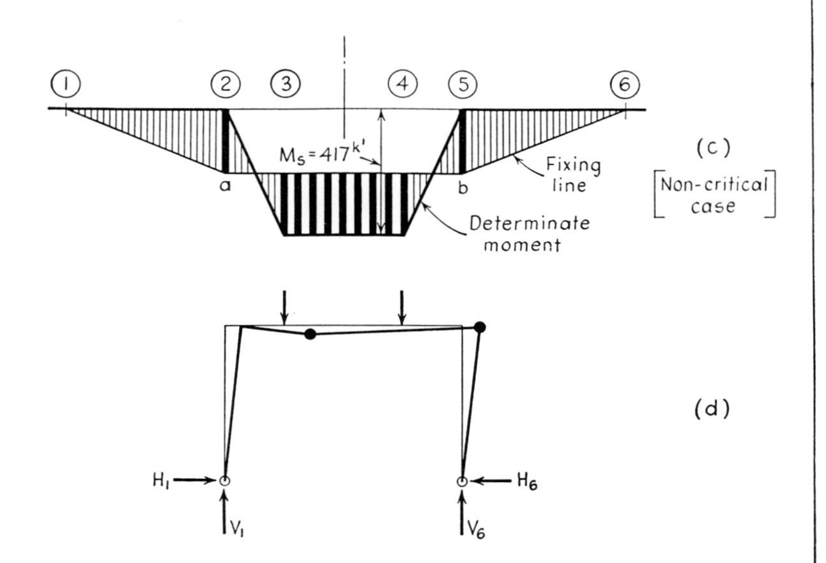

PLATE V (*Continued*)

(5) <u>Case II analysis</u>: (Table 7.3, Procedure 1)

Moment diagram: sketch (e) (Redundant = H_6)

$$M_2'\,(\text{Det.}) = W a L = \frac{w_u L}{10}\left(\frac{2L}{3}\right)$$

$$= \frac{w_u L^2}{15} = \frac{(2.80)(30)^2}{15} = 168.0^{k'}$$

$$M_s = \frac{PL}{4} = \frac{w_u L^2}{8} = \underline{315.6^{k'}}$$

Equilibrium at (3):

$$M_s + \tfrac{3}{4} M_2' = 2M_p$$

$$M_p = \frac{316 + 126}{2} = \underline{221^{k'}}$$

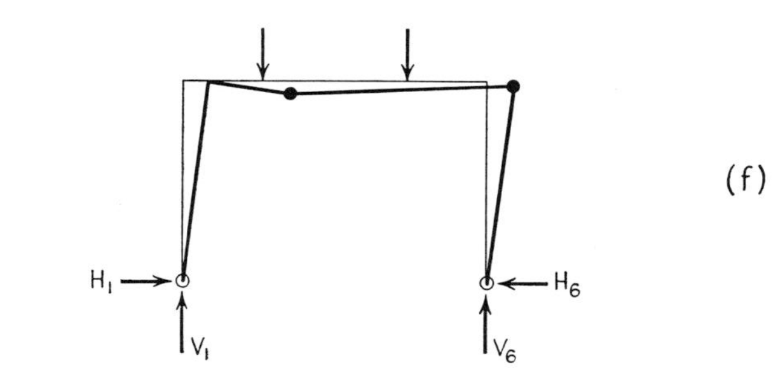

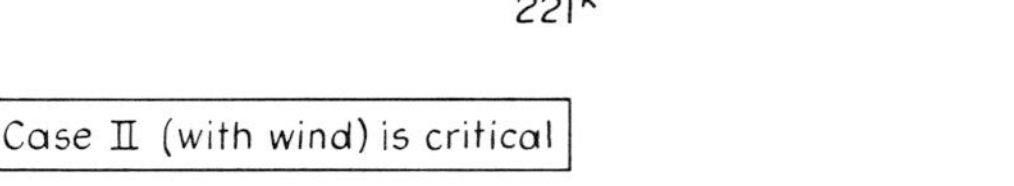

Mechanism: sketch (f)

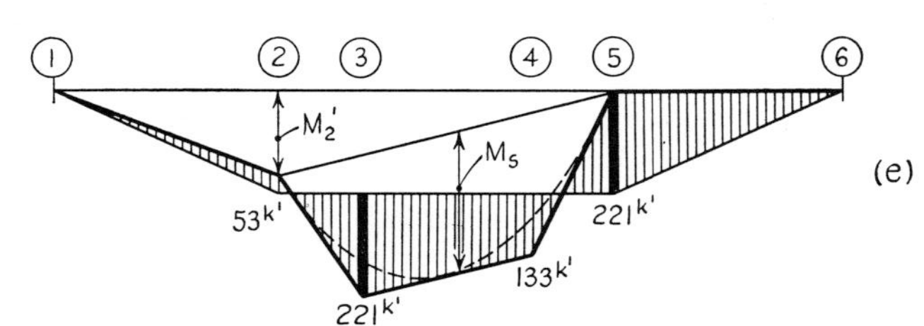

Reactions (Ultimate load)

$$H_6 = \frac{M_p}{aL} = \frac{221}{20} = 11.1^k$$

$$H_1 = H_6 - W = 11.1 - 8.4 = 2.7^k$$

$$V_1 = \frac{H_1 a L + M_p}{L/4} = \frac{84 + 221}{7.5} = 36.6^k$$

$$V_6 = 2P - V_1 = 84.0 - 36.6 = 47.4^k$$

Moment check (Table 7.3, Procedure 7)

$$M_2 = M_p - M_2' = 221 - 168 = 53^{k'}$$

$$M_4 = V_6\,(7.5) - H_6\,(20) = 133^{k'}$$ <u>ok</u>

(6) Selection of section (Table 7.1, Provision 5)

$$Z = M_p/\sigma_y = (221)(0.364) = 80.5 \text{ in.}^3$$

Use 16 WF 45
$Z = 82.0 \text{ in}^3$

$A = 13.24 \square''$	$w = 0.346''$
$d = 16.12''$	$t = 0.563''$
$b = 7.04''$	$S = 72.4 \text{ in.}^3$
$r_y = 1.52''$	$I = 583 \text{ in.}^3$

(7) Axial force (Right-hand column critical)
(Table 7.4, Guide 1)

$$\frac{P}{P_y} = \frac{V_6}{\sigma_y A} = \frac{55.5}{(33.0)(13.24)} = 0.125 < 0.15 \quad \text{ok}$$

(8) Shear force
(Table 7.4, Guide 2)

$$V_{max} = V_{45} = V_6 = 55.5^k$$

$$V_{allow.} = 18.0\, wd = 100.3^k > 55.5 \quad \text{ok}$$

(9) Cross-section proportions
(Table 7.8, Guide 1)

$b/t = 12.50$, $d/w = 46.6$ ok

PLATE V (*Continued*)

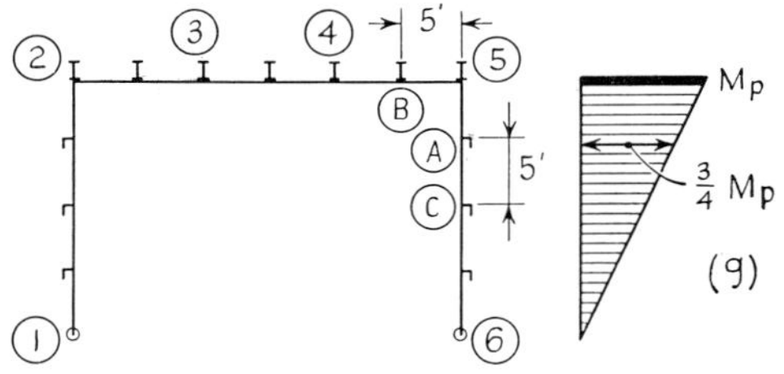

(10) Lateral bracing (Table 7.8, Guide 3)

Spacing: $\frac{L_B}{r_y} = \frac{(5)(12)}{1.52} = 39.5$

"Rafter" hinge OK. Last hinge forms in rafter

"Column" hinge $\frac{M}{M_p} = \frac{\frac{3}{4} M_p}{M_p} = 0.75$

Eq (4.28): $\left(\frac{L_B}{r_y}\right)_{cr} = 30 < 39.5$ ∴ More refined check is necessary as per Appendix I.

Evaluation of restraint coefficient:

$$C_f = 1.0 + 0.2 \left\{1 - \left(\frac{L}{L_1}\right)^2\right\} \left\{0.9 + \frac{1 - \left(\frac{L}{L_s}\right)^2}{1 - \left(\frac{L}{L_1}\right)^2}\right\} \quad \text{(A.1)}$$

Segment	Moment ratio	$\left(\frac{L}{r_y}\right)_{cr}$	L_{cr}
$L_B = 5\text{-}A$	$\frac{M}{M_p} = 0.75$	Eq 428: $\frac{L}{r_y} = 30$	$46'' = L_b$
$L_L = 5\text{-}B$	$g_L = -0.05$	Eq A2: $48 - 30(-0.05) = 49.5$	$75.3'' = L_s$
$L_R = A\text{-}C$	$g_R = 0.67$	Eq A3: $134/\sqrt{0.75} + 60(1 - 0.67) = 175$	$266'' = L_1$

$$C_f = 1.0 + 0.2 \left\{1 - \left(\frac{60}{266}\right)^2\right\} \left\{0.9 + \frac{1 - \left(\frac{60}{75.3}\right)^2}{1 - \left(\frac{60}{266}\right)^2}\right\} = 1.25$$

$(L_B)_{cr} = C_f L_b = (1.25)(46) = 58'' < 60''$ (Call adequate)

Bracing details

At sections 2 & 5 brace to inner (compression) corner from purlin

(11) Columns (Right-hand column critical, sketch d)
(Table 7.6, Guides 2 & 4)

$\frac{P}{P_y} = 0.125 < 0.15 \quad \therefore$ Full M_p is available

Weak axis:

$$\frac{L}{r_y} < \sqrt{\frac{8850}{\frac{P}{P_y}}}, \quad \sqrt{\frac{8850}{0.125}} = 268 > \frac{(5)(12)}{1.52} = 39.5 \quad \underline{ok}$$

(12) Connection detail
(Table 7.7, Guide 2)

$$t_s = \frac{\sqrt{2}}{b}\left(\frac{S}{d} - \frac{wd}{\sqrt{3}}\right) = \frac{\sqrt{2}}{7.04}\left(\frac{72.4}{16.12} - \frac{(0.346)(16.12)}{\sqrt{3}}\right) = 0.256 \text{ in}$$

See sketch (h)

NOTE: $\frac{7}{16}''$ PL needed to meet b/t requirement

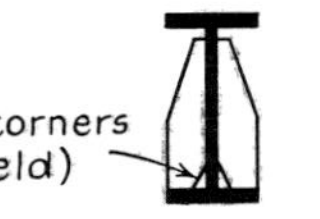

Use $3\frac{3}{8}'' \times \frac{7}{16}''$ PLs

(h)

(13) Splices — Provided as part of corner connection detail

PLATE V (Continued)

(14) Deflection at working load
(Table 7.3, Procedure 12)

Ultimate load,	$P_u = 42.0^k$
Moment diagram,	Sketch (e)
Mechanism,	Sketch (f)
Free body diagram,	Sketch (i)
Slope deflection eqs,	$(\theta_{32} = \theta_{35})$

$$\theta_{32} = \frac{\delta_{V3}}{l} + \frac{l}{3EI}\left(M_3 - \frac{M_2}{2}\right) = \frac{\delta_{V3}}{\frac{L}{4}} + \frac{\frac{L}{4}}{3EI}(-221 + 27)$$

$$\theta_{35} = \theta'_{35} + \frac{\delta_{V3}}{\frac{3L}{4}} + \frac{\frac{3L}{4}}{3EI}(+221 - 111)$$

$$\theta'_{35} = 0.028\frac{PL^2}{EI}$$

$$\delta_{V3} = \delta_u = 1.55''$$

$$\delta_w < \frac{\delta_u}{F} = \frac{1.55}{1.85} = 0.84'' \quad \underline{ok}$$

(i)

$M_2 = 53^{k'}$
$M_3 = 221^{k'}$
$M_5 = 221^{k'}$

(15) Frame layout

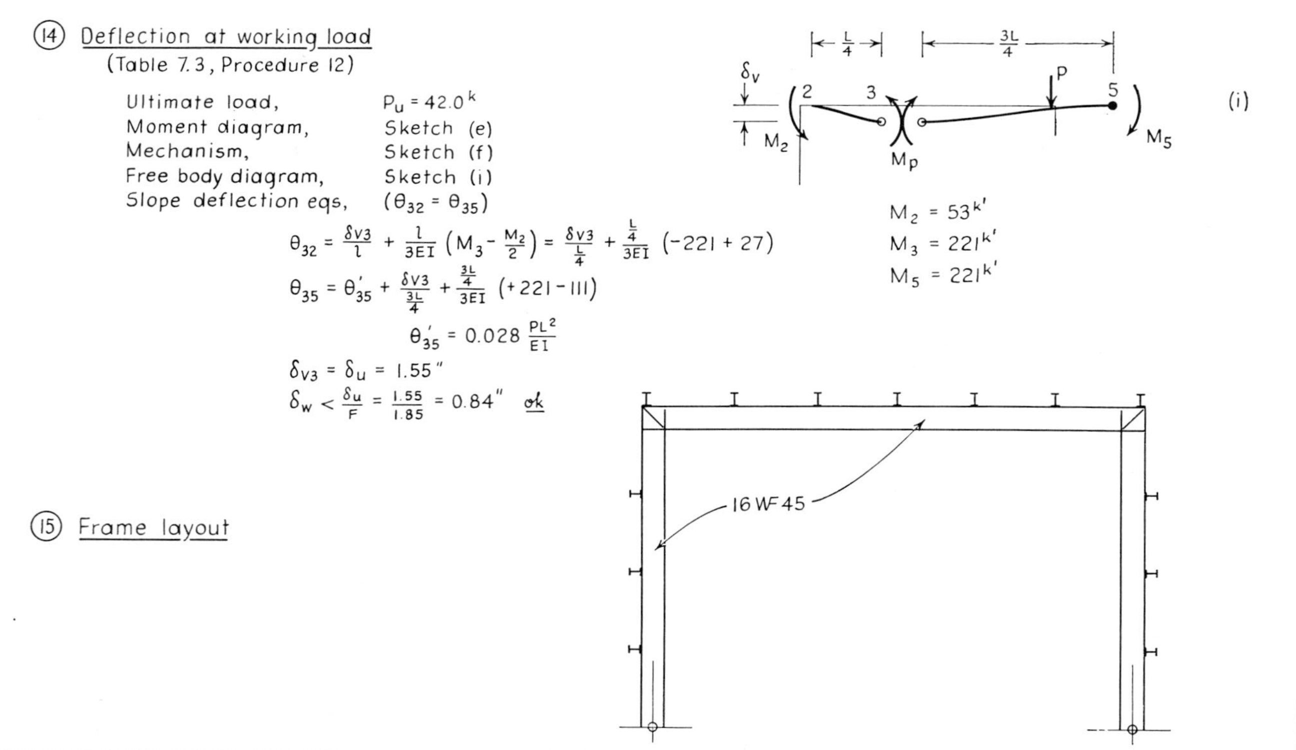

Step 2: In the absence of wind, $F = 1.85$. For the second loading condition a load factor of 1.40 is applied against all loads (live + dead + wind). For simplicity the uniform vertical load is replaced by concentrated loads at the quarter points (sketch *b*). (The solution of a problem in which the loading was actually assumed as distributed was given in Chapter 8.) The distributed horizontal load is replaced by a single load acting at the eaves line. The horizontal load W (called T elsewhere) produces an equivalent moment about the base; or,

$$W = \frac{(w_h aL)}{aL}\frac{(aL)}{2} = \frac{w_h aL}{2}$$

Step 3: In arriving at the preliminary choice of member sizes, it is assumed that a constant section will be used throughout. The important load is the vertical load and thus maximum restraining moments at the ends are desirable (Art. 7.2).

Step 4: The analysis is carried out by the statical method. The redundant is selected as H_6, the horizontal reaction at 6. The fixing line 1–a–b–6 is drawn in such a way that a mechanism forms as shown in sketch (*d*) with plastic hinges at sections 3 and 5. It happens that the moment at sections 2 and 4 is also equal to M_p, and thus the frame is overdeterminate at failure with hinges forming at sections 2, 5, and along 3–4. The required plastic moment is 209 kip-ft.

Step 5: Case II is next analyzed and it is found that the mechanism is the same as that for case I; it is shown in sketch *f*. The "complete" or composite moment diagram is that shown by the shaded portion of sketch (*e*). In that sketch, the heavy solid line is the determinate moment, the lighter line being the moment due to loading by the redundant H_6. The required plastic moment for case II is 221 kip-ft. Therefore this case controls the design.

Step 6: In selecting the section, a plastic modulus of 80.5 in.^{3} would be required. The 16WF45 supplies a value of Z equal to 82.0 in.^{3} In view of the fact that the analysis was carried out on the conservative basis of concentrated loads (sketch *b*) a somewhat lighter section certainly would be adequate. The dotted parabolic moment diagram reveals, in fact, that the required M_p-value is reduced from 221 to about 210 kip-ft, as determined by measurement.

Step 7: After the reactions are computed, the next step is to examine the secondary design considerations. In checking the axial force according to Design Guide 1 of Table 7.4, it is found that $P/P_y = 0.125$; the full value of M_p is thus available. It will be noted that the reaction V_6 for case I has been used, although case II controls the initial choice of member sizes. Had the resulting check for axial force required a modification (that is, $P/P_y, > 0.15$) it might have been worth while to examine the case I solution to see whether or not the assumed hinge at section 5 actually formed. Figure 4.19, case 2, shows that for $a > 0.5$ and $C = 0$ *, the first hinge forms at midspan and the last hinge forms at section 5. Thus if the case I loading were applied to a frame that was strong enough to support the case II loading, a hinge at section 5 could not form. In this case, Design Guide 1 of Table 7.4 would be too restric-

* C is the side load factor and is computed as $w_h/w_u(a + b)^2$.

tive. Of course, the question is academic in the present problem because $P/P_y < 0.15$.

Step 10: In checking for lateral bracing, it will be assumed that the purlin spacing is 5 ft. Between bracing points the ratio L/r_y is 39.5. This value is greater than the value $L/r_y = 30$ given in Eq. 4.28, but the latter value also assumes that some plastic rotation is required. If the last hinge forms in the rafter, then all that is required is that the section reach M_p in the rafter. Again, examining Fig. 4.19 with $C = w_h/w_u(a)^2 = (0.3)(\frac{2}{3})^2 = 0.133$, it is found that the last hinge is truly in the rafter. Therefore the 5-ft spacing is adequate.

Turning attention to the columns, the girt spacing is also assumed as 5 ft. Since the first hinge forms at 5, the selected spacing may or may not be adequate, depending on how much plastic rotation is required at 5 in order to develop the last hinge in the rafter, and how much the adjoining beam lengths restrain the "critical" segment. Although one could calculate the required plastic rotation at section 5,* it is usually quicker to check the restraint coefficient according to Appendix 1. For use in Eq. A.1 the braced segment L_B is segment 5–A, as shown in sketch (g). Segment L_L is 5–B, and segment L_R is A–C. Both segments L_L and L_B contain plastic zones, and therefore Eq. 4.28 is used to compute the critical slenderness ratio. For segment 5–A the moment ratio is $M_A/M_5 = 0.75$ therefore $L_B/r_y = 30$. For segment 5–B the moment ratio is -0.05 and thus $L_L/r_y = 49.5$. Segment A–C is elastic and the critical length ($C_f = 1.0$) is given by Eq. A.3, using a moment ratio $q_A = M_C/M_A = 0.67$. A value of 175 is determined for L_R/r_y. The critical lengths (with $C_f = 1.0$) are next computed; it is found that segment L_L is the shorter (L_s) and segment L_R is the longer (L_l). Upon substituting into Eq. A.1 a value of 1.25 is obtained for C_f. The "adjusted" critical length is thus 58″ which is sufficiently close to the selected value of the 60-in. purlin spacing.

Bracing details are suggested in the example. No stiffener was specified near mid-span of the rafter, because it was determined earlier that the last hinge forms there.

Step 11: Since the columns are loaded in case B loading (see Guide 2, Table 7.6) and since $P/P_y < 0.15$, the full plastic moment will be transmitted.

Step 12: The connection detail is shown in sketch (h), the thickness of diagonal stiffener being determined from Design Guide 2, Table 7.7. The local buckling requirement that b/t be less than 17 suggests a $\frac{7}{16}$-in. plate.

Step 14: Although the deflection of such a structure would probably not be computed, an "estimate" by Procedure 12 of Table 7.3 shows that the deflection at working load is less than 0.84. This is undoubtedly satisfactory since the usual rule-of-thumb value ($L/360$), gives one inch as the limit for live load deflection, taken alone.

This deflection analysis was completed without the necessity of making any preliminary trials because it was known in advance that the last hinge formed at section 3. Had this not been the case, then a trial calculation would also be

* The rotation angle has been calculated from Ref. 4.5 and found to give a value of $H_B/\phi_p L_B < 1.0$, indicating a small rotation-angle requirement.

necessary, assuming that the last hinge formed at section 5. The quantity θ_{35}' (rotation at end 3 of span 3–5 due to concentrated load P) was obtained with the aid of tables in Ref. 6.3. Of course, it could be computed by slope-deflection or moment areas.

PLATE VI Fixed-Base Rectangular Frame

A frame similar to that of Plate V, p. 268, will be designed except that the column bases will be fixed. Considerable expense is involved in providing sufficient rigidity to resist the overturning moments at column bases, and this factor should be considered carefully to see that the additional expense is warranted. There is less advantage to fixed column bases if the side loads are small (or absent altogether). At the opposite extreme is a structure designed to withstand blast load. In one instance [3.9] the capacity of a structure to resist externally applied side load was increased nine-fold simply by fixing the column bases and without changing member sizes whatsoever. Quite evidently there are instances in which the additional construction expense would be warranted in view of the improved load-carrying capacity of the structure. Tall buildings and industrial frames carrying relatively large cranes which might otherwise be sensitive to lateral deflections would constitute two other cases where fixed bases might be considered.

Step 1: The frame has a span of 48 ft, a column height of 16 ft, a uniformly distributed vertical load of 1.2 kip/ft, and a side (wind) load of 0.5 kip/ft, as shown in sketch (*a*) of Plate VI.

Step 2: Of the various methods for handling distributed loads, one which has been discussed (Art. 3.7) but not yet illustrated is to assume the purlin spacing at the outset and to analyze the frame on the basis of the resulting purlin loading. This method will be used here; the purlin load is found to be 13.32 kips for case I and 10.1 kips for case II.

Step 3: As in the previous example, the most economical design results from using a uniform member throughout; girder 2–4 should be restrained by the maximum possible restraining moment, and a member of equal moment capacity accomplishes this.

Step 4: The mechanism method of analysis is used in this problem in view of the greater redundancy of the structure when compared with Plate V. For case I with no side load, mechanism 1 will control and it is found that $M_p =$ 320 kip-ft. Note that if the actual distributed load had been used, then $M_p = w_u L^2/16 = 320$ kip-ft. This is precisely the same value as for concentrated load and is contrary to what would normally be expected. The reason is that the end purlin reacts directly on the column. Although the frame is redundant at failure $[I = X - (M - 1) = 3 - (3 - 1) = 1]$ the moment check is easily made by remembering that the elastic carry-over factor is one-half for cases such as members 2–1 and 4–5.

PLATE VI. FIXED-BASE RECTANGULAR FRAME

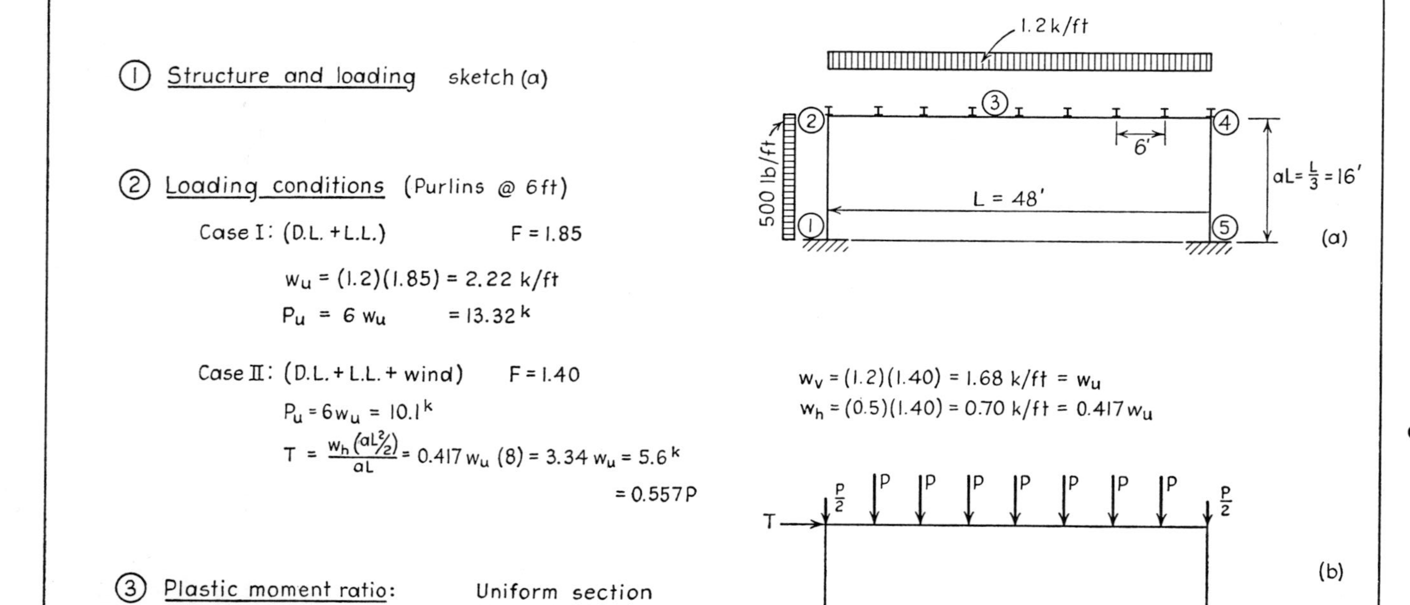

① Structure and loading sketch (a)

② Loading conditions (Purlins @ 6 ft)

Case I: (D.L. + L.L.) $F = 1.85$

$w_u = (1.2)(1.85) = 2.22$ k/ft

$P_u = 6\,w_u = 13.32^k$

Case II: (D.L. + L.L. + wind) $F = 1.40$

$w_v = (1.2)(1.40) = 1.68 \text{ k/ft} = w_u$

$w_h = (0.5)(1.40) = 0.70 \text{ k/ft} = 0.417\,w_u$

$P_u = 6w_u = 10.1^k$

$$T = \frac{w_h (aL^2/2)}{aL} = 0.417\,w_u\,(8) = 3.34\,w_u = 5.6^k = 0.557P$$

③ Plastic moment ratio: (Table 7.2, Procedure 1) Uniform section throughout

④ <u>Case I analysis</u> (Mechanism method)
(Table 7.3, Procedure 2)

Possible plastic hinges = 5 (sections 1, 2, 3, 4, 5)
Possible mechanisms = 2 ($n = N - X = 5 - 3 = 2$)

Independent : Nos. 1 & 2
Composite : No. 3

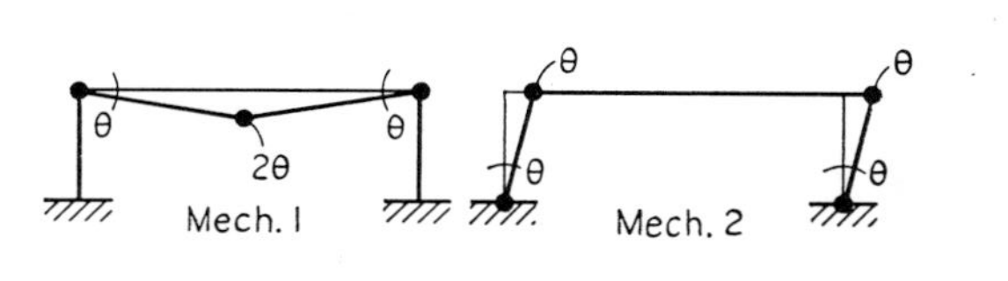

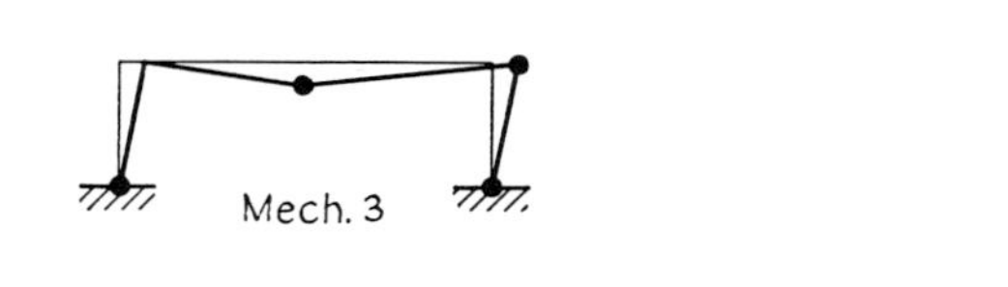

(c)

Solution for mechanism 1

$$M_p(\theta + 2\theta + \theta) = \frac{PL\theta}{8}\left(1 + 2 + 3 + \frac{4}{2}\right)2 = 2PL\theta$$

$$M_p = \frac{PL}{2} = \frac{(13.32)(48)}{2} = \underline{320^{k'}}$$

Moment check sketch (d)
(Table 7.3, Procedure 7)

$$M_1 = \frac{M_p}{2} = \frac{320}{2} = 160^{k'}$$

Sway equilibrium

$$M_5 = M_1 = 160^{k'} < M_p \quad \underline{ok}$$

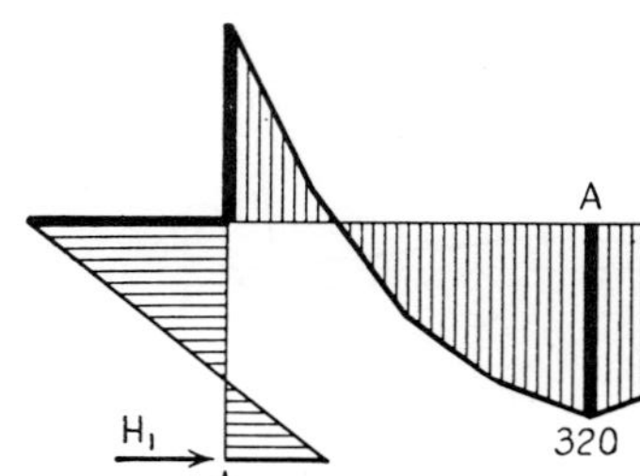

(d)

Reactions at ultimate load

$$V_1 = V_5 = 4P = 53.3^k$$

$$H_1 = H_5 = \frac{M_2 + M_1}{aL} = \frac{320 + 160}{16} = 30^k$$

PLATE VI (Continued)

⑤ <u>Case II analysis</u>

(Table 7.3, Procedure 2)

Hinges and mechanisms (See case I)

Solution for mech. 1 — sketch (c)

$$M_p = \frac{PL}{2} = (10.1)(24) = 242^{k'}$$

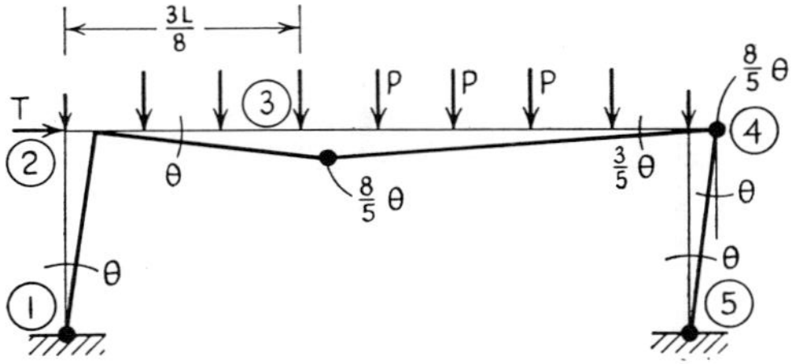

Solution for mech. 3 — sketch (e)

$$M_p\theta\left(1 + \frac{8}{5} + \frac{8}{5} + 1\right) = T\theta aL + \frac{PL\theta}{8}\left(1+2+3+\frac{3}{5}[1+2+3+4]\right)$$

$$5.2\,M_p = (5.6)(16) + \frac{3}{2}(10.1)(48)$$

$$M_p = \underline{157^{k}} < 242^{k'}$$

Moment check for mech. 1

(Table 7.3, Procedures 7 & 8)

Use "trial and error" method. Assume $M_5 = -M_p$

$$M_1 + M_2 + M_4 + M_5 + TaL = 0$$

$$M_1 = -TaL - M_2 - M_4 - M_5 = -(5.6)(16) - 242 + 242 + 242$$

$$= 242 - 90 = 152^{k'} < 242^{k'} \quad \underline{ok}$$

Case I (without wind) is critical

⑥ Selection of section

(Table 7.1, Provision 5)

$$Z = \frac{M_p}{\sigma_y} = (320)(0.364) = 116.5 \text{ in}^3$$

Use 18W60, Z = 122.6

$$\begin{cases} A = 17.64 & w = 0.416 \\ d = 18.25 & t = 0.695 \\ b = 7.56 & S = 107.8 \\ r_y = 1.63 & I = 984 \end{cases}$$

⑦ Axial force

(Table 7.4, Guide 1)

$$\frac{P}{P_y} = \frac{V_1}{\sigma_y A} = \frac{53.3}{(33)(17.64)} = 0.092 < 0.15 \quad \underline{ok}$$

⑧ Shear force

(Table 7.4, Guide 2)

$$V_{max.} = V_{34} = V_5 = 53.3^k$$

$$V_{allow.} = 18.0\,wd = 137^k > 53.3^{kips} \quad \underline{ok}$$

⑨ Cross section

(Table 7.8, Guide 1)

$$b/t = 10.9 < 17$$

$$d/w = 43.9 < 55 \quad \underline{ok}$$

PLATE VI (*Continued*)

(10) <u>Lateral bracing</u>

(Table 7.8, Guide 3)

Spacing check

Rafter at section 3: (Purlin spacing = 6.0 ft)

$$\frac{L_B}{r_y} = \frac{(6.0)(12.0)}{1.63} = 43.2 > 30$$ ok Last hinge is in rafter.

Column: (girt spacing = 5.0 ft from top)

$$\frac{L_B}{r_y} = \frac{(5.0)(12.0)}{1.63} = 36.8$$

$$\frac{M}{M_p} = \frac{140}{320} = +0.44$$

$$\text{Eq 4.28: } \left(\frac{L_B}{r_y}\right)_{cr} = 48 - 30\frac{M}{M_p} = 48 - 30\,(0.44) = 34.8 < 36.8$$

More refined check necessary

$$\text{Eq 4.29: } \left(\frac{L_B}{r_y}\right)_{cr} = 60 - 40\frac{M}{M_p} = 60 - 40\,(0.44) = 42.4 > 36.8$$ ok

Bracing details

At sections 2 & 4 brace to inner (compression) corners.

(11) Columns

(Table 7.6, Guides 2 & 4)

$\frac{P}{P_y} = 0.092 < 0.15$ $\therefore$ Full M_p is available

$\frac{L}{r_y} < \sqrt{\frac{8850}{\frac{P}{P_y}}}, \sqrt{\frac{8850}{0.092}} = 311 > \frac{(6)(12)}{(1.63)} = 43.2$ ok

(12) Connection detail

(Table 7.7, Guide 2)

$t_s = \frac{\sqrt{2}}{b}\left(\frac{S}{d} - \frac{wd}{\sqrt{3}}\right) = \frac{\sqrt{2}}{7.56}\left(\frac{107.8}{18.25} - \frac{(0.416)(18.25)}{\sqrt{3}}\right) = 0.286 \text{ in.}$

See detail h, Plate V

Use $3\frac{1}{2}'' \times \frac{7}{16}''$ Ⓟs

(13) Splices

Provide as part of corner connection detail.
(Beam is continuous across column top)

(14) Frame layout

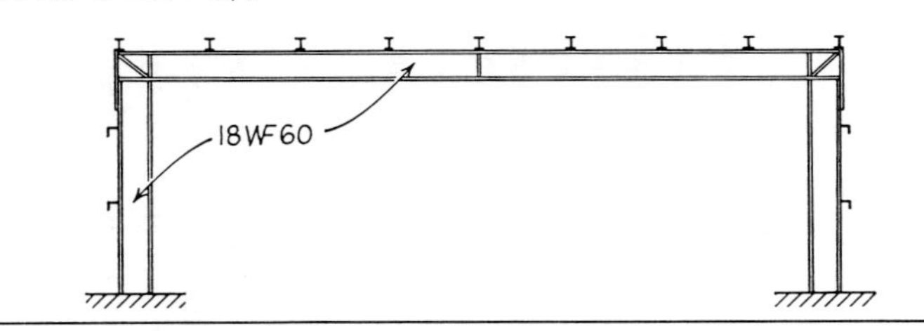

Step 5: Analyzing case II it is found that mechanism 1 still controls with a required M_p of 242 kip-ft. In this part of the problem two approaches were possible: (1) try the two most likely mechanisms (namely, mechanisms 1 and 3) or (2) try mechanism 1 and make a moment check. The former was done in this case, the moment check following for mechanism 1 when it was discovered that mechanism 1 controlled.

Actually it was not necessary in this problem to work out the beam mechanism (No. 1) a second time, because the required plastic moment will always be less for such a mechanism than for the larger case I loading. Nonetheless, for completeness, the calculation is shown.

The question might be asked: how was it known that the hinge would form under the third purlin from the left corner? Actually this was simply a guess. When the resulting moment was found to be substantially less than the required moment for mechanism 1, then the next logical step was to go directly to the moment check. If the hinge had been assumed at the center instead of at the third purlin, the corresponding plastic moment would have been 177 kip-ft.

To complete the moment check, the trial-and-error method is used. It is assumed that the moment at section 5 is the maximum possible value (namely, $-M_p$ with the minus sign denoting a counterclockwise moment). From sway equilibrium using an equation similar to Eq. 7.12 of Procedure 8, Table 7.3, it is found that the moment at section 1 is $+152$ kip-ft, which is less than M_p. Therefore, mechanism 1 is the correct mechanism for loading condition II.

Step 6: Case I is found to be critical with a required M_p of 320 kip-ft. The required plastic modulus is 116.5 in.^{3}; consequently an 18WF60 shape is specified (Z = 122.6 in.^{3}).

Step 10: The check according to Eq. 4.28 to determine the adequacy of the selected purlin spacing shows that the rafter is satisfactory in the vicinity of section 3 (segment A–B) since the last hinge forms there (see Fig. 4.19). However, the check of the selected girt spacing indicates that a more refined calculation is needed. In Appendix 1 the refinement was made by use of the restraint correction process described there and this same structure was used for illustration. In this plate, Eq. 4.29 is used to illustrate the alternate procedure; by it the selected spacing is adequate. Actually in using the alternate method, no trial by Eq. 4.28 would be needed, the check being made directly according to Eq. 4.29.

A deflection analysis was not made in this example. If one were desired, the procedure would be the same as is outlined in Art. 6.3. Since case I is the controlling condition, the deflection would be calculated on the basis that $\theta_{32} = 0$ (last hinge forms at the centerline).

PLATE VII Gabled Frame

As background for the design (to follow) of a gabled frame with haunched connections, a similar frame will be designed with straight corner connections.

The solution is shown in Plate VII. All of the steps are not shown, the problem only being carried to the point that the member size is selected.

Step 1: The loading upon the frame of 100-ft span is shown in sketch (*a*). Column bases are assumed as pinned.

Step 2: Only the case I loading condition is examined. It will be shown later Plate X, p. 311, that case I is, in fact, the controlling condition for this problem.

Step 4: The statical method of analysis is used. The determinate moment diagram, 2–*e*–6, is drawn first and represents the parabolic moment diagram due to uniform load. In plotting the redundant moment diagram the technique described in Procedure 1 of Table 7.3 is used to locate the "pole point." The redundant diagram will be correctly drawn when the moment at section 2 is equal to the moment in the rafter. Therefore the pole point, O, is located a distance x from the column (section 2), which distance is given by

$$x = \frac{a}{b} L_c \tag{7.8}$$

$x = 66\frac{2}{3}$ ft. A straight edge is rotated about that point until $M_2 = M_3 = M_p$. It is also observed that the maximum moment in the rafter will be close to the first purlin left of the crown. Thus the equilibrium equation at section 3 is formulated on this basis. The magnitude of H_1 is determined from Eqs. (2) and (3), and the required plastic moment M_p is solved from Eqs. 1 and 2.

PLATE VIII Gabled Frame with Haunched Connections

A single-span portal frame with gabled roof will be designed to resist vertical and side load. The example is the same as design problem No. 3 in Ref. 5.12. The frame has a span of 100 ft, the column height is 20 ft, and the rafter rise is 15 ft. Greatest economy of steel will be realized if haunches are used at the corners.* The problem is presented and solved in Plate VIII.

Step 2: The vertical distributed load is replaced by concentrated loads applied at the purlins (5-ft spacing). The horizontal distributed load is replaced by a single concentrated load T acting at the eaves, which produces the same moment about point 1. In other words, it is a concentrated load which produces an overturning moment equal to that of the uniformly distributed horizontal load.

Step 4: Since the frame is only redundant to the first degree, the statical method of analysis is used. The composite moment diagram is drawn roughly to scale in sketch (*b*) for case I, the parabola for uniform load being used for the determinate case. In drawing the moment diagram due to the redundant (line 1–*a*–*b*–*ε*–9) it is found that the maximum moment in the rafter is closest to the second purlin to the left of the crown. Therefore the equilibrium equations are formulated on that basis. Indeed, from this point on, the method is no longer

* Of course it does not necessarily follow that the total cost of the structure is also at a minimum.

PLATE VII. GABLED FRAME

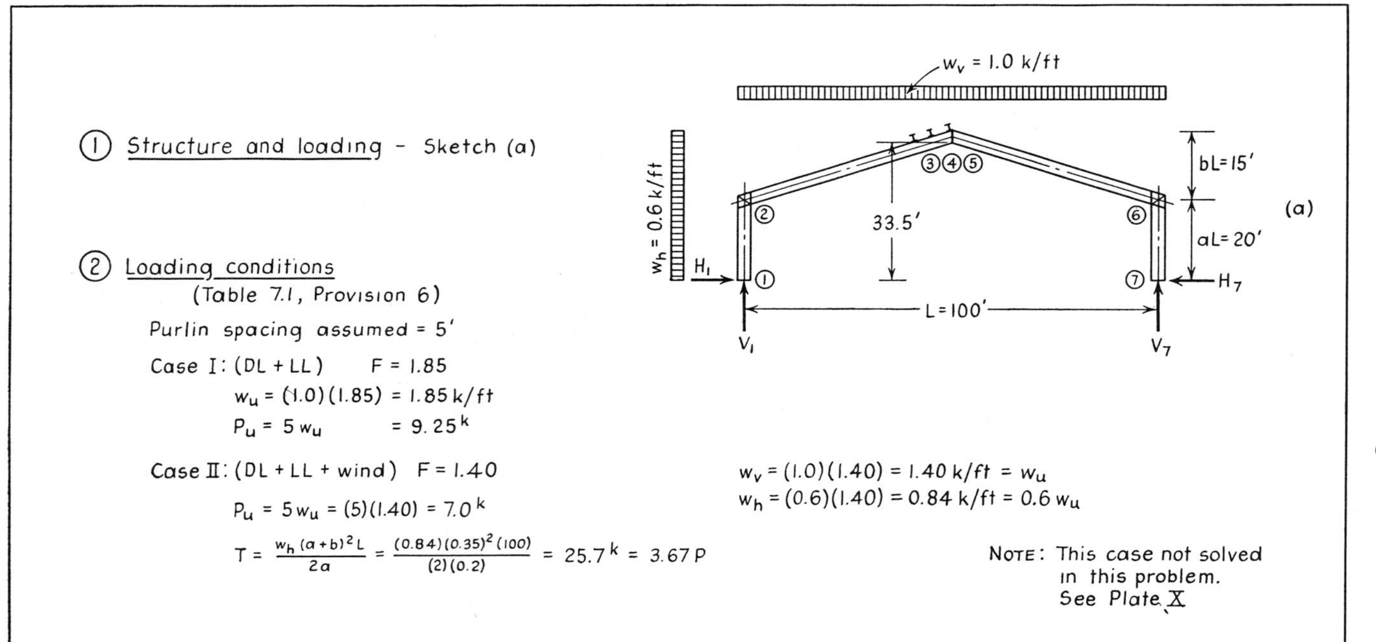

① Structure and loading - Sketch (a)

② Loading conditions
(Table 7.1, Provision 6)

Purlin spacing assumed = 5′

Case I: (DL + LL) $F = 1.85$

$w_u = (1.0)(1.85) = 1.85$ k/ft

$P_u = 5\,w_u = 9.25^k$

Case II: (DL + LL + wind) $F = 1.40$

$w_v = (1.0)(1.40) = 1.40 \text{ k/ft} = w_u$

$w_h = (0.6)(1.40) = 0.84 \text{ k/ft} = 0.6\,w_u$

$P_u = 5\,w_u = (5)(1.40) = 7.0^k$

$$T = \frac{w_h (a+b)^2 L}{2a} = \frac{(0.84)(0.35)^2(100)}{(2)(0.2)} = 25.7^k = 3.67\,P$$

NOTE: This case not solved in this problem. See Plate X

③ Plastic moment ratios Use constant section throughout

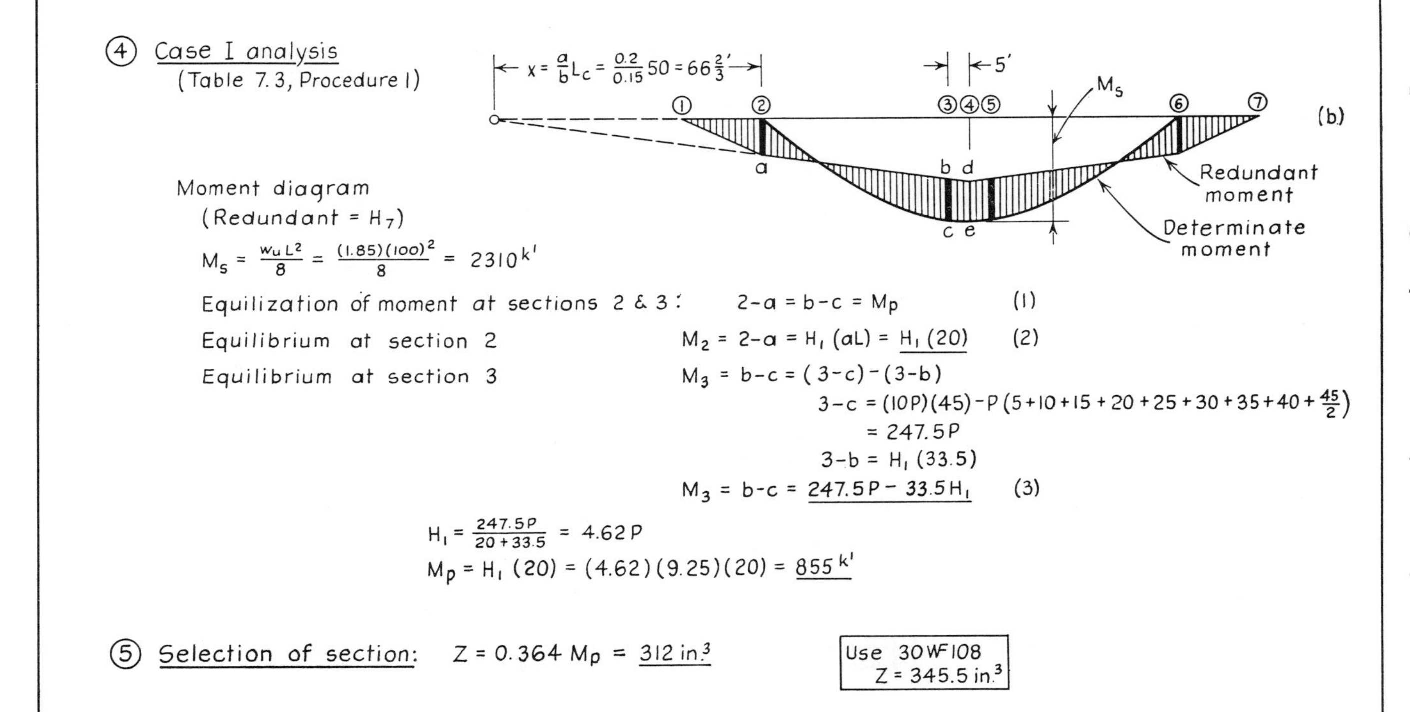

④ Case I analysis
(Table 7.3, Procedure I)

Moment diagram
(Redundant = H_7)

$M_s = \frac{w_u L^2}{8} = \frac{(1.85)(100)^2}{8} = 2310^{k'}$

Equilization of moment at sections 2 & 3: $2\text{-}a = b\text{-}c = M_p$ (1)

Equilibrium at section 2: $M_2 = 2\text{-}a = H_1 (aL) = \underline{H_1 (20)}$ (2)

Equilibrium at section 3: $M_3 = b\text{-}c = (3\text{-}c) - (3\text{-}b)$

$3\text{-}c = (10P)(45) - P\left(5 + 10 + 15 + 20 + 25 + 30 + 35 + 40 + \frac{45}{2}\right)$

$= 247.5P$

$3\text{-}b = H_1 (33.5)$

$M_3 = b\text{-}c = \underline{247.5P - 33.5H_1}$ (3)

$H_1 = \frac{247.5P}{20 + 33.5} = 4.62P$

$M_p = H_1 (20) = (4.62)(9.25)(20) = \underline{855^{k'}}$

⑤ Selection of section: $Z = 0.364\, M_p = \underline{312 \text{ in.}^3}$

Use 30WF108
$Z = 345.5 \text{ in.}^3$

PLATE VIII. GABLED FRAME WITH HAUNCHED CONNECTIONS

① Structure and loading Sketch (a)

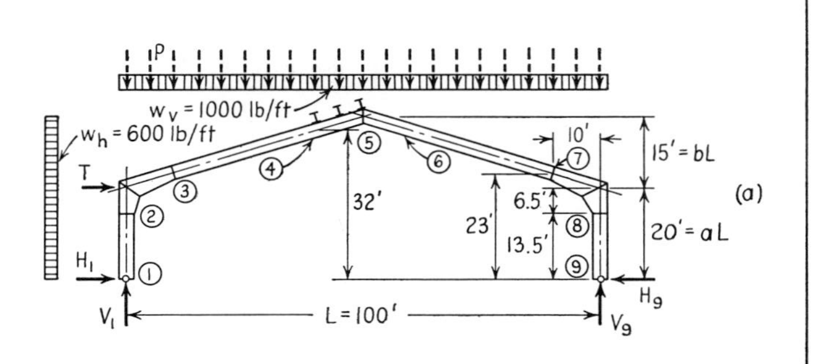

② Loading conditions (Table 7.1, Provision 6)

Purlin spacing assumed = 5′

Case I: (DL + LL) F = 1.85

$w_u = (1.0)(1.85) = 1.85\ k/ft$

$P_u = 5w_u = 9.25^k$

Case II (DL + LL + wind) F = 1.40

$w_v = (1.0)(1.40) = 1.40\ k/ft = w_u$

$w_h = (0.6)(1.40) = 0.84\ k/ft = 0.6\,w_u$

$P_u = 5w_u = (5.0)(1.40) = 7.0^k$

$$T = \frac{w_h(a+b)^2L}{2a} = \frac{(0.84)(35)^2}{(2)(20)} = 25.7^k = 3.67P$$

③ Plastic moment ratios:

Adjust M_p to suit the moment diagram – see below

(4) <u>Case I analysis</u> (Statical method)
(Table 7.3, Procedure I)

<u>Moment diagram</u>
(Redundant = H_9)

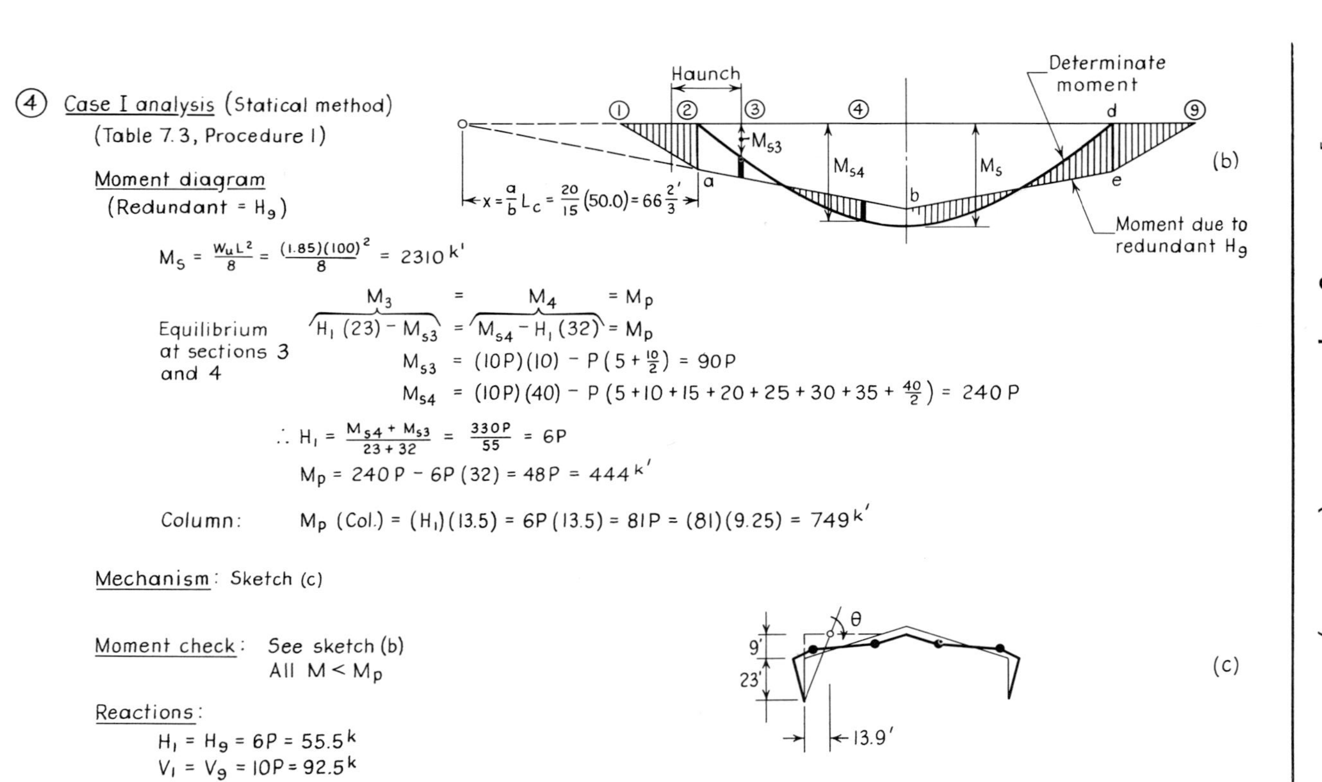

$$M_s = \frac{w_u L^2}{8} = \frac{(1.85)(100)^2}{8} = 2310^{k'}$$

Equilibrium at sections 3 and 4

$$\overbrace{H_1(23) - M_{s3}}^{M_3} = \overbrace{M_{s4} - H_1(32)}^{M_4} = M_p$$

$$M_{s3} = (10P)(10) - P\left(5 + \frac{10}{2}\right) = 90P$$

$$M_{s4} = (10P)(40) - P\left(5 + 10 + 15 + 20 + 25 + 30 + 35 + \frac{40}{2}\right) = 240P$$

$$\therefore H_1 = \frac{M_{s4} + M_{s3}}{23 + 32} = \frac{330P}{55} = 6P$$

$$M_p = 240P - 6P(32) = 48P = 444^{k'}$$

Column: $M_p\,(\text{Col.}) = (H_1)(13.5) = 6P(13.5) = 81P = (81)(9.25) = 749^{k'}$

<u>Mechanism</u>: Sketch (c)

<u>Moment check</u>: See sketch (b)
All $M < M_p$

<u>Reactions</u>:
$H_1 = H_9 = 6P = 55.5^k$
$V_1 = V_9 = 10P = 92.5^k$

PLATE VIII (Continued)

(5) <u>Case II analysis</u>: (Table 7.3, Procedure I)

<u>Moment diagram</u>:
(Redundant = H_9)

$M_s = 250\,P_u$

$M_{corner\,(det.)} = T(20) = 73.4P$

<u>Mechanism</u>: Sketch (f)

Equilibrium at Section 4:

$M_4 + H_9\,(30.5) = M_{s4}$

$M_{s4} = (0.65)(73.4P) + M_a$

$M_a = (10P)(35) - P\left(5+10+15+20+25+30+\frac{35}{2}\right)$

$M_a = 227.5P$

$M_4 = M_p = \frac{444}{7.0}\,P = 63.4P$

$H_9\,(30.5) = (0.65)(73.4P) + 227.5P - 63.4P$

$H_9 = \frac{211.8P}{30.5} = 6.95P$

Equilibrium at section 8:

$M_8 = H_9\,(13.5)$

$M_8 = (6.95)(13.5)(P) = 93.7P = (93.7)(7.0) = 656^{k'}$

$M_8 = 656 < 749$ (Case I)

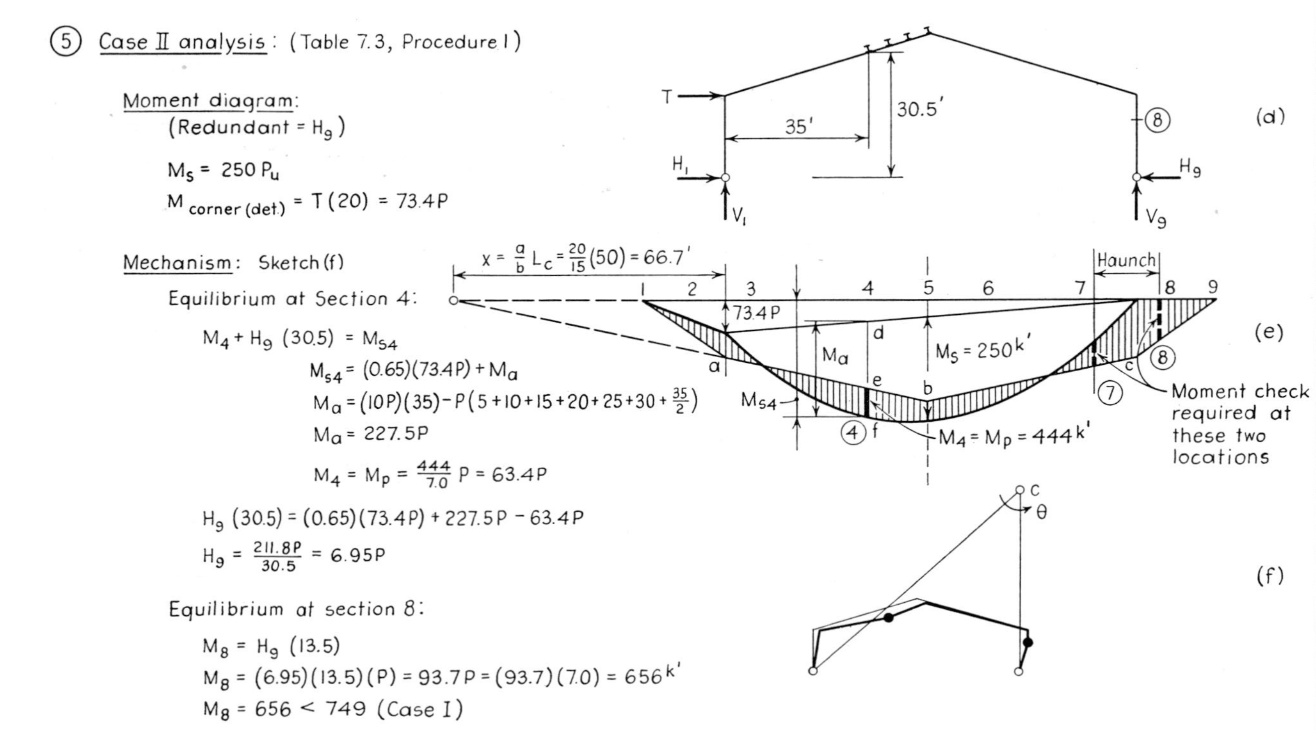

Moment check: Diagram plotted – sketch (e)

$$M_7 = H_9\,(23) - M_7\,(\text{Det.}) = (6.95P)(23) - \left[(10P)(10) - P\left(5 + \frac{10}{2}\right)\right]$$

$$M_7 = 62.7P = (62.7)(7.0) = 439^{k'} < 444^{k'} \quad \underline{ok}$$

(g)

Reactions

$$H_9 = 6.95P = 48.65^{k}$$

$$H_1 = H_9 - T = 48.65 - 3.67P = 48.65 - 25.7 = 23.7^{k}$$

$$V_1 = \frac{18P\,(50) + \frac{P}{2}\,(100) - T\,(20)}{100} \qquad V_1 = \frac{(900 + 50 - 73.4)P}{100} = 8.77P = 61.4^{k}$$

$$V_9 = 20P - V_1 = 140 - 61.4 = \underline{78.6}^{k}$$

(h)

Case I (without wind) is critical

NOTE: Case I reactions control the design.

⑥ Selection of section (Table 7.1, Provision 5)

Girder: $Z = \dfrac{M_p}{\sigma_y} = (444)(0.364) = 161.4$ in.^{3}

Use 21WF73 $Z = 172.1$

A = 21.46	r_y = 1.76
d = 21.24	z = 0.455
b = 8.295	S = 150.7
t = 0.740	I = 1600.3
w = 0.455	

Column: $Z = (749)(0.364) = 272.0$ in.^{3}

Use 27WF94 $Z = 277.7$

A = 27.65	w = 0.490
d = 26.91	z = 2.04
b = 9.99	S = 242.8
t = 0.747	I = 3266.7
r_y = 2.04	

PLATE VIII (Continued)

(7) Axial force (Table 7.4, Guide 1)

Column: $\frac{P}{P_y} = \frac{V_9}{\sigma_y A} = \frac{92.5}{(33)(27.65)} = 0.101 < 0.15$ ok

Girder: $\frac{P}{P_y} = \frac{H_9 \cos\theta + (V_9 - 2.5P)\sin\theta}{(33)(21.46)}$

$\theta = \tan^{-1}\frac{15}{50} = 16° \, 40'$

$\sin\theta = 0.288$

$\cos\theta = 0.960$

$\frac{P}{P_y} = \frac{(55.5)(0.96) + (75.0)(0.288)}{(33)(21.46)} = 0.106 < 0.15$ ok

(8) Shear force (Table 7.4, Guide 2)

V_{max} = Shear at end of girder

$V_{max} = V_9 - \frac{P}{2} = 92.5 - 9.25 = 83.2^k$

$V_{allow} = 18.0\ wd = (18.0)(0.455)(21.24) = 174.0^k > 83.2^k$ ok

(9) Cross-section proportions
(Table 7.8, Guide 1)

	Girder	Column		
b/t	= 11.21	13.37	< 17	ok
d/w	= 46.7	54.9	< 55	ok

⑩ <u>Lateral bracing</u>
(Table 7.8, Guide 3)

Spacing check

Girder: (Purlin spacing = 5′-0″)

$\frac{L_B}{r_y} = \frac{(5.0)(12.0)}{1.73} = \frac{60}{1.73} = 34.5$ "Center" position OK as hinges form there last. Check moment ratio correction on hinges at haunches.

Sketches (b) and (i):

$M_3 = M_p = 444^{k'}$

$M_A = -H_1(24.5) + V_1(15) - P\left(5 + 10 + \frac{15}{2}\right) = -180^{k'}$

$\therefore \frac{M_A}{M_p} = \frac{-180}{-444} = +0.406$

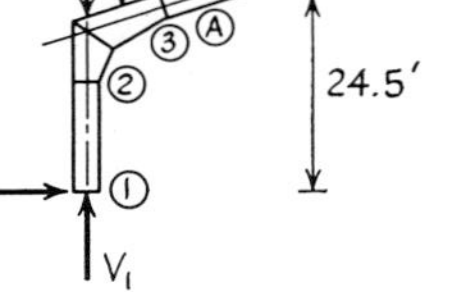

$\therefore$ Eq (4.28): $\left(\frac{L_B}{r_y}\right)_{cr} = 48 - 30\frac{M}{M_p} = 48 - 30(0.406) = 35.8 > 34.5$ ok

Column: Provide brace midway between end of haunch (section 8, sketch a) and column base.

$\frac{L_B}{r_y} = \frac{(6.75)(12)}{2.04} = 39.8$ OK because hinge rotation not required at Section 2

Bracing details:

Haunches: Provide bracing to inner (compression) flange at each end and at center.

Peak: Provide bracing to inner flange.

NOTE: Purlins must be adequately braced
In order to provide lateral support to rafters.

PLATE VIII (*Continued*)

⑪ Columns

(Table 7.6, Guides 2 & 4)

$$\frac{P}{P_y} = 0.101 < 0.15 \quad \underline{ok}$$

$$\text{weak axis}: \frac{L}{r_y} < \sqrt{\frac{8850}{\frac{P}{P_y}}} = \sqrt{\frac{8850}{0.101}} = 296 > 78.5 \quad \underline{ok}$$

⑫ Connection details

(Table 7.7, Guide 4)

Peak: Proportion stiffener to transmit flange thrust.

$$\sigma_y A_s = 2\sigma_y A_f \sin\theta$$

$$bt_s = 2bt \sin\theta$$

$$t_s = 2t \sin\theta$$

$$t_s = (2)(0.740)(0.288)$$

$$t_s = 0.426 \text{ in.}$$

θ

0.740″

Use $\frac{1}{2}$″ PL

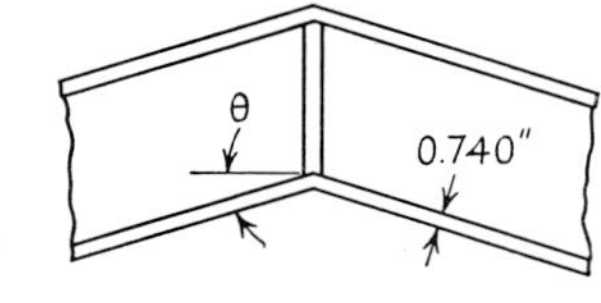

(j)

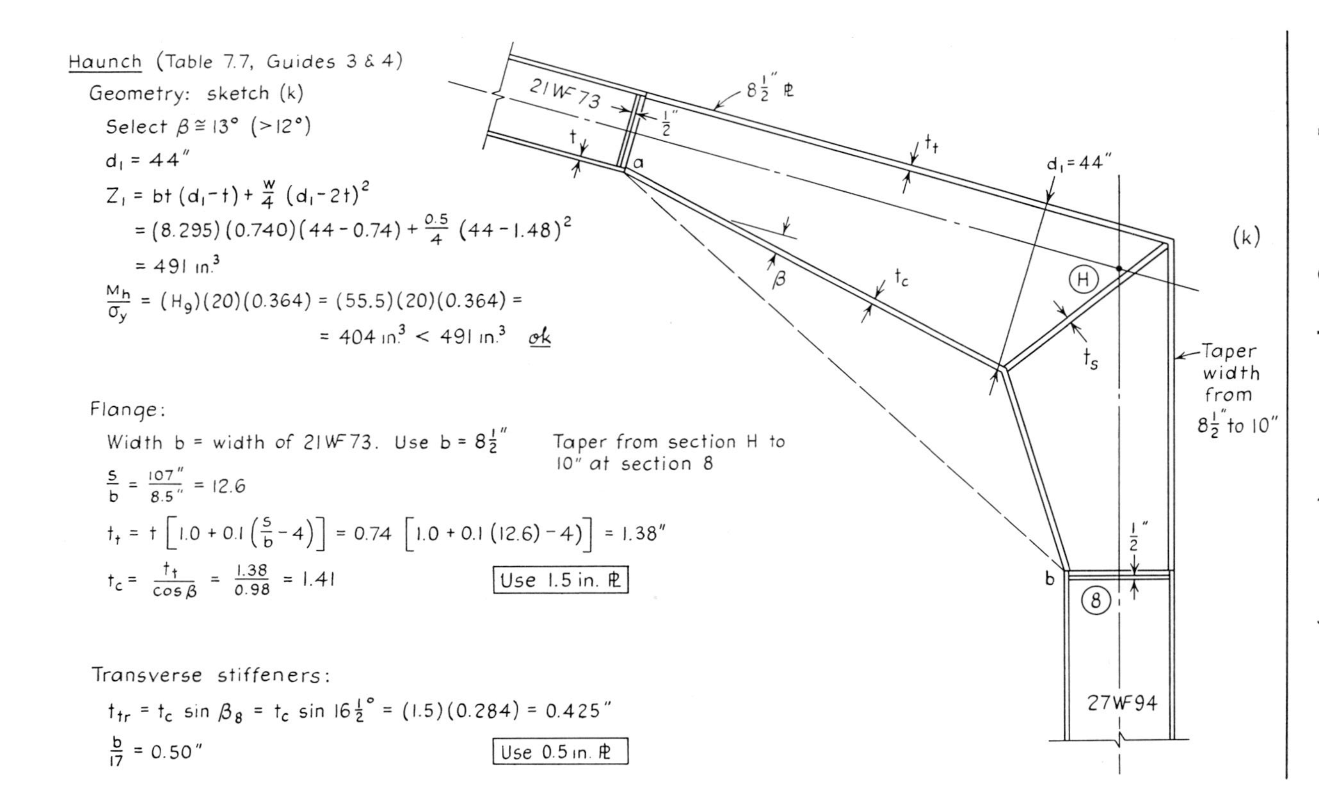

<u>Haunch</u> (Table 7.7, Guides 3 & 4)

Geometry: sketch (k)

Select $\beta \cong 13°$ $(>12°)$

$d_1 = 44''$

$$Z_1 = bt\,(d_1 - t) + \frac{w}{4}\,(d_1 - 2t)^2$$

$$= (8.295)(0.740)(44 - 0.74) + \frac{0.5}{4}\,(44 - 1.48)^2$$

$$= 491 \text{ in.}^3$$

$$\frac{M_h}{\sigma_y} = (H_g)(20)(0.364) = (55.5)(20)(0.364) =$$

$$= 404 \text{ in.}^3 < 491 \text{ in.}^3 \quad \underline{ok}$$

Flange:

Width b = width of 21WF73. Use $b = 8\frac{1}{2}''$

Taper from section H to 10″ at section 8

$$\frac{s}{b} = \frac{107''}{8.5''} = 12.6$$

$$t_t = t\left[1.0 + 0.1\left(\frac{s}{b} - 4\right)\right] = 0.74\left[1.0 + 0.1\,(12.6) - 4)\right] = 1.38''$$

$$t_c = \frac{t_t}{\cos\beta} = \frac{1.38}{0.98} = 1.41$$

Use 1.5 in. ℞

Transverse stiffeners:

$$t_{tr} = t_c \sin\beta_8 = t_c \sin 16\tfrac{1}{2}° = (1.5)(0.284) = 0.425''$$

$$\frac{b}{17} = 0.50''$$

Use 0.5 in. ℞

PLATE VIII (*Continued*)

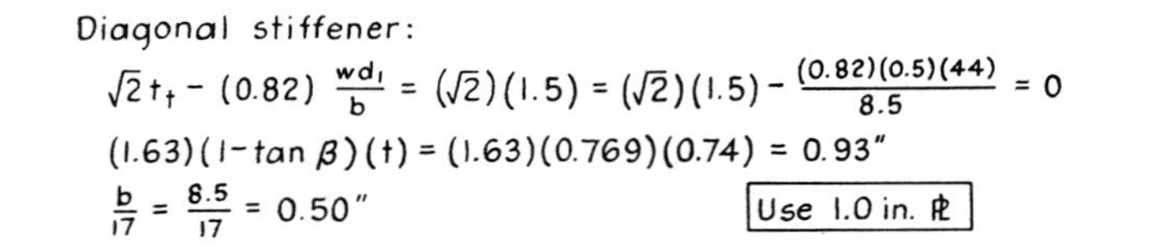

Diagonal stiffener:

$$\sqrt{2}\,t_t - (0.82)\,\frac{wd_1}{b} = (\sqrt{2})(1.5) = (\sqrt{2})(1.5) - \frac{(0.82)(0.5)(44)}{8.5} = 0$$

$$(1.63)(1 - \tan\beta)(t) = (1.63)(0.769)(0.74) = 0.93''$$

$$\frac{b}{17} = \frac{8.5}{17} = 0.50''$$

Use 1.0 in. ℙ

Web: Use $w = 0.50''$

(13) Splices

Provide as part of haunch and peak detail.

(14) Frame layout

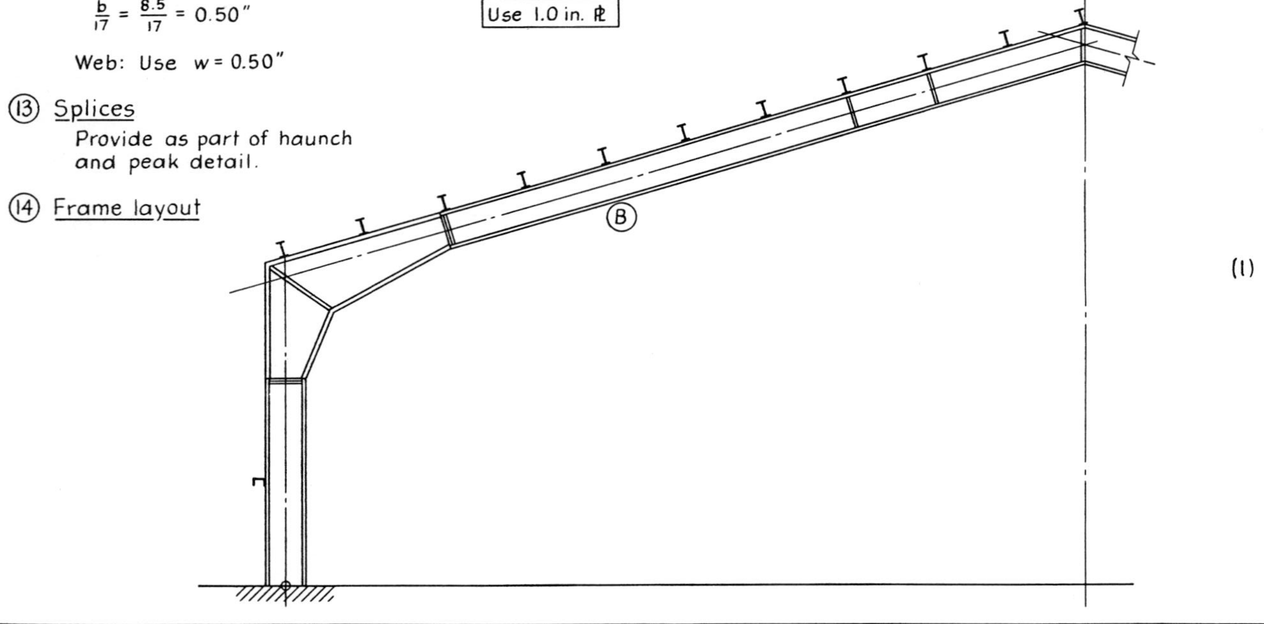

a graphical one because the equations of statics are used to compute M_p. (Of course, the magnitude of M_p could be picked off a graph drawn to sufficiently large scale.)

As in Plate VII (step 4), the correct position of line a–b on the redundant diagram is determined by rotating a straight-edge about a center at O (located a distance $x = a/bL_c = 66\frac{2}{3}$ ft from the column line) until the moment at section 3 is equal to the moment at section 4.

Solving the equilibrium equations for the condition that $M_3 = M_4$, gives a required plastic moment of 444 kip-ft.

The column is selected to have adequate strength to support the required moment at section 2. This value is computed by statics and the required plastic moment value is 749 kip-ft.

The mechanism of sketch (c) could have been used as the basis for satisfying the equilibrium condition. The position of the instantaneous center is first located, the coordinate being 32 ft vertically and (10)(32)/23 = 13.9 ft horizontally from the column base. The mechanism angle at hinge 3 is therefore $(\frac{9}{23} + 1)\theta = \frac{32}{23}\theta$. The angle at section 4 is equal to θ and the virtual work equation may next be written. The external work for one half of the frame is

$$W_E = P\theta\left(1.1 + 6.1 + 11.1 + 16.1 + 21.1 + 26.1 + 26.1 + \frac{26.1}{2}\right)$$

$$- P(\tfrac{9}{23}\theta)(5.0 + 10.0)$$

$$W_E = 114.9P\theta$$

The internal work is

$$W_I = M_p\theta(\tfrac{32}{23} + 1) = 2.39M_p\theta$$

Thus

$$M_p = \frac{114.9P}{2.39} = \frac{(114.9)(9.25)}{2.39} = 444 \text{ kip-ft}$$

which is identical with the answer obtained before.

Step 5: Analyzing for the case II loading, the redundant is selected as H_9 (sketch d). The moment diagram for the determinate structure is shown by the solid line in sketch (e). Rather than solve case II as a new problem, it will only be determined whether or not the member selected for case I is adequate for the case II loading. Therefore, the redundant moment diagram in sketch (e) (line 1–a–b–c–9) is drawn in such a way that the rafter moment is equal to 444 kip-ft, which is the value obtained in case I. The construction is made by locating point O and rotating line O–b until this desired condition is reached. The moment is maximum closest to the third rafter to the left of the crown and therefore section 4 is located at that point.

At this stage it is not yet known whether the plastic moment condition has been violated in the frame. If a hinge formed at section 8 (M = 749 kip-ft), the mechanism would be as shown in sketch (f). Another possibility is that the moment at section 7 is greater than 444 kip-ft. Sketch (g) would be the resulting mechanism. From equilibrium at section 4, H_9 is determined ($6.95P$); hence by equilibrium at section 8 it is found that the required value of M_p equals 656 kip-ft. Since this is less than the value of 749 kip-ft, the case I loading appears to control.

The moment check must be continued, and, while satisfactory, the moment at section 7 is very close to the maximum available M_p (439 kip-ft as compared with 444 kip-ft). The design of details should therefore be done on the basis that a plastic hinge could form at 4, 7, and 8 (and by symmetry, at 2, 3, and 6).

Case I is thus found to be critical; the reactions for this case are also the greatest.

Step 7: In checking for axial force, it is found that the P/P_y ratios are greater than in the previous problems but are still less than 0.15. As a matter of fact, the axial force ratio is higher in the girder than in the column because the member is lighter and due to the sloping roof, both the horizontal and vertical reactions at the column base produce a thrust component in the rafter.

Step 10: With a purlin spacing of 5 ft, the L_B/r_y ratio is 34.5. In the center hinge positions (sections 4 and 6) this slenderness ratio is satisfactory even though the moment diagram is "flat," because the corresponding plastic hinges will be the last to form. With regard to the hinges that form in the rafter adjacent to the haunch, a consideration of the moment ratio ($M/M_p = 0.406$) shows that the resulting allowable slenderness ratio is 35.8, which is greater than the value of 34.5 supplied. Therefore, no further check is necessary.

Concerning the column, the member was proportioned simply to provide strength and not to participate in mechanism action. Therefore a single brace between the end of the haunch and the column base would be adequate.

With regard to the bracing details, support is required on the inner (compression) side at all points on the haunch where the flange force changes in direction. Similar bracing should be provided at the peak.

Step 12: In calling for a $\frac{1}{2}$-in. plate at the peak (sketch j), it is assumed that the web will carry no thrust, and a plastic analysis is carried out to proportion the vertical plate stiffener. The analysis is the same as that for the end stiffener of a haunched connection (Table 7.7, Guide 4, Eq. 5.29) except that the thrust is doubled.

Step 12 (Haunch): The design of the haunch details is controlled in part by the initial choice of dimensions. At the outset it was decided that the haunch would extend 10 ft into the girder span and $6\frac{1}{2}$ ft down the column as shown in sketch (a). To a certain extent the remaining geometry is open to choice. For a geometry similar to that selected in this example, it has been shown that the angle β shown in sketch (k) should be greater than 12° [5.13] in order to assure that the plastic moment strength at the common intersection point will not be less than required from the applied moment diagram. An angle of $\beta = 13°$ was therefore selected. This gives a depth d_1 of 44 in. and provides a reasonable appearance to the haunch.

The width of this flange is made uniform at $8\frac{1}{2}$ in. along the girder portion of the haunch and then is gradually tapered to meet the 10-in. width of the 27WF94 flange. A web thickness of $\frac{1}{2}$ in. is selected, which is about the same as that of the rolled members joined.

The next step is to check the plastic modulus supplied at the transverse section of the common intersection point of the two tapered segments. Z_1 is found to be 491 in.^{3}, which is more than the value required by the moment at the haunch point H.

With a flange width of $8\frac{1}{2}$ in., the s/b ratio is 12.6, which is greater than the critical value of 4.0. According to Design Guide 3 (Table 7.7) three alternatives are open. Two intermediate braces could be supplied, the depth d_1 could be increased substantially, or the flange thickness can be increased. The first one complicates fabrication. The second (although the least expensive) tends to destroy the esthetic value, since it would be necessary to use an outline shown by the dotted line a–b in sketch (k). The third procedure is used in this example and a flange thickness of $1\frac{1}{2}$ in. is specified according to Eq. 5.30.

Step 13: This completes the design of the frame which is shown in sketch (l). With regard to splices, the columns and haunches could be shop-assembled with a field splice at section 7. Alternatively, a splice for full moment using high-strength bolts could be made at section 7 or bolted splices for less than full moment strength could be supplied at a section near the point of inflection (section B of sketch l).

It is of interest to compare the results of this design with the elastic solution and with the plastic solution for the case where no haunch has been used. The following table provides such a comparison, the elastic solutions being those of design problem No. 3 of Ref. 5.12. (The only

	Conditions			Elastic Design	Plastic Design
1	No haunch	Uniform section throughout	Girder Column	30WF124	30WF108
2	Haunch	Uniform section (in plastic design)	Girder Column	24WF94 30WF108	24WF94
3	Haunch	Different sections	Girder Column	24WF94 30WF108	21WF73 27WF94

difference is that the haunch extends $6\frac{1}{2}$ ft down the column instead of 6 ft.) Not only is there a considerable saving in each plastic design, but the example shows that possible weight savings may be achieved in plastic design as well as elastic design when a haunch is specified. Of course, the haunch-fabrication expense must be borne in mind when making comparison of over-all costs.

PLATE IX Fixed-Base Gabled Frame with Crane Loading

In Plate VI a fixed-base frame was designed in which the roof was flat; the mechanism method of analysis was used. In the present problem of a fixed-base gabled frame (Plate IX) the statical method will be used. As was pointed out earlier in this book, it is somewhat a matter of choice as to which method will be most suitable.

Step 1: This problem is similar to that included in p. 904 of Ref. 9.1, except that the column bases are fixed. (In Ref. 9.1, a shape weighing 40 lb/ft was required.) The side load due to wind is concentrated at the eaves, in a force T_1 that produces equivalent overturning moment. Horizontal crane thrusts T_2 are applied to each crane rail.

Step 2: The worst loading condition, insofar as the crane is concerned, is with the maximum load located on the windward side.

Step 3: The design is commenced assuming constant section throughout. Later in a routine design, it might be necessary to adjust column sizes, etc.

Step 4: For ease of construction of the determinate moment diagram, a table of moments is prepared (sketch *c*). In row 1 is shown the moment at each section due to the vertical roof loads. For example, at section 6 the roof load moment is given by

$$M_{s(6)} = V_1(9.54) - P_2(5.12) - P_1(9.54)$$

$$= (27.33)(9.54) - (6.85)(5.12) - (3.33)(9.54)$$

$$= 192 \text{ kip-ft}$$

Row (2) contains the moments introduced by the vertical crane loads. Row 3 gives the moments due to forces T_2. Finally, a sum is taken (row 4), giving the total statical (determinate) moment at each section. The sign convention is that positive moment produces tension on the lower fiber (inside) of the frame. (Sign conventions are discussed in Arts. 3.4 and 7.3.)

The process of drawing the determinate moment diagram is straight-forward and it is shown as the heavy solid line in sketch (*e*). Since there are three redundants in this problem, it is by no means clear how the redundant diagram is to be combined with the determinate diagram to complete the composite construction. So, what is done is to assume a mechanism and see if the resulting moment diagram satisfies the plastic moment condition. In other words, the process is a combination of the *statical* and the *mechanism* methods, carried out on a trial and error basis.

The method is simplified, somewhat, in that the redundant moment diagrams may be specified not only in terms of the unknowns but also in terms of M_p. Four possible mechanisms are shown in sketch (*d*). Mechanism 1 would be encountered only for high side loads. In fact, the solution for this mechanism can be determined directly as

$$T_1 aL + 2T_2(13.6) = 4M_p$$

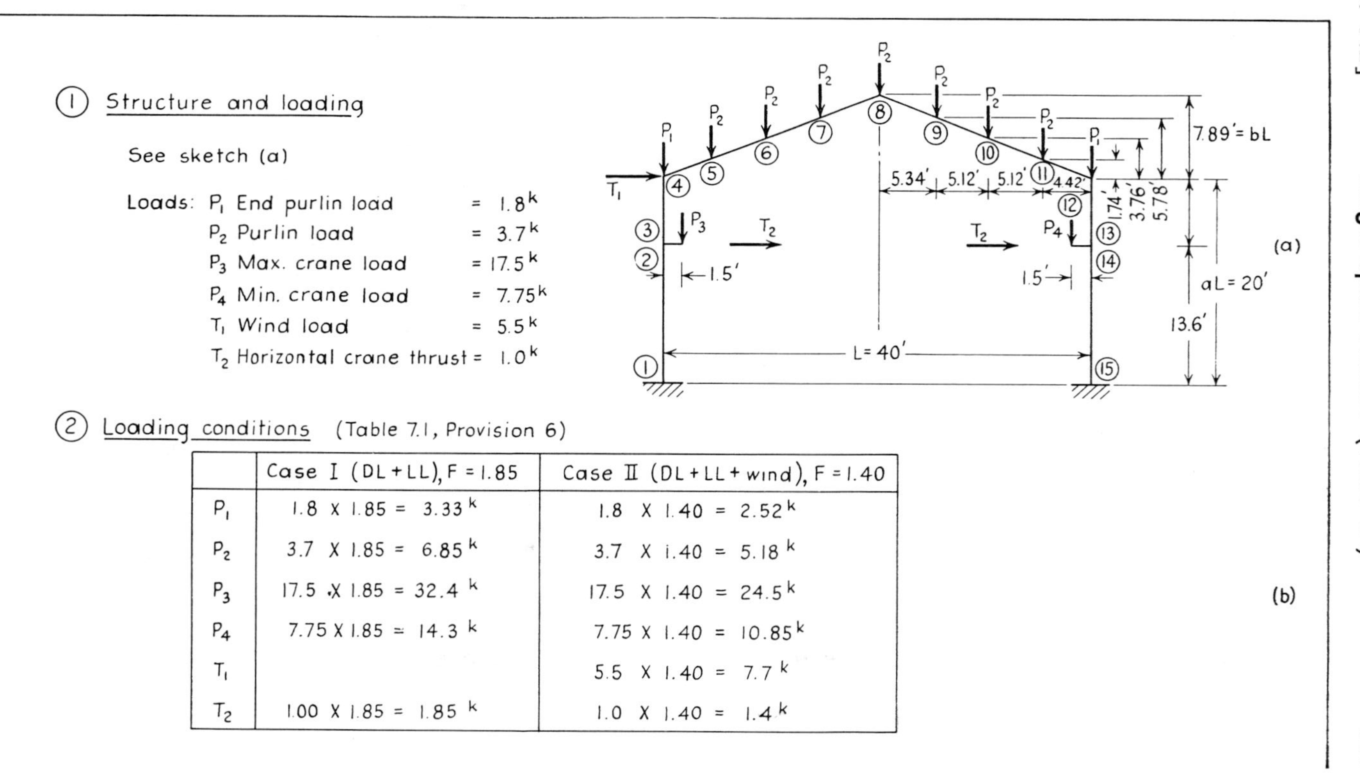

PLATE IX. FIXED-BASE GABLED FRAME WITH CRANE LOADING

① Structure and loading

See sketch (a)

Loads: P_1 End purlin load = 1.8^k
P_2 Purlin load = 3.7^k
P_3 Max. crane load = 17.5^k
P_4 Min. crane load = 7.75^k
T_1 Wind load = 5.5^k
T_2 Horizontal crane thrust = 1.0^k

② Loading conditions (Table 7.1, Provision 6)

	Case I (DL + LL), F = 1.85	Case II (DL + LL + wind), F = 1.40
P_1	1.8 X 1.85 = 3.33^k	1.8 X 1.40 = 2.52^k
P_2	3.7 X 1.85 = 6.85^k	3.7 X 1.40 = 5.18^k
P_3	17.5 X 1.85 = 32.4^k	17.5 X 1.40 = 24.5^k
P_4	7.75 X 1.85 = 14.3^k	7.75 X 1.40 = 10.85^k
T_1		5.5 X 1.40 = 7.7^k
T_2	1.00 X 1.85 = 1.85^k	1.0 X 1.40 = 1.4^k

(b)

PLATE IX (*Continued*)

(3) Plastic moment ratio   Try constant section throughout

(4) Case I analysis (Table 7.3, Procedure I)

Moment diagram: For construction of determinate moment diagram, use following data. (Redundants = M_1, M_{15}, H_{15})

Loading	1	2	3	4	5	6	7	8	9	10	11	12	13	14	15
(1) Roof load	0	0	0	0	106	192	247	265	247	192	106	0	0	0	0
(2) Vert. crane	0	0	48	48	49	45	42	38	34	30	27	24	24	0	0
(3) Horiz. crane	0	50	50	62	60	57	55	52	41	31	21	12	0	0	0
(4) Total	0	50	98	110	215	294	344	355	322	253	154	36	24	0	0

(c)

$$V_{1(1)} = P_1 + 3.5P_2 = 3.33 + (6.85)(3.5) = 27.33^k$$

$$V_{1(2)} = \frac{P_3(38.5) + P_4(1.5)}{40} = \frac{(32.4)(38.5) + (14.3)(1.5)}{40} = 31.7^k$$

$$V_{1(3)} = 2T_2 \frac{(13.6)}{40} = 1.26^k$$

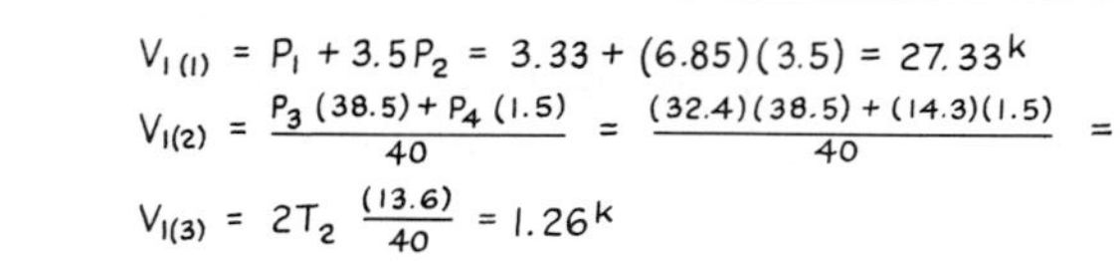

Possible mechanisms:

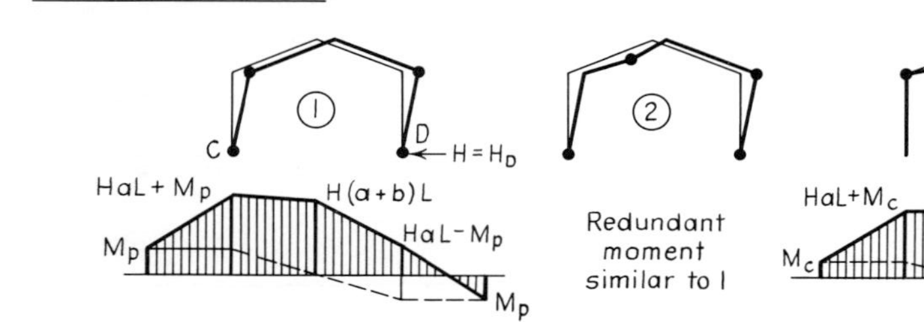

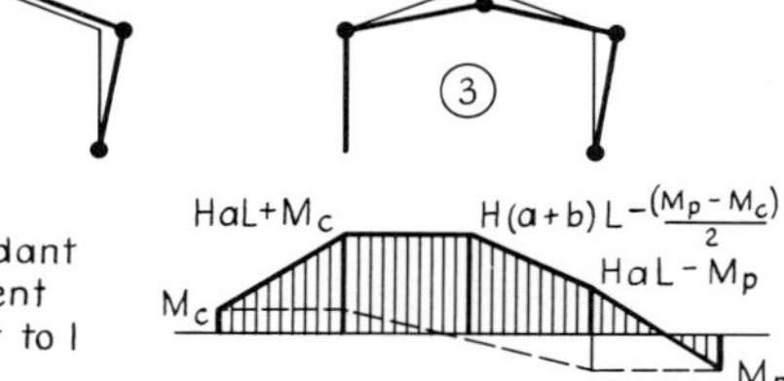

(d)

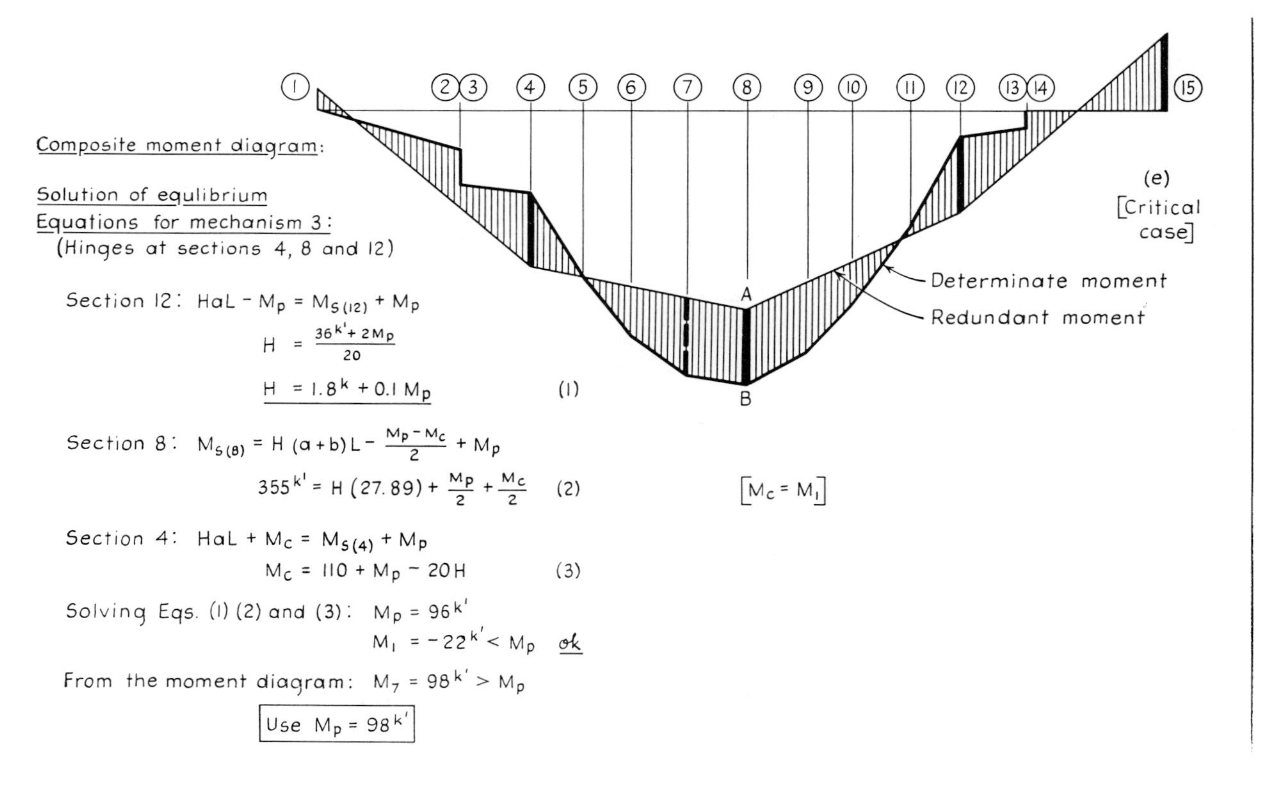

Composite moment diagram:

Solution of equlibrium
Equations for mechanism 3:
(Hinges at sections 4, 8 and 12)

Section 12: $HaL - M_p = M_{s(12)} + M_p$

$$H = \frac{36^{k'} + 2M_p}{20}$$

$$H = 1.8^{k} + 0.1\,M_p \qquad (1)$$

Section 8: $M_{s(8)} = H\,(a+b)L - \frac{M_p - M_c}{2} + M_p$

$$355^{k'} = H\,(27.89) + \frac{M_p}{2} + \frac{M_c}{2} \qquad (2)$$

Section 4: $HaL + M_c = M_{s(4)} + M_p$

$$M_c = 110 + M_p - 20H \qquad (3)$$

Solving Eqs. (1) (2) and (3): $M_p = 96^{k'}$

$M_1 = -22^{k'} < M_p$ ok

From the moment diagram: $M_7 = 98^{k'} > M_p$

Use $M_p = 98^{k'}$

PLATE IX (Continued)

(5) Case II analysis

Moment diagram (Determinate): (Redundants = M_1, M_{15}, H_{15})

Loading	1	2	3	4	5	6	7	8	9	10	11	12	13	14	15
(1) Roof load	0	0	0	0	80	145	187	201	187	145	80	0	0	0	0
(2) Vert. crane	0	0	36	36	37	34	32	29	26	23	20	18	18	0	0
(3) Horiz. crane	0	38	38	47	45	43	42	39	31	23	16	9	0	0	0
(4) Wind	0	105	105	154	137	117	97	77	56	37	17	0	0	0	0
(5) Total	0	143	179	237	299	339	358	346	300	228	133	27	18	0	0

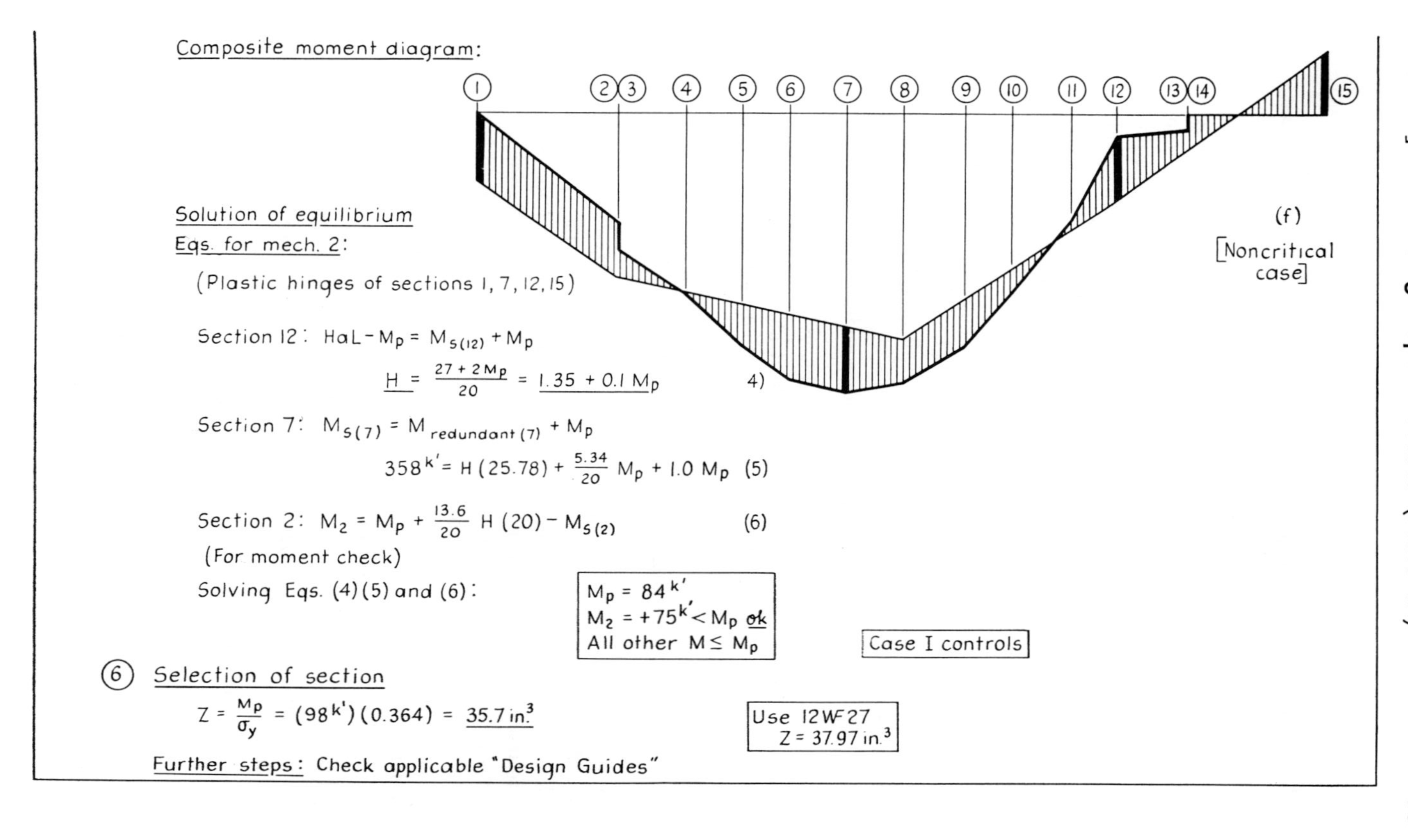

Composite moment diagram:

Solution of equilibrium

Eqs. for mech. 2:

(Plastic hinges of sections 1, 7, 12, 15)

Section 12: $HaL - M_p = M_{s(12)} + M_p$

$$\underline{H} = \frac{27 + 2M_p}{20} = \underline{1.35 + 0.1\,M_p} \qquad 4)$$

Section 7: $M_{s(7)} = M_{redundant\,(7)} + M_p$

$$358^{k'} = H\,(25.78) + \frac{5.34}{20} M_p + 1.0\,M_p \qquad (5)$$

Section 2: $M_2 = M_p + \frac{13.6}{20} H\,(20) - M_{s(2)} \qquad (6)$

(For moment check)

Solving Eqs. (4) (5) and (6):

$M_p = 84^{k'}$
$M_2 = +75^{k'} < M_p$ ok
All other $M \leq M_p$

Case I controls

(6) Selection of section

$$Z = \frac{M_p}{\sigma_y} = (98^{k'})(0.364) = \underline{35.7\text{ in.}^3}$$

Use 12WF27
$Z = 37.97\text{ in.}^3$

Further steps: Check applicable "Design Guides"

or

$$M_p = \frac{20T_1 + 27.2T_2}{4}$$

A solution to this equation gives the minimum required plastic moment. (For this problem, then, $M_p \geqq (20T_1 + 27.2T_2)/4 = 181/4 = 45.3$ kip-ft.)

Mechanism 2 would be a frequent case for relatively high side load. Mechanism 3 will be the most likely case for vertical load alone. Mechanism 4 represents another and somewhat unusual form that is possible when intermediate load is applied to the columns. The redundant moment diagrams for mechanisms 1 and 2 are identical in form, and one is shown beneath mechanism 1. It represents the summation of moments due to redundants H_D (called H hereafter), $M_D = M_p$, and $M_C = M_p$. Similarly, the redundant moment diagram is identical for mechanisms 3 and 4 and the diagram is shown beneath the former. (It is left as an exercise to obtain these diagrams.)

If the side loads are small, then, referring to sketch (*d*), the moment M_C in mechanism 3 might lie below the line instead of above it. If the sign convention previously described is used, the sign of M_C will be correct when the solution is completed.

Using the first trial as an illustration, the procedure is as follows:

(1) Draw determinate moment diagram (heavy solid line, sketch *e*).

(2) Assume a mechanism. For vertical load alone, either mechanisms 3 or 4 probably would control, and it was assumed that mechanism 3 would be the correct one.

(3) Roughly sketch the redundant moment diagram upon the determinate one. This need not be done to scale; it is a guide for writing the equilibrium equations.

(4) Write the equilibrium equation at the sections at which plastic hinges form. As shown beneath sketch (*e*) there are three sections (12, 8, and 4) at which equilibrium equations are written. Illustrating for section 8, the determinate moment at 8 (distance 8–B in sketch *e*) is equal to M_p (distance A–B) plus the redundant moment 8–A. The value 8–A may be obtained from sketch (*d*), mechanism 3, and is equal to $H(a + b)L - (M_p - M_C)/2$. (In this problem $M_C = M_1$.) Thus Eq. 2 is obtained. Similarly, at section 4 the equilibrium equation is

$$M_{\text{redundant}} = H(aL) + M_C = M_{s4} + M_p$$

(5) Finally the three resulting equilibrium equations, may be solved for the three unknowns (H, M_p, M_C) and a value M_p obtained (96 kip-ft in this example). The minus sign for the moment at section 1 indicates that the moment is opposite to that assumed in mechanism 3.

(6) The precise moment diagram is now drawn to make sure that $M \leq M_p$ throughout. It was found that the plastic moment condition was violated at section 7. So the correct hinge position is at section 7 instead of section 8.

(7) The next step is to make a new trial if an incorrect answer was obtained. In this particular case $96 < M_p < 98$ and the latter value is chosen for the required M_p for case I.

Step 5: Case II was next analyzed by the same technique. The new chart of determinate moments is conveniently prepared by multiplying the moments for rows (1), (2) and (3) in the table of sketch (*c*) by the ratio 1.40/1.85. The moment due to wind load must be added, of course.

The first trial mechanism was mechanism 2. Using the same procedure as before, it is found that $M_p = 84$ kip-ft. Since completion of the moment diagram shows that the plastic moment condition is not violated, then this is the correct answer for case II.

It should be noted that whenever mechanisms 1 or 2 control the design, then only two equilibrium equations are needed to solve the problem, since there are only two unknowns: M_p and H. Thus in the trial solution for mechanism 2 in step 5, only Eqs. 4 and 5 were needed to solve for M_p. Eq. 6 was set up to make the moment check at section 2.

Step 6: Case I controls the design. A 12WF27 shape, supplying a Z-value of 37.97 in.^{3} will be adequate.

The problem is not completely solved in Plate IX, since all of the appropriate design guides have not been checked, modifying the section requirements for axial force, shear, etc., where necessary. These are left as exercises.

9.3 SIMPLIFIED PROCEDURES FOR SINGLE-SPAN FRAMES

The possible advantages of the use of "short-cut" methods in plastic design were outlined in Art. 8.5. Two approaches are possible in simplifying the procedure for the solution of single-span frames. The virtual work equations can be expressed as formulas which would reflect both the frame geometry and the loading conditions. Alternatively curves may be prepared which present the solution in chart form and Ref. 7.4 represents an outstanding contribution in that direction. It enables the engineer to determine the required plastic moment of a single-span frame with the aid of charts in a fraction of the time required in a "routine" plastic analysis. The method of derivation and some examples are contained in Ref. 9.2. Reference 7.1 makes use of both the "formula" and the "chart" approach in presenting numerous practical examples; the chart solutions are based substantially on Ref. 7.4. Space here only permits an indication of the approach together with a few illustrations. Attention will be restricted to single-story structures of uniform plastic moment throughout.

Fig. 9.1 shows a gabled frame with uniformly distributed vertical and horizontal loading. For simplicity the horizontal distributed load is replaced by a concentrated load, acting at the eaves, such that it produces the same overturning moment about the base at section 1.

Since this overturning moment is given by

$$M = \frac{w_h(a + b)^2L^2}{2}$$

then

$$T = \frac{w_h(a + b)^2L}{2a} \tag{9.1}$$

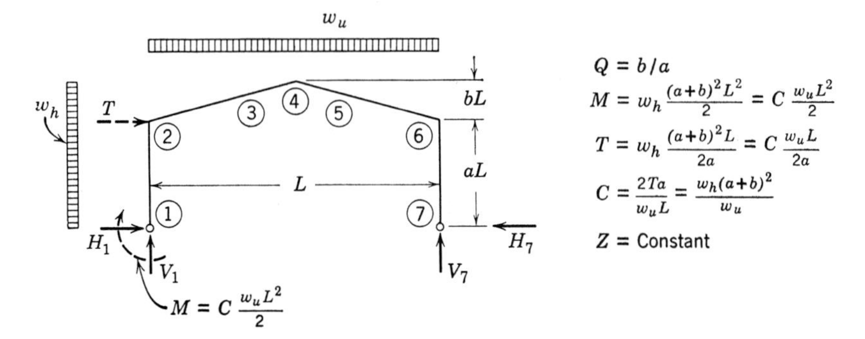

Fig. 9.1. Loading and geometrical functions involved in the analysis of gabled frames.

In order to simplify the form of the solution, a parameter C is introduced which is a function of the magnitude of the overturning moment. It is determined from

$$M = C\frac{w_uL^2}{2}$$

and thus

$$C = \frac{2Ta}{w_uL} = \frac{w_h}{w_u}(a + b)^2 \tag{9.2}$$

Consider, now, the mechanism shown in Fig. 9.2. (Of course there are other possible mechanisms but in most practical cases, the mechanism shown will be the one to form.) Using instantaneous centers, the rotation at each of the plastic hinges may be computed and then, by use of the mechanism method, the required plastic moment may be determined in terms of the variables w_u, L, Q, C, and x. The following equation results:

$$M_p = \frac{wL^2}{4}\left[\frac{\left(1 - \frac{x}{L}\right)\left(C + \frac{x}{L}\right)}{\sqrt{(1 + Q)(1 - QC)}}\right] \tag{9.3}$$

where x is given by

$$\left.\begin{aligned} x &= \frac{L}{Q}[\sqrt{(1+Q)(1-QC)} - 1] & (Q > 0) \\ x &= L\frac{1-C}{2} & (Q = 0) \end{aligned}\right\} \tag{9.4}$$

and is computed by the methods already discussed.

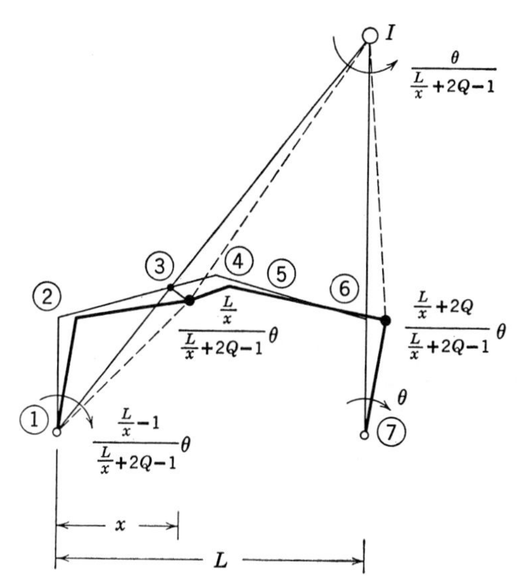

Fig. 9.2. A frequently encountered gabled frame mechanism and corresponding mechanism angles.

The only remaining problem is to determine the range of variables for which the mechanism shown in Fig. 9.2 is in fact the correct solution. Figure 9.3 summarizes the applicable formulas for the pinned-base, single-span, single-story frame. Similar solutions may be developed for other loading conditions and for fixed bases.

In Ref. 7.4 are presented all possible solutions to the single-span, single-story frame in the form of two charts—one which gives the value of M_p/wL^2 as influenced by C and Q, and one which gives the distance x to the plastic hinge in the rafter (also a function of C and Q). These two charts are indicated in Fig. 9.4 and for this major range of variables, they are simply representations of Eqs. 9.3 and 9.4. Their use will be indicated by the example which follows.

Vertical load alone

$$V_1 = V_7 = \frac{w_u L}{2}$$

$$H_1 = H_7 = \frac{M_p}{aL}$$

For $Q = 0$:

$$M_p = \frac{w_u L^2}{16}, \qquad x = \frac{L}{2}$$

For $Q > 0$:

$$M_p = \frac{w_u L^2}{4}\left[\frac{\frac{x}{L}\left(1 - \frac{x}{L}\right)}{\sqrt{1 + Q}}\right]$$

$$x = \frac{L}{Q}[\sqrt{1 + Q} - 1]$$

Vertical and horizontal load

$$V_1 = \frac{w_u L}{2}(1 - C) \qquad V_7 = \frac{w_u L}{2}(1 + C)$$

$$H_1 = w_h(a + b)L - H_7 \qquad H_7 = M_p/aL$$

For $C > \dfrac{1}{1 + Q}$ (panel mechanism):

$$M_p = \frac{w_u L^2}{4} C, \qquad x = 0$$

For $C < \dfrac{1}{1 + Q}$ (combined mechanism):

$$Q = 0{:} \quad M_p = \frac{w_u L^2}{16}(1 + C)^2, \qquad x = L\frac{(1 - C)}{2}$$

$$Q > 0{:} \quad M_p = \frac{w_u L^2}{4}\left[\frac{\left(1 - \frac{x}{L}\right)\left(C + \frac{x}{L}\right)}{\sqrt{(1 + Q)(1 - QC)}}\right]$$

$$x = \frac{L}{Q}[\sqrt{(1 + Q)(1 - QC)} - 1]$$

Fig. 9.3. Formulas for the solution of pinned-base frames.[7.1]

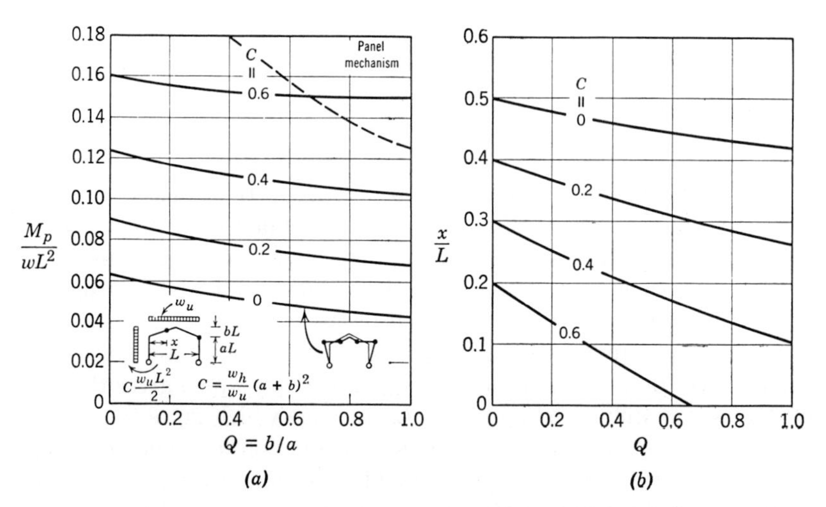

Fig. 9.4. Charts for the solution of pinned base frames.[7.4]

PLATE X. GABLED FRAME WITH STRAIGHT CONNECTIONS

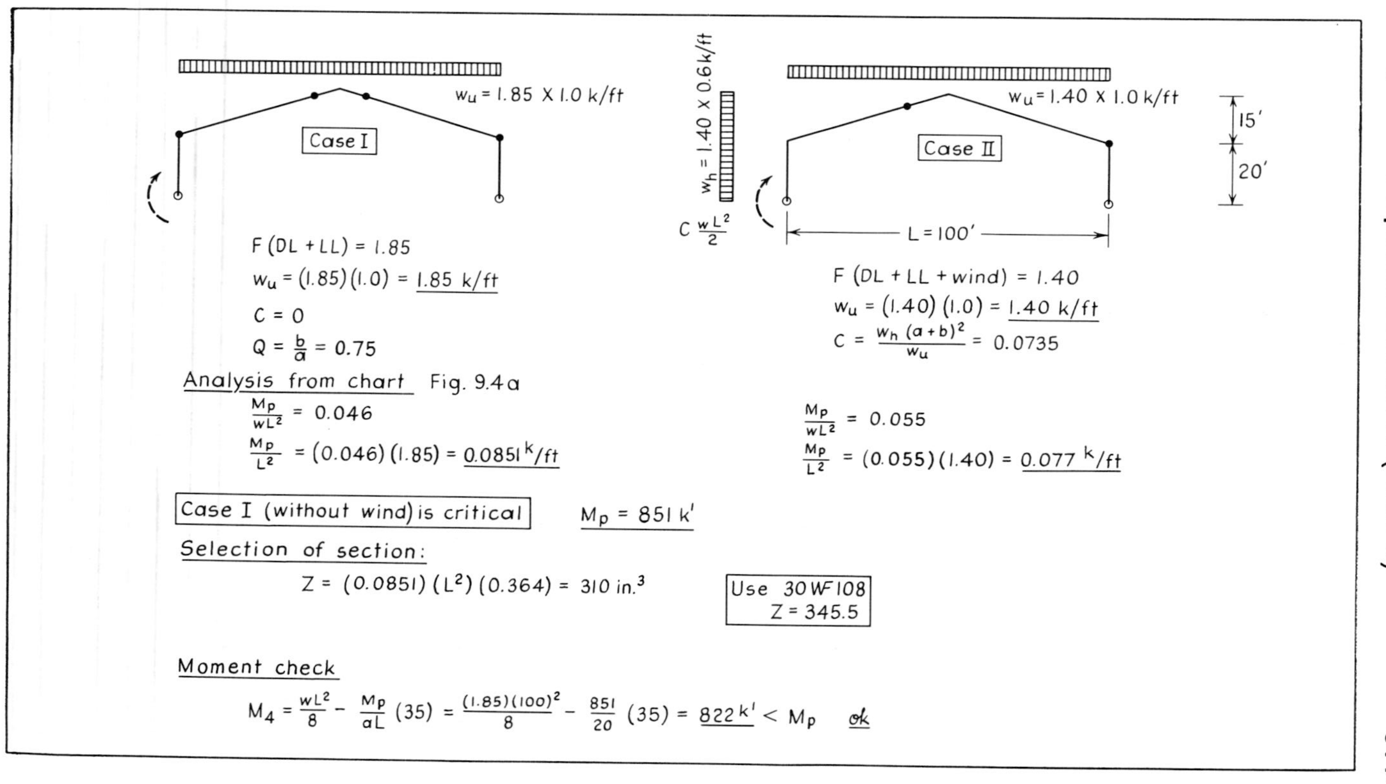

$w_u = 1.85 \times 1.0$ k/ft
Case I
$w_h = 1.40 \times 0.6$ k/ft
$C \frac{wL^2}{2}$
$w_u = 1.40 \times 1.0$ k/ft
Case II
15′
20′
L = 100′
F (DL + LL) = 1.85
$w_u = (1.85)(1.0) = 1.85$ k/ft
C = 0
$Q = \frac{b}{a} = 0.75$
Analysis from chart Fig. 9.4a
$\frac{M_p}{wL^2} = 0.046$
$\frac{M_p}{L^2} = (0.046)(1.85) = 0.0851^k/ft$
F (DL + LL + wind) = 1.40
$w_u = (1.40)(1.0) = 1.40$ k/ft
$C = \frac{w_h (a+b)^2}{w_u} = 0.0735$
$\frac{M_p}{wL^2} = 0.055$
$\frac{M_p}{L^2} = (0.055)(1.40) = 0.077^k/ft$
Case I (without wind) is critical
$M_p = 851\ k'$
Selection of section:
$Z = (0.0851)(L^2)(0.364) = 310$ in.^{3}
Use 30 WF 108
Z = 345.5
Moment check
$M_4 = \frac{wL^2}{8} - \frac{M_p}{aL}(35) = \frac{(1.85)(100)^2}{8} - \frac{851}{20}(35) = 822\ k' < M_p$ ok

PLATE X Gabled Frame with Straight Connections

This example is the same as that given in Plate VII, p. 286. The two loading conditions are as shown at the top of Plate X. The distributed load acting horizontally on the frame produces an overturning moment from which C may be computed (Eq. 9.2). The values of C are thus determined as zero for case I and 0.0735 for case II. Knowing that $Q = a/b = 0.75$, all the needed information is available for entering the chart of Fig. 9.4a.

For case I, M_p/wL^2 is equal to 0.046 and for case II it is 0.055. To determine the critical or controlling case, it is sufficient to compare M_p/L^2 ratios since L is the same in both cases. On this basis, case I is found to be critical. A 30WF108 member is specified, an answer that agrees with the value obtained in Plate VII, p. 286.

The moment check shows that the plastic moment condition is not violated and thus the answer is correct. The secondary design conditions would next be checked.

9.4 MULTI-SPAN FRAMES

PLATE XI Two-Span Rectangular Frame

An industrial frame will be designed to carry a vertical load of 1.2 k/ft and a horizontal load of 0.6 k/ft. The mechanism method will be used to analyze the various loading conditions. Distributed load will be treated as such, although the loads will actually be applied to the structure through purlins. Opportunity will be afforded in this two-span frame to illustrate the "preliminary design" procedures for estimating plastic moment ratios.

Step 1: As shown in sketch (a) the column height is 15 ft, the left span is 30 ft, and the right span is 60 ft. For the time being the value M_p is assigned to the left rafter, the value k_1M_p to the right rafter, and k_2M_p to the interior column.

Step 2: There are two possible loading conditions. For case I with dead load and live load, the load factor of safety is 1.85. The distributed load becomes 2.22 kips/ft. For case II (dead load plus live load plus wind) the load factor of safety is 1.40 and the vertical load is 1.68 k/ft, the horizontal load being half this value.

Step 3: In order to determine a suitable plastic moment ratio for the rafters, the beams are considered as fixed ended as shown in sketch (b). The value k_1 is thus determined as 4.0. For this special condition, the minimum possible plastic moment values would be determined, the joints being fixed against rotation but the frame theoretically free to sway. The resulting ratio is therefore the basis for later analysis of the frame. For greatest economy, the end columns should provide full restraint to the beams, and therefore the plastic moment values are made equal to the appropriate beam values. The value

k_2 for the interior column may be determined by considering equilibrium of joint 6–7–8 (sketch *c*).* A value of k_2 equal to 3 is obtained, and M_7 acts in a counter clockwise direction on the joint.

Step 4: The structure is now analyzed for case I loading, an analysis that actually was completed in the previous step. There are seven possible plastic hinges. The frame is redundant to the third degree ($X = 3$). Therefore, there are four possible independent mechanisms and these are shown as mechanisms 1 through 4, sketches (*b*), (*c*), and (*d*).

The solution for mechanism 1 is made on the basis that mechanisms 1 and 2 form simultaneously. Consequently M_p is determined as 125 kip-ft and k_1M_p is equal to 500 kip-ft.

The moment check shown in sketch (*e*) reveals that the moment is nowhere greater than M_p or k_1M_p as the case may be, and thus the solution is correct for this loading condition. The other two mechanisms need not be checked because the three necessary conditions (equilibrium, mechanism, plastic moment) have been satisfied. The computation of reactions at ultimate load completes the first analysis.

Step 5: The analysis for case II is now performed to see whether or not the plastic moment values determined will be adequate. The same plastic moment ratios, k_1 and k_2, will be used.

The solution for mechanism 1 for case II loading will always be less than that for case I in the ratio of the two load factors. Thus, it could never control the design. M_p is found to be 94.5 kip-ft. The solution for mechanism 4 (sketch *d*) shows that M_p is so much less than the value determined for mechanism 1 that no further consideration of it is necessary.

The solution for mechanism 5, which is a combination of mechanisms 1, 2, 3, and 4, shows a required M_p value of 99.3 kip-ft. Incidentally, the solution for this mechanism assumed that hinges formed in mid-span. Thus, x_1 in sketch (*d*) was made equal to $L_1/2$ and $x_2 = L$. Although M_p is less than the value 125 kip-ft for case I, it is close enough so that the moment check must be made. In the first place, we must expect that the plastic moment condition will be violated in each of the rafters because of our initial assumption that plastic hinges formed at mid-span (sections 5 and 9). This is only the correct position when the end moments are equal. The moment check is completed by using the equilibrium equations, and these are shown above sketch (*f*). Using the equation for beam 4–6 it is found that M_4 equals 80 kip-ft which is less than M_p, and similarly for span 8–10. By the joint equilibrium equation for 6–7–8, it is found that the moment in the column top is also less than k_2M_p.

Thus sufficient information is available for drawing the moment diagram, and it is plotted to scale in sketch (*f*). As expected, the moment is greater than the plastic moment value near the center of the two rafters. To the left of section 5, $M = 100$ kip-ft as compared with $M_p = 99$ kip-ft. To the left of section 9, $M = 400$ kip-ft as compared with $4M_p = 397$ kip-ft. We may therefore conclude that mechanism 5 is the correct one, M_p being slightly larger than 99 kip-ft. Case I ($M_p = 125$ kip-ft) therefore controls the design.

* The sign convention is that clockwise moments are positive.

PLATE XI. TWO-SPAN RECTANGULAR FRAME

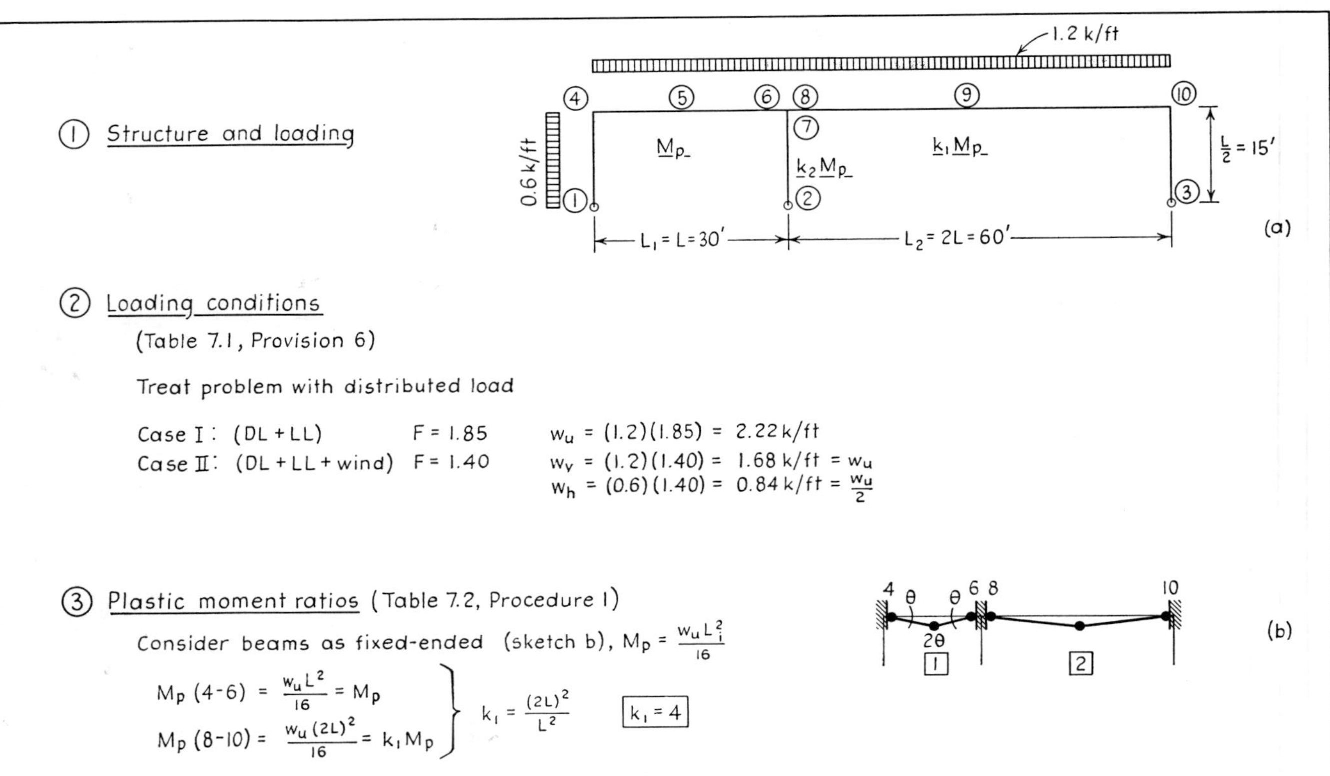

① Structure and loading

② Loading conditions

(Table 7.1, Provision 6)

Treat problem with distributed load

Case I: (DL + LL) F = 1.85 $\quad w_u = (1.2)(1.85) = 2.22$ k/ft

Case II: (DL + LL + wind) F = 1.40 $\quad w_v = (1.2)(1.40) = 1.68 \text{ k/ft} = w_u$

$w_h = (0.6)(1.40) = 0.84 \text{ k/ft} = \frac{w_u}{2}$

③ Plastic moment ratios (Table 7.2, Procedure 1)

Consider beams as fixed-ended (sketch b), $M_p = \frac{w_u L_i^2}{16}$

$$\left.\begin{aligned} M_p\,(4\text{-}6) &= \frac{w_u L^2}{16} = M_p \\ M_p\,(8\text{-}10) &= \frac{w_u (2L)^2}{16} = k_1 M_p \end{aligned}\right\} \quad k_1 = \frac{(2L)^2}{L^2} \qquad \boxed{k_1 = 4}$$

End columns provide full restraint to beams

$M_p\,(1\text{-}4) = M_p\,, \quad M_p\,(3\text{-}10) = k_1 M_p = 4M_p$

Interior column to provide joint equilibrium: sketch (c)

$M_6 = -M_p \quad M_8 = +k_1 M_p$

M_7

3

(c)

$M_6 + M_7 + M_8 = 0$

$M_7 = -M_6 - M_8 = +M_p - (+k_1 M_p) = M_p\,(1-k_1) = -3M_p\,, \quad \boxed{k_2 = 3}$

④ Case I analysis (Mechanism method)
(Table 7.3, Procedure 2)

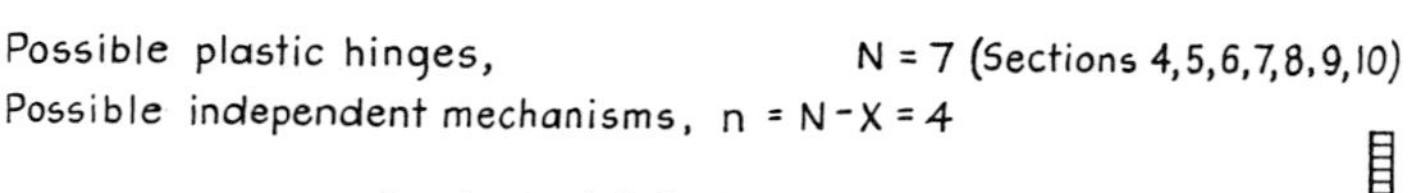

Possible plastic hinges, $N = 7$ (Sections 4,5,6,7,8,9,10)

Possible independent mechanisms, $n = N - X = 4$

mechanisms 1 & 2: beam

mechanism 3: joint

mechanism 4: panel

mechanism 5: composite

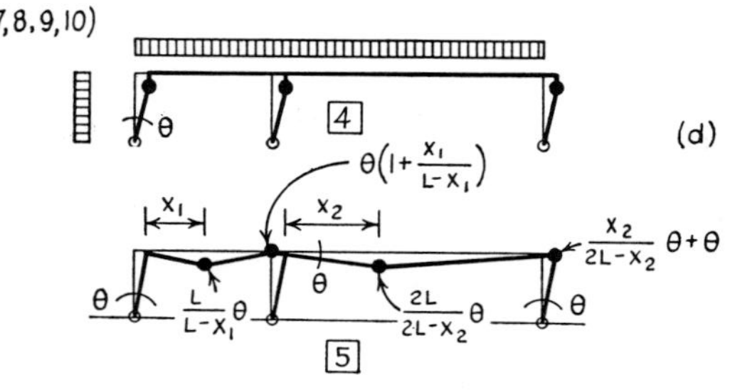

PLATE XI (Continued)

Solution for mechanism I

$$M_p(\theta + 2\theta + \theta) = w\frac{L\theta}{2}\cdot\frac{L}{2}$$

$$M_p = \frac{w_u L^2}{16} = \frac{(2.22)(30)^2}{16} = \underline{125^{k'}} \qquad \text{(Member 4-6)}$$

$$k_1 M_p = (4)(125) = \underline{500^{k'}} \qquad \text{(Member 8-10)}$$

Moment check: sketch (e)
(Table 7.3, Procedure 7)

Joint equilibrium at 6-7-8: $\underline{k_2 M_p = 375^{k'}}$ (Member 2-7)

All $M < kM_p$ ok

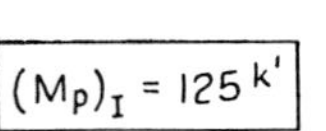

$$(M_p)_I = 125^{k'}$$

Reactions at ultimate load

$$H_1 = \frac{M_p}{\frac{L}{2}} = \frac{125}{15} = 8.3^k \qquad V_1 = \frac{w_u L}{2} = \frac{(2.22)(30)}{2} = 33.3^k$$

$$H_2 = \frac{k_2 M_p}{\frac{L}{2}} = \frac{375}{15} = 25.0^k \qquad V_2 = \frac{w_u L}{2} + \frac{w_u(2L)}{2} = 1.5\,w_u L = 100^k$$

$$H_3 = \frac{k_1 M_p}{\frac{L}{2}} = \frac{500}{15} = 33.3^k \qquad V_3 = \frac{w_u(2L)}{2} = 66.7^k$$

⑤ Case II analysis

(Table 7.3, Procedure 2)

Hinges and mechanisms: See Case I

Solution for mechanism 1: $M_p = (125)\,\frac{1.40}{1.85} = \underline{94.5^{k'}}$

Solution for mechanism 4: (sketch d)

$$M_p\,\theta + (k_2 M_p)\,\theta + (k_1 M_p)\,\theta = w_h\,\theta\,\frac{L}{2}\cdot\frac{\frac{L}{2}}{2}$$

$$M_p\,(1+3+4) = \frac{0.5\,w_u}{2}\left(\frac{L^2}{4}\right) = \frac{(1.68)(900)}{16}\,,\quad M_p = \underline{11.8^{k'}} \ll M_p \text{ case I}$$

Solution for mechanism 5 (1+2+3+4): sketch (d)

NOTE: Assume hinges form at mid-span ($x_1 = \frac{L}{2}$, $x_2 = L$, $\theta_5 = \theta_6 = \theta_9 = \theta_{10} = 2\theta$)

$$M_p\,\theta\,(2+2+2\times 4+2\times 4) = w_u\left(\frac{L}{2}\theta\right)(L)\left(\frac{1}{2}\right) + w_u\,(\theta L)(2L)\left(\frac{1}{2}\right) + \frac{w_u}{2}\,\frac{L}{2}\left(\theta\frac{\frac{L}{2}}{2}\right)$$

$$20\,M_p = w_u L^2\left(\frac{1}{4}+1+\frac{1}{16}\right) = \frac{21}{16}\,(1.68)(30)^2;\quad M_p = \underline{99.3^{k'}}$$

"Upper bound" solution
∴ check moment

PLATE XI (*Continued*)

Moment check: (sketch f)

(Table 7.3, Procedures 7 & 8)

Beam (4-6): $M_5 = \frac{M_4}{2} - \frac{M_6}{2} + \frac{w_u L^2}{8}$

$$M_4 = 2M_5 + M_6 - \frac{w_u L_1^2}{4} = 2M_p + M_p - \frac{w_u L^2}{4} = 3(99.3) - \frac{(1.68)(900)}{4} = -80^{k'} < 99.3 \quad \text{ok}$$

Beam (8-10): $M_8 = 2M_9 + M_{10} - \frac{w_u L_2^2}{4} = (2)(4M_p) + 4M_p - \frac{w_u (2L)^2}{4}$

$$M_8 = (12)(99.3) - (1.68)(3600)/4 = -321^{k'} < 397^{k'} \quad \text{ok}$$

Joint (6-7-8): $M_7 = -M_6 - M_8 = +99 - 321 = -222^{k'} < 3M_p = 298^{k'}$ ok

Case I (without wind) is critical

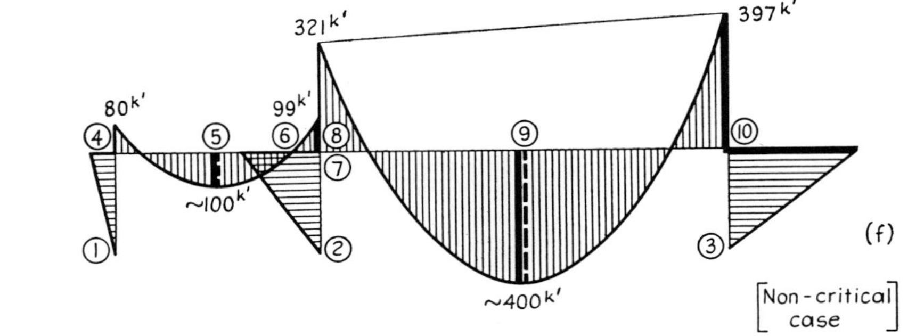

⑥ Selection of sections — Controlling moment diagram: sketch (e), $M_p = 125^{k'}$
(Table 7.1, Provision 5)

Left beam, Left column: $Z_{4\text{-}6} = \frac{M_p}{\sigma_y} = (125)(0.364) = 45.5 \text{ in.}^3$

Use 14 WF 30, $Z = 47.1 \text{ in.}^3$ — $A = 8.81$, $d = 13.86$, $b = 6.73$, $t = 0.383$; $w = 0.270$, $S = 41.8$, $I = 289.6$, $r_y = 1.41$

Right beam, Right column: $Z_{8\text{-}10} = 4Z_{4\text{-}6} = 182.0 \text{ in.}^3$

Use 24 WF 76, $Z = 200.1 \text{ in.}^3$ — $A = 22.37$, $d = 23.91$, $b = 8.99$, $t = 0.68$; $w = 0.440$, $S = 175.4$, $I = 2096.4$, $r_y = 1.85$

Center column: $Z_{2\text{-}7} = 3Z_{4\text{-}6} = 136.0 \text{ in.}^3$

Use 21 WF 62, $Z = 144.1 \text{ in.}^3$ — $A = 18.23$, $d = 20.99$, $b = 8.24$, $t = 0.615$; $w = 0.400$, $S = 126.4$, $I = 1326.8$, $r_y = 1.71$

⑦ Axial force
(Table 7.4, Guide 1)

Left column (14 WF 30): $\frac{P}{P_y} = \frac{V_1}{\sigma_y A_{1\text{-}4}} = \frac{33.3}{(33)(8.81)} = 0.115 < 0.15$ ok

Interior column (21 WF 62): $\frac{P}{P_y} = \frac{V_2}{\sigma_y A_{2\text{-}7}} = \frac{100}{(33)(18.23)} = 0.166 > 0.15$ [Modification req'd.]

$$Z_{req.} = Z_{trial}\left(\frac{P}{P_y} + 0.85\right)$$
$$= 136.0\,(0.166 + 0.85) = 138 \text{ in.}^3 < 144.1 \text{ in.}^3 \quad \text{ok}$$

PLATE XI (*Continued*)

(8) Shear force
(Table 7.4, Guide 2)

Left beam: (14WF30)
$V_{max.} = V_1 = 33.3^k$
$V_{allow.} = 18\,wd = (18)(0.270)(13.86) = 67.3 > 33.3$ ok

Right beam: (24WF76)
$V_{max.} = V_3 = 66.7^k$
$V_{allow.} = 18\,wd = (18)(0.440)(23.91) = 189 > 66.7$ ok

(9) Cross-section proportions
(Table 7.8, Guide 1)

Shape	$\frac{b}{t}(<17)$	$\frac{d}{w}(<55)$		
14WF30	17.58	51.3	ok	NOTE 1
21WF62	13.40	52.5	ok	NOTE 2
24WF76	13.2	54.3	ok	

NOTE 1: $\frac{b}{t}$ for 14WF30 within 3% of specified value – adequate

NOTE 2: $\frac{d}{w} \leq 70 - 100\frac{P}{P_y} = 70 - 16.6 = 53.4 > 52.5$ ok

(10) Lateral bracing Purlin spacing = 5 ft, girt spacing = 5 ft
(Table 7.8, Guide 3)

Spacing Left rafter: (14 WF 30) Sec. 5 $\frac{L_B}{r_y} = \frac{(5.0)(12.0)}{1.41} = 42.5$ ok (Last hinge in rafter)

Left column: (14 WF 30) $\frac{L_B}{r_y} = \frac{(5.0)(12.0)}{1.41} = 42.5$ $\left[\frac{M}{M_p} = \frac{\frac{2}{3}M_p}{M_p} = 0.67\right]$

Eq 4.28: $\left(\frac{L_B}{r_y}\right)_{cr} = 30 < 42.5$ $\therefore$ More refined check necessary

Restraint coefficient

Segment	Moment ratio	$\left(\frac{L}{r_y}\right)_{cr}$	L_{cr}
L_B	$\frac{M}{M_p} = \frac{\frac{2}{3}M_p}{M_p} = 0.67$	Eq (4.28): $\frac{L}{r_y} = 30$	$42.3'' = L_b$
L_L	$g_L = \frac{\frac{1}{3}M_p}{\frac{2}{3}M_p} = 0.50$	Eq (A2): $\frac{134}{\sqrt{0.67}} + 60\,(1-0.50) = 194$	$274'' = L_l$
L_R	$g_R = \frac{-13}{125} = -0.10$	Eq (A4): $48 - 30\,(-0.10) = 51$	$72'' = L_s$

Eq (A1): $C_f = 1.0 + 0.2\left\{1-\left(\frac{60}{274}\right)^2\right\}\left\{0.9 + \frac{1-\left(\frac{60}{72}\right)^2}{1-\left(\frac{60}{274}\right)^2}\right\} = 1.23$

$(L_B)_{cr} = C_f L_b = (1.23)(42.3) = 52'' < 60''$

[Provide additional brace to column]

Bracing details: (1) Provide vertical welded plates at center purlins of both rafters.
(2) At sections 4, 7 and 10 brace to inner (compression) corners.

PLATE XI (*Continued*)

(11) Columns
(Table 7.6, Guide 2)

Left column: (14 WF 30) $\frac{P}{P_y} = 0.115 < 0.15$ ∴ Full M_p available

Right column: (21 WF 73) $\frac{P}{P_y} = \frac{V_3}{\sigma_y A_{3-10}} = \frac{66.7}{(33)(21.46)} = 0.094 < 0.15$ ∴ Full M_p available

Interior column: (21 WF 62)

$\frac{P}{P_y} = 0.166 > 0.15$

$\frac{L}{r_y} = \frac{(15)(12)}{8.53} = 21.2$

Simple check for axial force is adequate. ∴ Original design is OK

(12) Connection details (Use straight connections w/o haunches)

Connection 4: (14 WF 30) (Table 7.7, Guide 2)

Try diagonal stiffener equal to flange thickness = 0.383″

For local buckling: $t \leq \frac{b}{17} = \frac{6.73}{17} = 0.394''$

See detail h, Plate V

Use $3\frac{1}{4}'' \times \frac{7}{16}''$ ℄

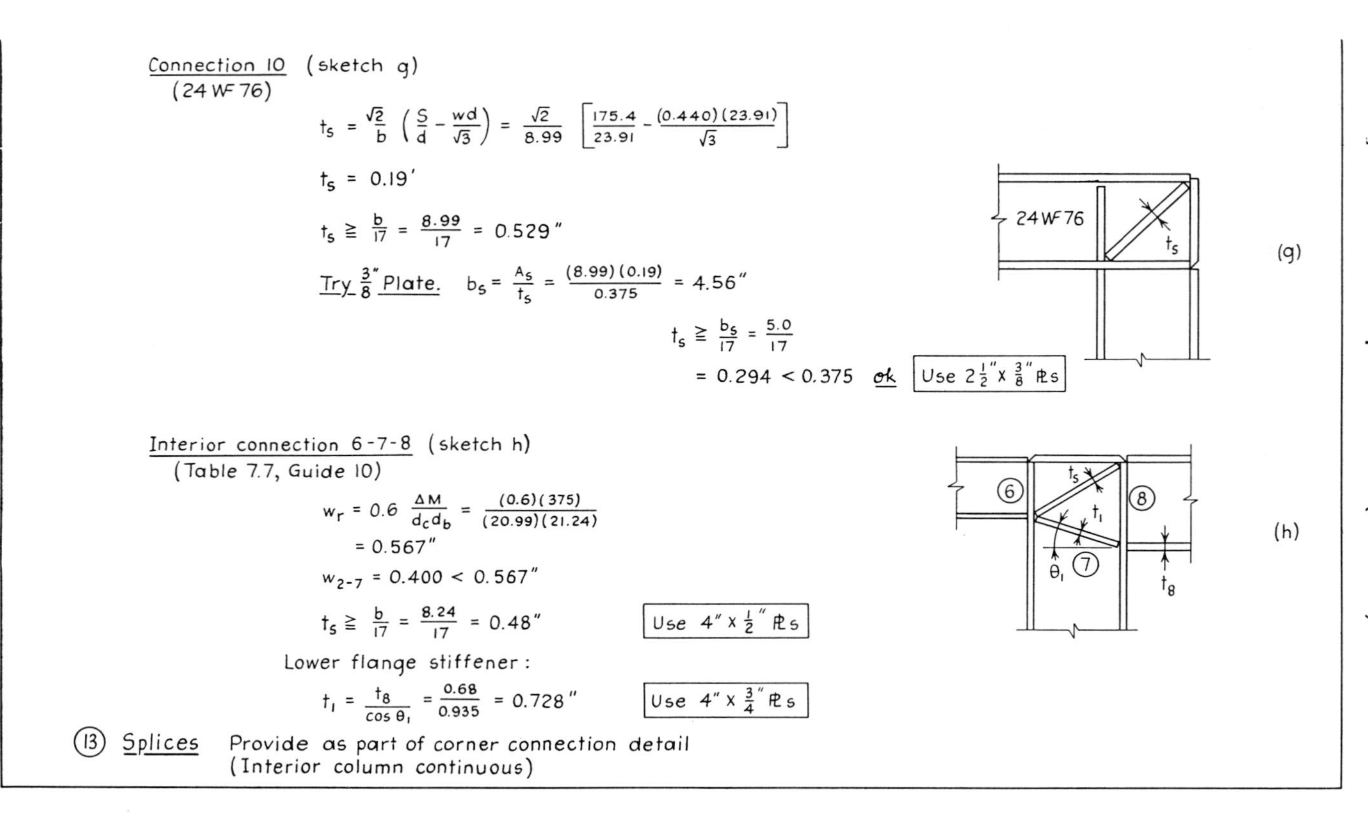

<u>Connection 10</u> (sketch g)
(24 WF 76)

$$t_s = \frac{\sqrt{2}}{b}\left(\frac{S}{d} - \frac{wd}{\sqrt{3}}\right) = \frac{\sqrt{2}}{8.99}\left[\frac{175.4}{23.91} - \frac{(0.440)(23.91)}{\sqrt{3}}\right]$$

$$t_s = 0.19'$$

$$t_s \geqq \frac{b}{17} = \frac{8.99}{17} = 0.529''$$

Try $\frac{3}{8}''$ Plate. $b_s = \frac{A_s}{t_s} = \frac{(8.99)(0.19)}{0.375} = 4.56''$

$$t_s \geqq \frac{b_s}{17} = \frac{5.0}{17}$$

$$= 0.294 < 0.375 \quad \text{ok}$$

Use $2\frac{1}{2}'' \times \frac{3}{8}''$ PLs

<u>Interior connection 6-7-8</u> (sketch h)
(Table 7.7, Guide 10)

$$w_r = 0.6\,\frac{\Delta M}{d_c d_b} = \frac{(0.6)(375)}{(20.99)(21.24)}$$

$$= 0.567''$$

$$w_{2\text{-}7} = 0.400 < 0.567''$$

$$t_s \geqq \frac{b}{17} = \frac{8.24}{17} = 0.48''$$

Use $4'' \times \frac{1}{2}''$ PLs

Lower flange stiffener:

$$t_1 = \frac{t_8}{\cos\theta_1} = \frac{0.68}{0.935} = 0.728''$$

Use $4'' \times \frac{3}{4}''$ PLs

(13) <u>Splices</u> Provide as part of corner connection detail
(Interior column continuous)

If it had been desirable to analyze mechanism 5 and determine the precise location of plastic hinges this could either be done graphically, by trial and error, or by maximizing the required M_p value expressed in terms of the distances x_1 and x_2. The following equation in terms of x_1 and x_2 would be differentiated partially with respect to x_1 and with respect to x_2, and would be set equal to 0; and the resulting two equations would be solved simultaneously for the x_1 and x_2 values.

$$M_p\theta\left(\frac{L}{L-x_1}+\frac{x_1}{L-x_1}\right)+k_1M_p\theta\left(\frac{2L}{2L-x_2}+\frac{x_2}{2L-x_2}\right)$$
$$= w_u\frac{L_1}{2}\theta x_1 + w_u\frac{L_2}{2}\theta x_2 + w_h\frac{L/2}{2}\left(\frac{L}{2}\right)\theta \qquad (9.5)$$

No reactions were computed for the case II loading because it is obvious that they would be less than for case I which controls the selection of member sizes.

Step 6: Since case I (without wind) is the critical condition, the selection of required section will be made on the basis of the M_p values thus determined: $M_p = 125$ kip-ft, $4M_p = 500$, and $3M_p = 375$ kip-ft.

Step 7: In checking Design Guide 1 of Table 7.4 for axial force in the members, it is found that the center column has a P/P_y value of 0.166, which is greater than 0.15. Using the recommended formula (Eq. 4.11) it is found that the original choice was satisfactory since the Z value actually furnished is greater than the modification factor requires. No check is necessary of the right-hand column, since the center column is satisfactory; and the beams are adequate because the horizontal thrusts are less than the vertical ones.

Step 9: In evaluating the cross-section proportions it is found that the 14WF30 beam has a b/t ratio of 17.58 (greater than 17). Since it is within 3% of the specified value it is considered satisfactory. The 21WF62 with a P/P_y ratio of 0.166 is checked for the limiting value of d/w allowable for this condition. The existing value is found to be satisfactory ($52.5 < 53.4$).

Step 10: Concerning the matter of lateral bracing, the purlin and girt spacing is selected as 5 ft. The left rafter will be the most critical since it has the smallest r_y value. A slenderness ratio of 42.5 will be adequate since the plastic hinge in the center of the rafter will be one of the last to form. A preliminary check of the left-hand column (14WF30) shows that a more refined examination is required. A consideration of the restraint coefficient improved the situation somewhat (compare the critical length of 52 in. with the value of 60 in. that exists in the structure). Either an additional brace could be placed part way down the column (tied to the eaves purlin) or one could check the hinge rotation at section 4 to see if the requirement was as severe as assumed in Design Guide 3. Alternatively a new shape could be selected for the left column so that $(L_B)_{cr} \geq 60$ in. Thus

$$r_y = \frac{60}{C_f(L/r_y)_{cr}} = \frac{60}{(1.23)(30)} = 1.63''$$

A 12WF40 ($Z = 57.6$ in.^{3}, $r_y = 1.94$ in.) would be suitable.

Step 11: In checking the columns, it is found that only the interior column (21WF62) requires special consideration. With a P/P_y ratio of 0.166 and the very low value of slenderness ratio in the strong direction (21.2), Fig. 4.23 indicates that a modification by Eq. 4.11 is all that is required. Since this modification has been made previously, the member is adequate.

Step 12: In proportioning the diagonal stiffener for connection 4 (14WF30), the member is so light that the initial choice will be based on a diagonal with thickness equal to that of the rolled section flange. In checking for local buckling of this element a slightly greater thickness is required ($t = 0.394$ in.). Therefore a $\frac{7}{16}$-in. plate is specified.

A similar situation arises for connection 10, except that the local buckling provision becomes more critical. In this case, Eq. 5.12 for t_s was used, resulting in a required value of 0.19 in. The buckling provision requires a thickness of 0.53 in. if the full flange width is maintained. Use of the flange thickness (0.615 in.) would have been adequate, and the example suggests that this rule-of-thumb guide is probably the best one to use in design where light members are involved. Actually a $\frac{3}{8}$-in. plate was specified and this required that the stiffener width be 5-in.

With regard to interior connection 6–7–8, since the full moment capacity of the 24WF76 member need not be transmitted into the column, the existing web thickness may be adequate. Equation 5.49 was used as a check and it was found that the required thickness is 0.567 in., the required thickness to be compared with that furnished by the 21WF62 shape. The web is inadequate on this basis ($w = 0.400$ in.) and therefore stiffening is required. In view of the fact that the local buckling requirement had controlled the previous design, and since the column web nearly meets the requirement, the $b/t < 17$ rule will be used to select the stiffener thickness t_s. A one-half-in. plate is therefore specified.

PLATE XII Two-Span Gabled Frame with Fixed Base

A two-span gabled frame will now be designed with fixed column bases and will illustrate the application of the mechanism method. The design is incomplete since all of the applicable design guides have not been examined; these are left as exercises. It is emphasized that problems of this type encountered in design would be solved most rapidly by using the simplified procedures referred to in Art. 9.5. The purpose of presenting Plate XII in this form is to show the basic method. More complicated problems, for which charts might not be available, could then be solved.

Step 1: The frame is symmetrical throughout, with individual spans of 80 ft, column height of 20 ft, and roof rise of 20 ft. The roof loading is concentrated at the quarter points of the rafters and might be thought of as an approximation to a uniformly distributed load of 1.0 kip/ft. Similarly, the side load ($T = 16$ kips) produces the same overturning moment about the base as that of a uniformly distributed horizontal load of 0.4 kip/ft acting on the vertical projection of the structure.

PLATE XII. TWO-SPAN GABLED FRAME WITH FIXED BASE

① Structure and loading:

$P = 20^k$
$T = 16^k$
Column bases fixed

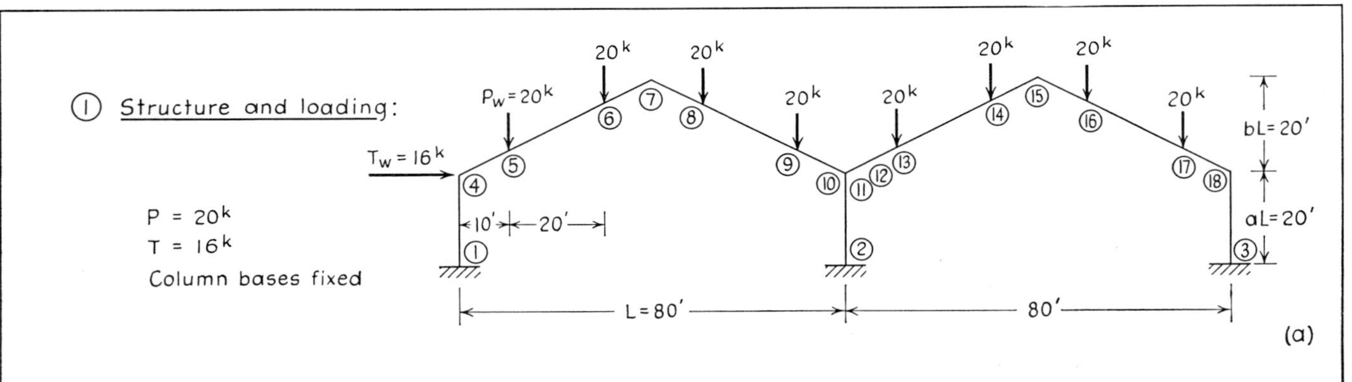

(a)

② Loading conditions (Table 7.1, Provision 6)

Load	Case I (DL+LL), F = 1.85	Case II (DL + LL + wind), F = 1.40
P_u	$(20)(1.85) = 37^k$	$(20)(1.40) = 28^k$
T_u	———	$(16)(1.40) = 22.4^k = 0.8P$

③ Plastic moment ratio Try constant section throughout

④ Independent mechanisms

(Table 7.3, Procedures 5 & 6)

Possible plastic hinges, N = 18 [Numbered sections in sketch (a)]

Redundants X = 6 [Remove supports at sections 2 & 3]

Number of independent mechanisms, n = N − X = 12

Mechanisms 1-4 : beam mechanisms

Mechanisms 5-8 : beam mechanisms

Mechanism 9 : panel mechanism

Mechanisms 10-11: gable mechanisms

Mechanism 12 : joint mechanism

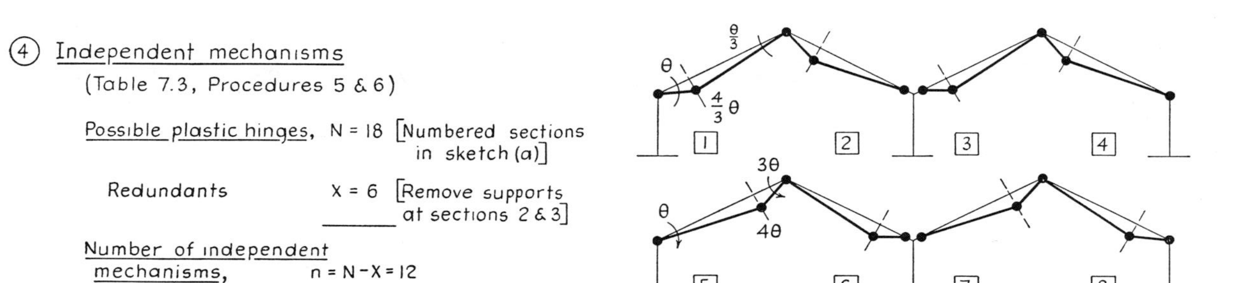

PLATE XII *(Continued)*

CASE I SOLUTION

(5) Mechanism analysis (Table 7.3, Procedure 2)

1	2	3	4	5
No.	Mechanism	Internal work $\left(W_I/M_p\theta\right)$	External work $\left(W_E/PL\theta\right)$	$\frac{M_p}{P_uL}$
1-4	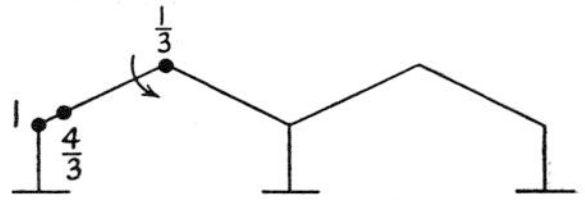	$1+\frac{4}{3}+\frac{1}{3}=\frac{8}{3}$	$\frac{1}{8}+\left(\frac{1}{8}\right)\left(\frac{1}{3}\right)=\frac{1}{6}$	$\frac{1}{16}$
5-8	(Similiar. see sketch b)	$1+4+3=8$	$\frac{1}{8}+\frac{3}{8}=\frac{1}{2}$	$\frac{1}{16}$
9	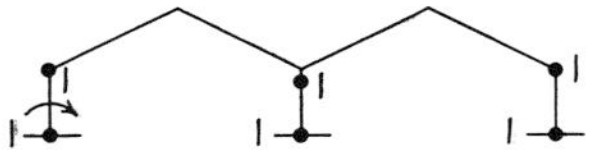	$1+1+1+1+1+1=6$	0	0
10 11	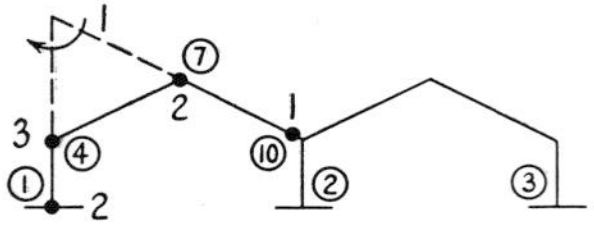	$2+3+2+1=8$	$\frac{1}{8}+\frac{3}{8}+\frac{3}{8}+\frac{1}{8}=1$	$\frac{1}{8}$
13 (9+10)	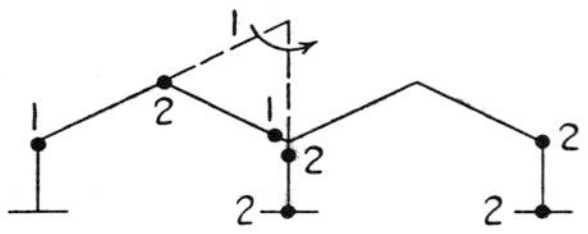	$2+2+1+2+1+2+2=12$	$\frac{1}{8}+\frac{3}{8}+\frac{3}{8}+\frac{1}{8}=1$	$\frac{1}{12}$

13a	Solution by summation of mechanism solutions	Two times Mech. 9 12 Mech. 10 8 Cancelled −8 Total 12	0 1 — 1	$\frac{1}{12}$
14 (11+12+13)	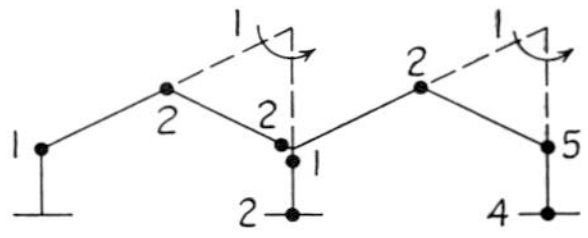	$2+4+1+2+2+1+2+5=19$	$\frac{1}{8}(1+3+3+1)(2)$	$\frac{2}{19}$
14a	Solution by summation	Mech. 11 8 Mech. 13 12 Mech. 12 3 Cancelled (section 11, 12) −4 Total 19	1 1 0 — 2	$\frac{2}{19}$

⑥ Moment check for $M_p = \frac{PL}{8}$ (Mech. 10 & 11)

Beam 7–10: $M_8 = \frac{3}{4}M_7 - \frac{M_{10}}{4} + \frac{PL}{8}$

$= +\frac{3}{4}M_p - \frac{M_p}{4} + M_p$

$= 1.5\,M_p$ [Violates]

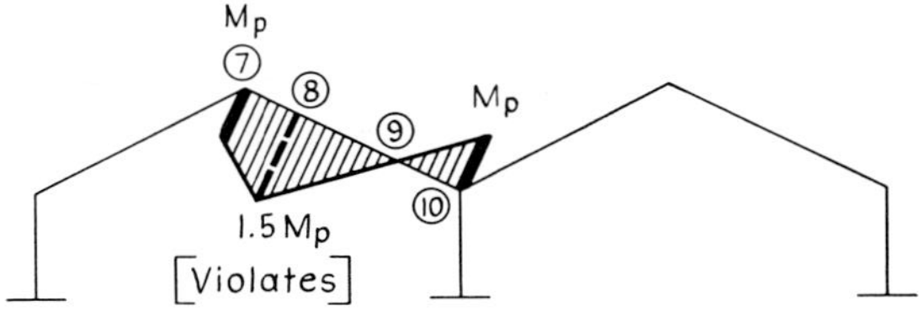

(c)

PLATE XII (Continued)

(7) Mechanism solution: (Additional)

15 (2+7+10+11)	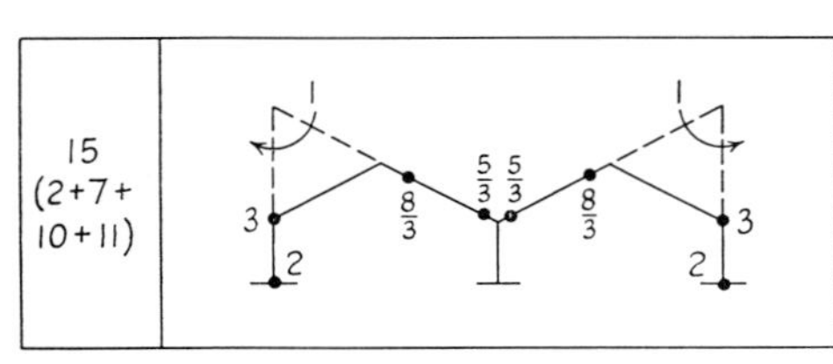	$2+3+\frac{8}{3}+\frac{5}{3}=\frac{28}{3}$	$\frac{1}{8}+\frac{3}{8}+\frac{5}{8}+\left(\frac{1}{8}\right)\left(\frac{5}{3}\right)=\frac{4}{3}$	$\frac{1}{7}$
		(Due to symmetry only one-half of frame is solved)		

(8) Moment check for mech. 15: $\left(M_p = \frac{1}{7}P_uL\right)$

Beam 7-10: $M_8 = \frac{3}{4}M_7 - \frac{1}{4}M_{10} + \frac{PL}{8}$

$$M_7 = \frac{4}{3}M_8 + \frac{1}{3}M_{10} - \frac{PL}{6}$$

$$= \frac{4}{3}M_p + \frac{1}{3}M_p - \frac{7}{6}M_p$$

$$M_7 = +\frac{M_p}{2}$$

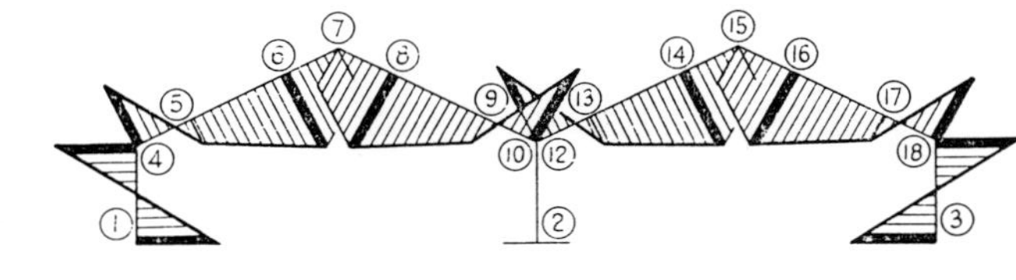

(d)

Beam 4-7: $M_6 = \frac{1}{4}M_4 - \frac{3}{4}M_7 + \frac{PL}{8}$

$$= -\frac{1}{4}M_p + \frac{3}{8}M_p + \frac{7}{8}M_p, \quad \underline{M_6 = +M_p}$$

All $M \leq M_p$

$$M_p = \frac{P_uL}{7} = \frac{(37)(80)}{7} = \underline{423^{k'}}$$

CASE II SOLUTION

(9) Mechanism solutions

9 (II)	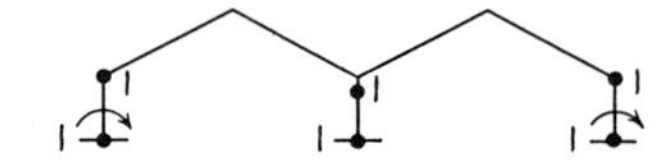	$1+1+1+1+1+1=6$	$(0.8)\left(\frac{1}{4}\right)(1)=0.2$	$\frac{1}{30}$
16 (9+10+11 +12+5+7)	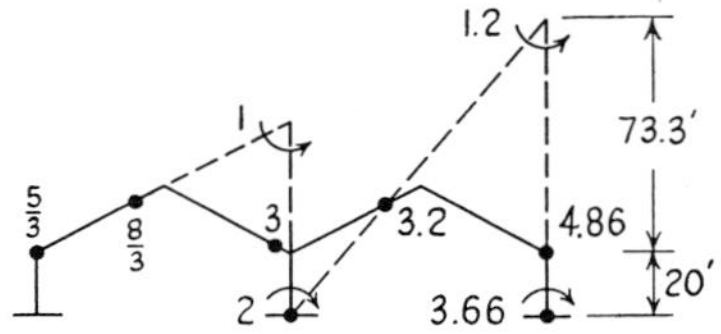	$2+3.66+\frac{5}{3}+\frac{8}{3}+3+3.2+4.86=21.06$	$\left(\frac{1}{8}\right)\left(\frac{5}{3}\right)+\left(\frac{3}{8}\right)\left(\frac{5}{3}\right)+\frac{3}{8}+\frac{1}{8}$ $+\left(\frac{1}{8}\right)(2)+\left(\frac{3}{8}\right)(2)+\left(\frac{3}{8}\right)(1.2)$ $+\left(\frac{1}{8}\right)(1.2)=2.93$	0.139.

(10) Moment check for mech. 16 $(M_p)_{II} = 0.139\,P_uL$

Beam 4-7: $M_6 = \frac{M_4}{4} - \frac{3M_7}{4} + \frac{PL}{8}$

$$M_7 = \frac{M_4}{3} - \frac{4}{3}M_6 + \frac{PL}{6} = -\frac{M_p}{3} - \frac{4}{3}M_p + \frac{M_p}{(6)(0.139)}$$

$M_7 = -0.13\,M_p < M_p$ ok

(e)

Sway: $TaL + M_1 + M_2 + M_3 + M_4 + M_{11} + M_{18} = 0$

$TaL + M_1 - M_p - M_p + M_p + 0 - M_p = 0$

$$M_1 = -TaL + 2M_p = -\frac{(0.8)M_p(20)}{(0.139)(80)} + 2M_p = +0.56\,M_p \quad ok$$

$(M_p)_{II} = 0.139\,P_uL = (0.139)(28)(80) = 312^{k'} < 423^{k'}$ Case I controls

PLATE XII (*Continued*)

(11) Reactions for case I

$$H_3 \times 20 = 2M_p$$

$$H_3 = \frac{M_p}{10} = 42.3^k$$

$$H_1 = H_3 = 42.3^k$$

$$V_1 = V_3 = 2P = 40^k$$

$$V_2 = 4P = 80^k$$

(12) Selection of section

$$Z = \frac{M_p}{\sigma_y} = (423)(0.364) = 154 \text{ in.}^3$$

Use 21 WF 68 $Z = 159.8$ in.^{3}

(13) Axial force

Center column: $\frac{P}{P_y} = \frac{V_2}{\sigma_y A} = \frac{80}{(33)(20.02)} = 0.12 < 0.15$ ok

Further steps:

Continue examination of "Design Guides"

Step 4: The 12 possible independent mechanisms are shown in sketch (*b*). Incidentally, the problem worked here is nearly identical with the procedure that would be followed if the column bases were pinned instead of fixed. In that case, X equals 3, but there are three fewer plastic hinges (sections 1, 2, and 3) and therefore 12 independent mechanisms still exist; these mechanisms are exactly the same as illustrated in sketch (*b*), except that there are no hinges at the bases.

Step 5: The mechanism solutions are conveniently developed in tabular form. Column 1 is the mechanism number. Column 2 contains a sketch of the mechanism; since the deformed shape was shown in sketch (*b*), this feature is not repeated in the small sketches.* In column 1 (for the composite mechanisms) the independent mechanisms are noted that were combined. The internal work is computed in column 3. To facilitate checking, it is desirable to list the work done at each hinge in the same sequence as the numbering given in sketch (*a*). Column 4 contains the computation of external work, listing the work done by each load in the same sequence. M_p in terms of P_uL is given in column 5.

To illustrate, for mechanism 10, the mechanism angles at sections 1, 4, 7, and 10 are 2θ, 3θ, 2θ, and 1θ, respectively. Thus the total internal work is $8M_p\theta$, or $W_I/M_p\theta = 8$ (see column 3). Using the instantaneous center, the load at 5 does work equal to $(P)[(L/8)\theta]$; that at 6 equal to $P[(3L/8)\theta]$. Segment 7–10 rotates about 10 through the angle θ, and therefore the work done by the load at 8 equals $(P)[(3L/8)\theta]$ and that done by the load at 9 equals $(P)[(L/8)\theta]$. Then $(\frac{1}{8} + \frac{3}{8} + \frac{3}{8} + \frac{1}{8})PL\theta = 1PL\theta$, or, as shown in column 4, $W_E/PL\theta = 1$. Equating W_I to W_E (column 5), one obtains

$$8M_p\theta = 1PL\theta$$

$$M_p = \frac{PL}{8}$$

Turning now to possible combinations, these are made in such a way as to eliminate plastic hinges, because only by this means can the ratio M_p/P_uL be increased. (It will be remembered that the largest possible value of M_p is being sought.) Mechanism 13 is formed by combining mechanisms 9 and 10. Hinges will be eliminated at sections 1 and (in part) at 4 only if $\theta_9 = 2\theta_{10}$, and mechanism 13 is sketched accordingly. The result is still less than mechanism 10 alone.

Mechanism 13*a* is the same as mechanism 13, except that the solution is obtained by summation of work equations for the independent mechanisms as described in Chapter 3. The combination eliminates mechanism angles of 2θ at sections 1 and 4 of mechanism 10 and of 2θ at the same sections of mechanism 9 ("cancel $8M_p$"). The same answer is thus obtained as by the first method.

Step 6: Following the analysis of mechanism 14, a moment check was made to see if mechanism 10 was critical. For beam 7–10 it was found to be the in-

* For convenience the number indicating the magnitude of the mechanism angle is shown on the side of the number deformed in tension. See also step 4.

correct answer, because the plastic moment value is $1.5M_p$ at section 8. This suggests a combination of mechanisms 10, 11, 2, and 7, and the resulting mechanism 15 does, in fact, give the correct answer as shown in sketch *d*, step 8. The required M_p value is 423 kip-ft.

Step 9: Case II loading requires the recalculation of only one of the independent mechanisms, namely mechanism 9. The resulting value of M_p is comparatively very low.

Mechanism 16 is next investigated, consisting of a combination of mechanisms 9, 10, 11, 12, 5, and 7. The moment check then follows in step 10.

Step 10: Since the frame is determinate at failure ($I = X - (M - 1) = 6 - (7 - 1) = 0$), a possible equilibrium moment diagram may be obtained without difficulty (sketch *e*). It was obtained by plotting the known M_p values (sections 2, 3, 4, 6, 10, 14, and 18) and solving first for the moment at section 7 ($M_7 = -0.13M_p$ in segment 4–7). Since $M_{10} = M_p$, the moment at section 8 also equals M_p, so the moment diagram may be completed for rafter 4–7–10. If the trial-and-error method is used (Art. 3.8, p. 85) one can assume $M_{11} = 0$. Hence $M_{12} = M_p$, and the moment diagram for the right-hand span would be identical to the left-hand span. Making use of the sway equilibrium equation (similar to Eq. 7.12), the moment at section 1 is $0.56\ M_p$. Therefore $M \leq M_p$ throughout and the value $M_p = 0.139PL$ is correct for case II loading.

Since $(M_p)_{II} < (M_p)_I$, the case I loading controls the design and $M_p = 423$ kip-ft.

Step 11: The horizontal and vertical reactions at the column bases are computed by statics, advantage being taken of symmetry in computing the vertical reactions.

Step 12: A 21WF68 shape supplies the needed plastic modulus.

Step 13: The center column must support the greatest axial force, even though a plastic hinge does not form there under case I loading. Since $P/P_y < 0.15$, neither it, nor the other two columns are critical.

The problem would be completed with an examination of the remaining appropriate design guides.

PLATE XIII Three-Span Unsymmetrical Gabled Frame

An unsymmetrical three-span frame with gabled roofs will now be designed, making use of the mechanism method of analysis.

Step 1: The three-span structure is shown in sketch (*a*). The column bases are pinned in this example, but the procedures would be quite similar had they been fixed, and a somewhat smaller size of member would result. The center span is the largest and most heavily loaded and therefore will require the largest size of member. The left span is 80 ft long and the right is 60 ft in length. All column heights are the same (20 ft) but the roof rises vary to give a more uniform appearance of the structure.

The concentrated loads of 40, 50, and 30 kips, are in proportion to the corresponding span lengths. A side load of 20 kips (due to wind) is assumed to act at the eaves at section 5.

Step 2: For unsymmetrical frames, there are three loading conditions: dead load plus live load and dead load plus live load in combination with wind acting first from the left and then from the right.

As a preliminary start to a selection of the proper relationship between plastic moments for the different spans, the plastic moment ratios are selected to be in proportion to the square of the span lengths. The reason for this is that the loads are in proportion to the span lengths, and therefore the bending moments will be in proportion to the square of these lengths. The exterior columns are equal to the corresponding rafter sizes and the inside columns are made equal to the difference in plastic moment values of the adjoining rafters. By expressing the moment capacity in terms of k_1M_p, k_2M_p, and k_3M_p, respectively, it will be seen later that the ratios may be adjusted conveniently to improve the over-all design of the frame.

Since the loading case for vertical load alone usually controls the design of multi-span frames, whatever ratios are found suitable for case I will be used for analyzing cases II and III.

Step 4: The twelve independent mechanisms for this frame are shown in sketch (*b*). Mechanism 10, a "gable-panel" mechanism is a consequence of the failure of span 2. The gable mechanism for this span cannot form without one of the interior columns pushing outward.

Step 5: The mechanism solutions are carried out in the tabular form shown in the Plate. In computing the expressions for internal work, the ratios k_1, k_2, and k_3 are retained for use in subsequently obtaining a more effective design.

In order to simplify the sketches in this table, the term θ has been omitted, it being understood that the virtual mechanism angle in every case is equal to the number written by the hinge times θ. For example in mechanism 1, the mechanism angles are 1θ, 2θ, and 1θ, respectively. In this problem, the deformed shape of the structure is not shown, but is represented by the arrows indicating the direction of rotation of a segment about the appropriate hinge position or the instantaneous center. For convenience the mechanism angle at the instantaneous center is selected as 1θ. As before, mechanism angles are shown on the tension side of the hinges.

To illustrate entries made in the tabulation, the calculation of the work equation for mechanism 10 is given in detail as follows. The instantaneous center at I is first located, from which the various mechanism angles may be computed. The internal work done at each hinge is next recorded in the sequence of joint numbers (sketch *a*). The internal work done at sections 11, 13, 15, 16, and 21 is equal to the summation of the products of the particular plastic moment value times the corresponding mechanism angle, or

$$W_I = (k_2M_p)(\theta) + (k_2M_p)(2\theta) + (k_2M_p)(\theta) + (k_1M_p)(2\theta) + (k_3M_p)(2\theta)$$

or as shown in the tabulation,

$$\frac{W_I}{M_p\theta} = k_2(1 + 2 + 1) + k_1(2) + k_3(2)$$

PLATE XIII. THREE-SPAN UNSYMMETRICAL GABLED FRAME

① Structure and loading

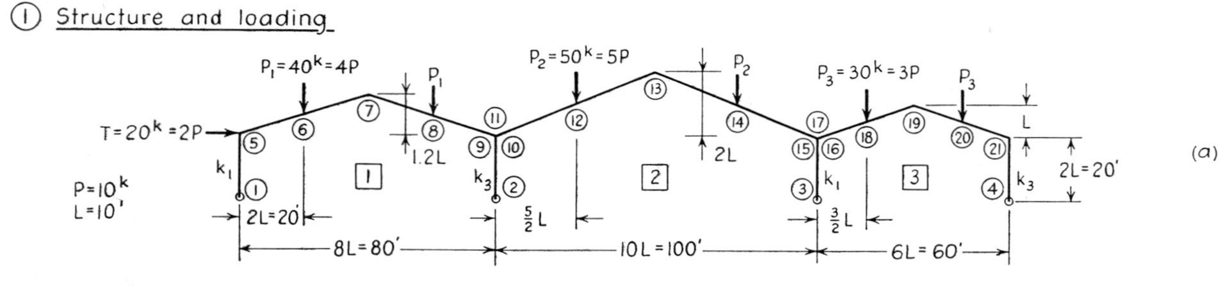

② Loading conditions (Table 7.1, Provision 6)

Load	Case I (DL + LL) F = 1.85	Case II (DL + LL + wind from left) Case III (DL + LL + wind from right) } F = 1.40
P_u $T_u = 2P_u$	10 X 1.85 = 18.5 ———	10 X 1.40 = 14.0 20 X 1.40 = 28.0

③ Plastic moment ratios (Table 7.2, Procedure 1)

Try plastic moments in ratio of square of span length, based on span 3

Span 1: $k_1 = \left(\frac{80}{60}\right)^2 = 1.78$

Span 2: $k_2 = \left(\frac{100}{60}\right)^2 = 2.78$

Span 3: $k_3 = 1.00$

Column 2-10 : $k = k_2 - k_1 = 2.78 - 1.78 = k_3$

Column 3-16 : $k = k_2 - k_3 = 2.78 - 1.00 = k_1$

NOTE Revisions may be made later as per failure mechanism

(4) Independent mechanisms (Solution by mechanism method) (Table 7.3, Procedure 2)

Possible plastic hinges, N = 17 (All numbered sections except 1-4)

Redundants X = 5 (Remove two supports each at 2 and 3 and remove H_4

Number of independent mechanisms, n = N − X = 12

Beam mechanisms	1-6
Panel mechanism	7
Gable mechanisms	8, 9
Gable-panel mechanism	10
Joint mechanisms	11, 12

(b)

PLATE XIII (*Continued*)

CASE I SOLUTION

⑤ Mechanism solutions

No.	Mechanism	Internal work $\left(W_I/M_p\theta\right)$	External work $\left(W_E/PL\theta\right)$	$\frac{M_p}{P_uL}$
1, 2	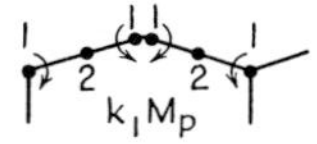 k_1M_p	$k_1(1+2+1) = 4k_1$ $= 4(1.78) = 7.12$	$(4)(2) = 8$	1.12
3, 4	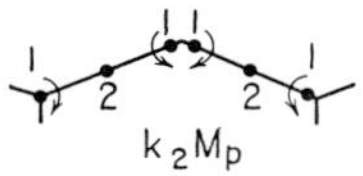 k_2M_p	$k_2(1+2+1) = 4k_2$ $= (4)(2.78) = 11.11$	$(5)(2\frac{1}{2}) = 12.5$	1.12
5, 6	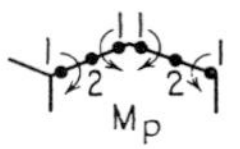 M_p	$k_3(1+2+1) = 4k_3 = 4.00$	$(3)(1\frac{1}{2}) = 4.5$	1.12
8	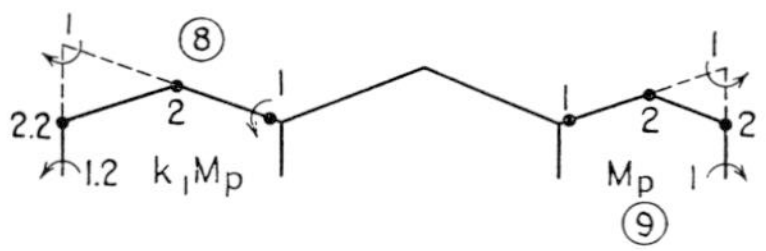 ⑧ k_1M_p	$k_1(2.2+2+1) = 5.2k_1$ $= (5.2)(1.78) = 9.25$	$(4)(2)+(4)(2) = 16$	1.73
9	⑨ M_p	$1+2+2 = 5$	$(3)(1\frac{1}{2})+(3)(1\frac{1}{2}) = 9$	1.80
10	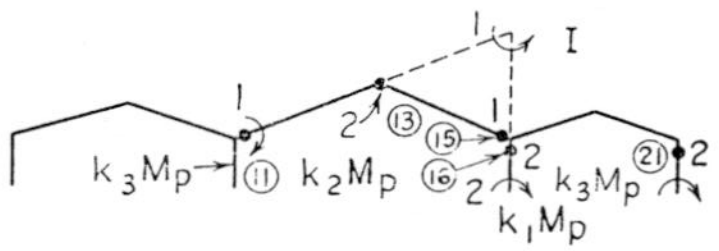 k_3M_p, k_2M_p, k_3M_p, k_1M_p	$k_2(1+2+1)+k_1(2)+k_3(2) =$ $4k_2+2k_1+2k_3 = 4(2.78)+2(1.78)+2$ $= 16.68$	$(5)(2\frac{1}{2})+(5)(2\frac{1}{2}) = 25$	1.50

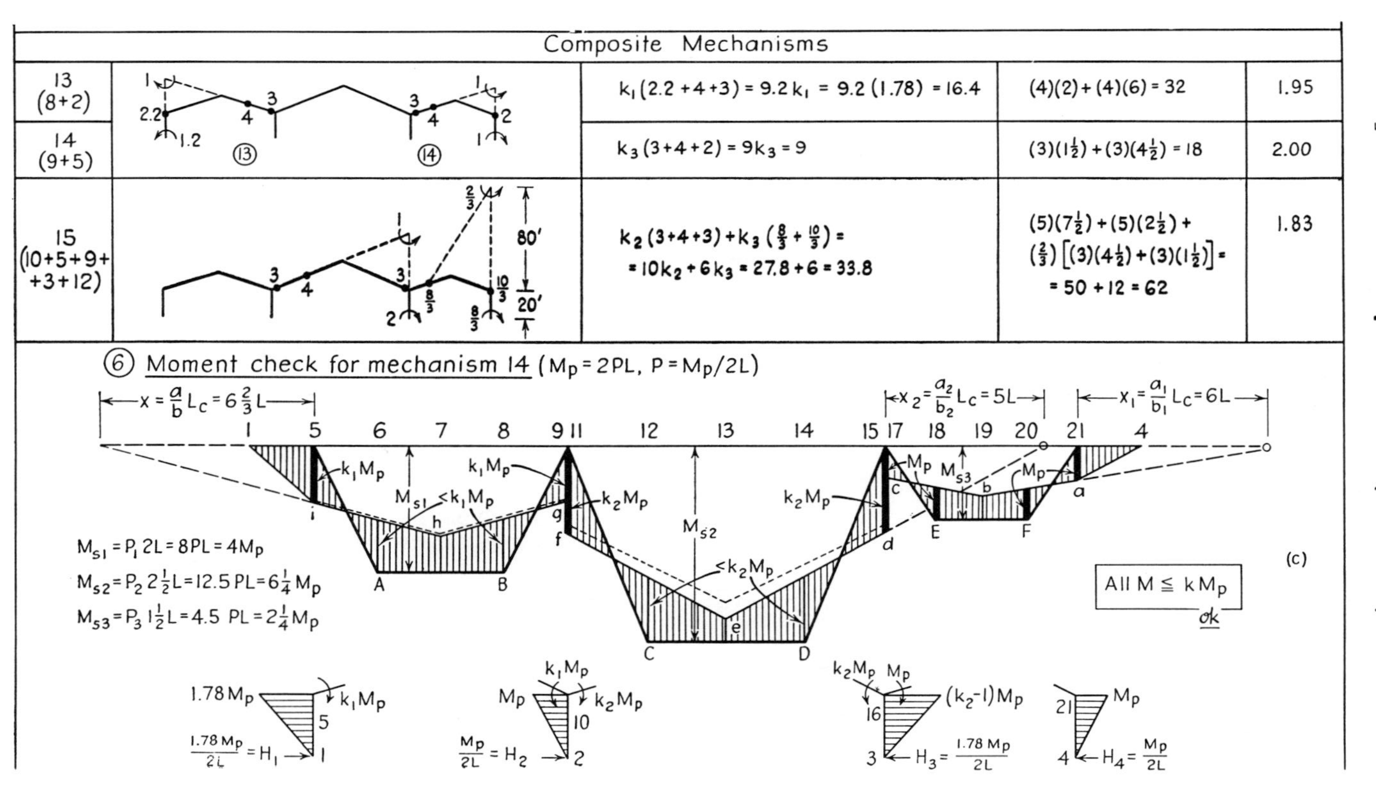

Composite Mechanisms				
13 (8+2)	⑬	$k_1(2.2+4+3) = 9.2k_1 = 9.2(1.78) = 16.4$	$(4)(2)+(4)(6) = 32$	1.95
14 (9+5)	⑭	$k_3(3+4+2) = 9k_3 = 9$	$(3)(1\frac{1}{2})+(3)(4\frac{1}{2}) = 18$	2.00
15 (10+5+9+ +3+12)		$k_2(3+4+3)+k_3(\frac{8}{3}+\frac{10}{3}) =$ $= 10k_2+6k_3 = 27.8+6 = 33.8$	$(5)(7\frac{1}{2})+(5)(2\frac{1}{2})+$ $(\frac{2}{3})[(3)(4\frac{1}{2})+(3)(1\frac{1}{2})] =$ $= 50+12 = 62$	1.83

PLATE XIII (*Continued*)

(7) Reformulation of M_p ratios

Select k values such that $P_{13} = P_{14} = P_{15}$, with $\underline{k_3 = 1.00}$

Mech. 13:	$9.2 k_1 M_p = 32 PL$	(1)
Mech. 14:	$9 k_3 M_p = 18 PL$	(2)
	$\therefore 9.2 k_1 M_p = 16 k_3 M_p$, $\underline{k_1 = 1.74}$	
Mech. 15:	$10 k_2 M_p + 6 M_p = 62 PL$	(3)
Solving (2)&(3):	$10 k_2 M_p + 6 M_p = \frac{(62)(9)(1.0)}{18} M_p = 31 M_p$	
	$10 k_2 = 25$, $\underline{k_2 = 2.50}$	

(8) Revised mechanism solutions

13*a*	*Same as* 13, but $k_1 = 1.74$	$(9.2)(1.74) = 16$	32	2.00
15*a*	*Same as* 15 except new values of k. $k_1 = 1.74$, $k_2 = 2.50$	$k_2(10) + k_3(6) = 25 + 6 = 31$	62	2.00 ok

(9) Moment check for mechanisms 13 & 15 (Revised)($M_p = 2PL$) Hinges *at* sections 5, 8, 9, 11, 12, 15, 18, 19

NOTE: Reference may be made to moment *diagram* of sketch (c). The following method is a straight forward use of equilibrium Eqs. Dotted line in spans 1 & 2 shows *equivalent graphical* solution.

Column 1-5: $H_1 = M_5/2L = \underline{1.74\,M_p/2L = H_1}$

Column 4-21: $H_4 = M_{21}/2L = \underline{M_p/2L = H_4}$

Beam 7-9: $M_7 = 2M_8 + M_9 - M_5 = 2k_1 M_p + k_1 M_p - \frac{(4P)(4L)}{2} = 3k_1 M_p - 4M_p = +1.22\,M_p$ ok

Beam 5-7: $M_6 = k_1 M_p$ (by symmetry with beam 7-9)

Beam 11-13: $M_{12} = \frac{M_{11}}{2} - \frac{M_{13}}{2} + M_5, \quad M_{13} = M_{11} - 2M_{12} + \frac{(5P)(5L)}{2} = -k_2 M_p - 2k_2 M_p + \frac{25}{4} M_p = -0.75\,M_p$ ok

Beam 13-15: $M_{14} = M_p$ (Symmetrical with beam 11-13)

Joint 9-10-11: $M_9 + M_{10} + M_{11} = 0, \quad M_{10} = -M_9 - M_{11} = k_1 M_p - k_2 M_p = -0.76\,M_p$ ok

Column 2-10: $H_2 = \frac{M_{10}}{2L} = \underline{0.38}\,\frac{M_p}{L} = \underline{H_2}$

$\Sigma H = 0$: $H_1 + H_2 = H_3 + H_4, \quad H_3 = H_1 + H_2 - H_4 = M_p/L\,(0.87 + 0.38 - 0.5) = \underline{0.75}\,\frac{M_p}{L} = \underline{H_3}$

Column 3-16: $H_3(2L) = M_{16}, \quad M_{16} = 0.75\,\frac{M_p}{L}(2L) = \underline{1.50\,M_p = M_{16}}$

Joint 15-16-17: $M_{17} = -M_{15} - M_{16} = -1.50\,M_p + 2.50\,M_p = +M_p$ ok

Beam 17-19: $M_{18} = M_{17}/2 - M_{19}/2 + M_5, \quad M_{19} = M_{17} - 2M_{18} + \frac{(3P)(3L)}{2} = -3M_p + \frac{9}{4} M_p = -0.75\,M_p$ ok

Beam 19-21: $M_{20} = M_p$ (Symmetry)

All $M \le k M_p$. $(M_p)_I = 2PL = 2(18.5)(10) = 370^{k'}$
$k_1 M_p = 644^{k'}$
$k_2 M_p = 925^{k'}$

PLATE XIII (*Continued*)

CASE II SOLUTION

Independant mechanisms: see step 4

(10) Mechanism solutions

7	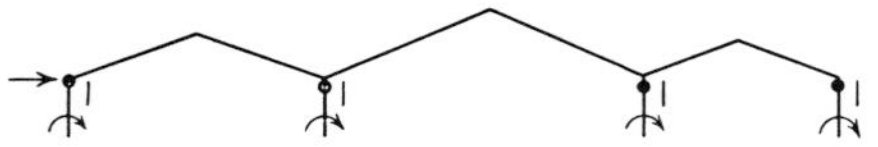	$k_1 + k_3 + k_1 + k_3 =$ $2(1\ 74) + 2 = 5.48$	$(2)(2) = 4$	0.73
16	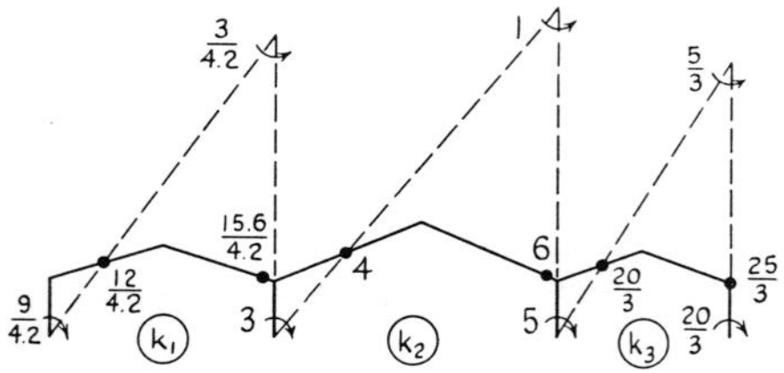	$k_1\left(\frac{12}{4.2} + \frac{15.6}{4.2}\right) + k_2\,(4+6) +$ $+ k_3\left(\frac{20}{3} + \frac{25}{3}\right) =$ $(1.74)\left(\frac{27.6}{4.2}\right) + (2.50)(10) + 15 = 51.4$	$\left(\frac{9}{4.2}\right)(2)(2)$ $\frac{3}{4.2}\,(4)(6+2) +$ $+1\,(5)\left(7\frac{1}{2} + 2\frac{1}{2}\right) +$ $+\left(\frac{5}{3}\right)(3)\left(4\frac{1}{2} + 1\frac{1}{2}\right) = 111.4$	2.17

Critical mechanism: No. 16, $M_p = 2.17\,PL$

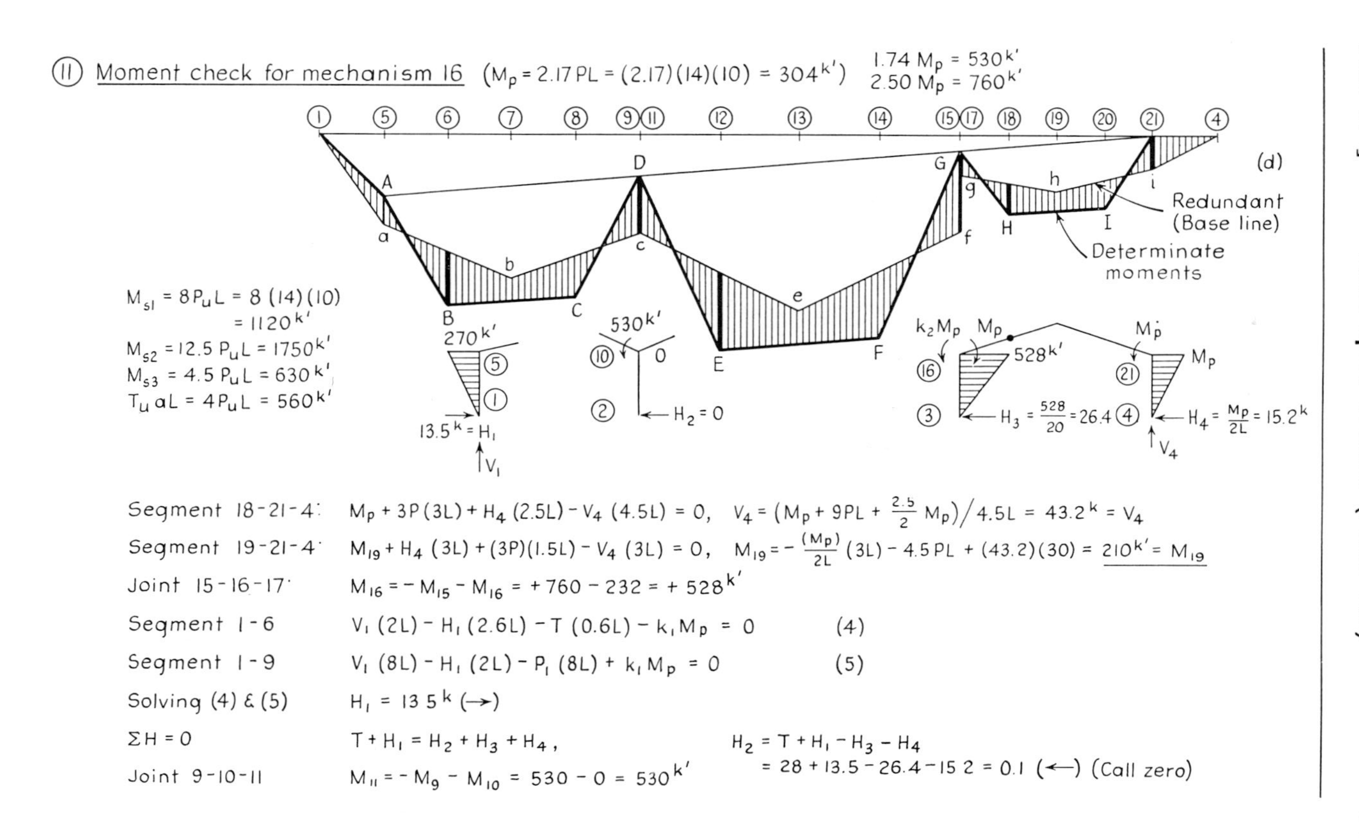
(11) Moment check for mechanism 16 ($M_p = 2.17PL = (2.17)(14)(10) = 304^{k'}$)
$1.74\,M_p = 530^{k'}$
$2.50\,M_p = 760^{k'}$
1 5 6 7 8 9 11 12 13 14 15 17 18 19 20 21 4
A B C D E F G H I
a b c e f g h i
(d)
Redundant (Base line)
Determinate moments
$M_{s1} = 8P_uL = 8\,(14)(10) = 1120^{k'}$
$M_{s2} = 12.5\,P_uL = 1750^{k'}$
$M_{s3} = 4.5\,P_uL = 630^{k'}$
$T_u aL = 4P_uL = 560^{k'}$
$270^{k'}$
$13.5^k = H_1$
V_1
$530^{k'}$
10 0 2
$H_2 = 0$
k_2M_p M_p $528^{k'}$
16 3
$H_3 = \frac{528}{20} = 26.4$
M_p 21 4
$H_4 = \frac{M_p}{2L} = 15.2^k$
V_4
Segment 18-21-4: $M_p + 3P(3L) + H_4(2.5L) - V_4(4.5L) = 0$, $V_4 = (M_p + 9PL + \frac{2.5}{2}M_p)/4.5L = 43.2^k = V_4$
Segment 19-21-4: $M_{19} + H_4(3L) + (3P)(1.5L) - V_4(3L) = 0$, $M_{19} = -\frac{(M_p)}{2L}(3L) - 4.5PL + (43.2)(30) = 210^{k'} = M_{19}$
Joint 15-16-17: $M_{16} = -M_{15} - M_{16} = +760 - 232 = +528^{k'}$
Segment 1-6 $V_1(2L) - H_1(2.6L) - T(0.6L) - k_1M_p = 0$ (4)
Segment 1-9 $V_1(8L) - H_1(2L) - P_1(8L) + k_1M_p = 0$ (5)
Solving (4) & (5) $H_1 = 13\,5^k\,(\rightarrow)$
$\Sigma H = 0$ $T + H_1 = H_2 + H_3 + H_4$, $H_2 = T + H_1 - H_3 - H_4 = 28 + 13.5 - 26.4 - 15\,2 = 0.1\,(\leftarrow)$ (Call zero)
Joint 9-10-11 $M_{11} = -M_9 - M_{10} = 530 - 0 = 530^{k'}$

PLATE XIII (*Continued*)

Moment check complete for case II. mechanism 16 controls.

$$(M_p)_{II} = 2.17 P_u L = (2.17)(14)(10) = \boxed{304^{k'} = (M_p)_{II}}$$

CASE III SOLUTION

(12) Mechanism solution (Wind from right)

17	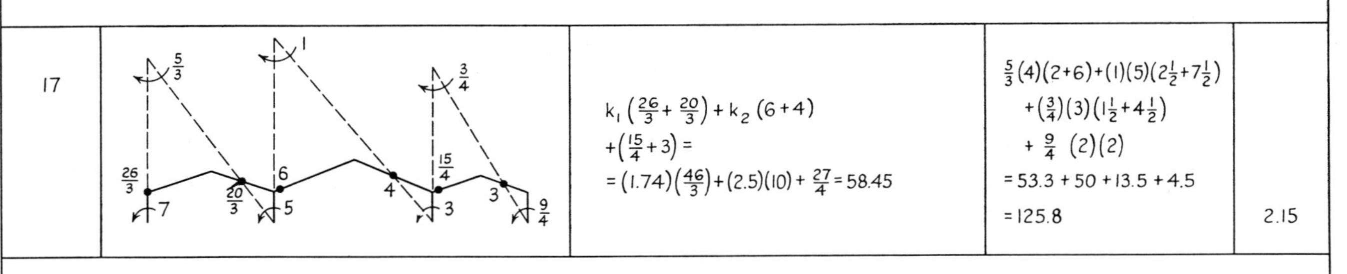	$k_1\left(\frac{26}{3}+\frac{20}{3}\right)+k_2(6+4)$ $+\left(\frac{15}{4}+3\right)=$ $=(1.74)\left(\frac{46}{3}\right)+(2.5)(10)+\frac{27}{4}=58.45$	$\frac{5}{3}(4)(2+6)+(1)(5)(2\frac{1}{2}+7\frac{1}{2})$ $+(\frac{3}{4})(3)(1\frac{1}{2}+4\frac{1}{2})$ $+\frac{9}{4}(2)(2)$ $=53.3+50+13.5+4.5$ $=125.8$	2.15

Moment check for mechanism 17

This procedure would be the same as in step 5. (*Not repeated here. It is found not to be the controlling case.*)

(13) <u>Reactions for case I</u> $(M_p = 370^{k'})$

$$H_1 = \frac{1.74\,M_p}{2L} = 0.087\,M_p = (0.087)(370) = 32.2^k$$

$$H_2 = \frac{0.38\,M_p}{L} = 0.038\,M_p = (0.038)(370) = 14.1^k$$

$$H_3 = \frac{0.75\,M_p}{L} = 0.075\,M_p = 28.5^k$$

$$H_4 = \frac{M_p}{2L} = 0.05\,M_p = 18.5^k$$

Since the moment diagram for every span is symmetrical, the vertical reactions for each span are symmetrical.

$$V_1 = P_1 = 4P_u = (4)(18.5) = 74^k$$

$$V_2 = P_1 + P_2 = 4P_u + 5P_u = 166.5^k$$

$$V_3 = P_2 + P_3 = 5P_u + 3P_u = 148^k$$

$$V_4 = P_3 = 3P_u = 55.5^k$$

PLATE XIII (*Continued*)

(14) Selection of sections

(Table 7.1, Provision 5)

$k = k_1$ — Span 1

Left column, Rafter: $Z_1 = \frac{1.74 M_p}{\sigma_y} = (1.74)(370)(0.364) = 234 \text{ in.}^3$

Column 3-16: $Z = Z_1 = 234 \text{ in.}^3$

Use 24WF94, $Z = 253 \text{ in.}^3$

$A = 27.63$
$d = 24.29$
$b = 9.06$
$t = 0.872$
$w = 0.516$

$k = k_2$. Span 2

Rafter: $Z_2 = \frac{2.5 M_p}{\sigma_y} = (2.50)(370)(0.364) = 336 \text{ in.}^3$

Use 30WF108, $Z = 344.5 \text{ in.}^3$

$A = 31.77$
$d = 29.82$
$b = 10.48$
$t = 0.76$
$w = 0.548$

$k = k_3$ — Span 3

Rafter, Right column: $Z_3 = \frac{M_p}{\sigma_y} = (370)(0.364) = 135 \text{ in.}^3$

Column 2-10: $Z = Z_3 = 135 \text{ in.}^3$

(Revised later to 21WF68. Step 15)

Use 21WF62, $Z = 144.1 \text{ in.}^3$

$A = 18.23$
$d = 20.99$
$b = 8.24$
$t = 0.615$
$w = 0.400$

(15) <u>Axial force</u>

(Table 7.4, Guide 1)

Column 1-5: (24 WF 94)

$$\frac{P}{P_y} = \frac{V_1}{\sigma_y A_1} = \frac{74}{(33)(27.63)} = 0.081 < 0.15 \quad \underline{ok}$$

Column 2-10. (21 WF 62)

$$\frac{P}{P_y} = \frac{V_2}{\sigma_y A_3} = \frac{166.5}{(33)(18.23)} = 0.277 > 0.15 \quad \text{[Modification required]}$$

$$Z = Z_t \left[\frac{P}{P_y} + 0.85\right]$$

$$= 135\ (0.277 + 0.85) = 152 \text{ in}^3$$

Use 21WF68 $Z = 159.8 \text{ in.}^3$

Column 3-16: (24 WF 94)

$$\frac{P}{P_y} = \frac{V_3}{\sigma_y A_1} = \frac{148}{(33)(27.63)} = 0.163 > 0.15 \quad \text{[Modification may be required]}$$

$$Z = Z_t \left[\frac{P}{P_y} + 0.85\right]$$

$$= 234\ (0.163 + 0.85) = 236 < 253 \quad \underline{ok}$$

Column 4-21: (21 WF 62)

$$\frac{P}{P_y} = \frac{V_4}{\sigma_y A_3} = \frac{55.5}{(33)(18.23)}$$

$$= 0.0925 < 0.15 \quad \underline{ok}$$

PLATE XIII (*Continued*)

(16) Connection detail (Joint 15-16-17)
(Table 7.7, Guide 10)

$$w = \frac{0.6\,\Delta M}{d_c\, d_b} \qquad (5.49)$$

$$\Delta M = M_{15} + M_{17} = -925 + 370 = 555^{k'}$$

$$d_c = d_{16} = 24.29''$$

$$d_b = d_{15} = 29.82''$$

$$w = \frac{(0.6)(555)}{(24.29)(29.82)} = 0.460 \qquad < 0.516'' \text{ (Web of column)} \quad \underline{ok}$$

Further steps: Check remaining "Design Guides"

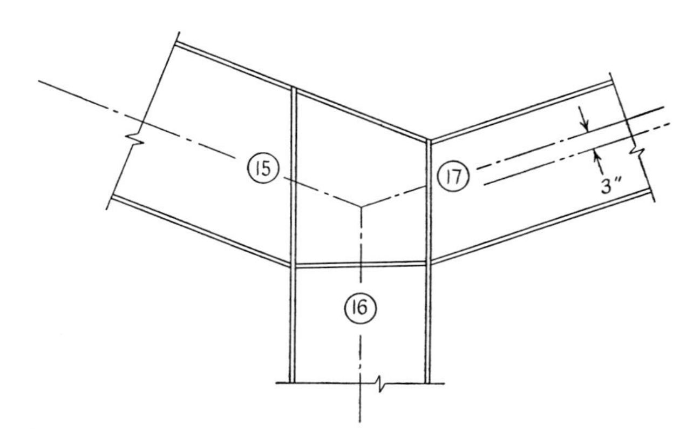

The external work is equal to the load at 12 times the displacement at 12 plus the load at 14 times the displacement at 14, or

$$W_E = (5P)(\tfrac{5}{2}L)(\theta) + (5P)(\tfrac{5}{2}L)(\theta)$$

or as shown in the tabulation

$$\frac{W_E}{PL\theta} = (5)(\tfrac{5}{2}) + (5)(\tfrac{5}{2})$$

Using the k values assumed in step 3 and equating W_I to W_E there is obtained

$$16.68M_p\theta = 25P_uL\theta$$

$$M_p = 1.50P_uL$$

Mechanisms 13, 14, and 15 are composite mechanisms. Since mechanism 14 requires the largest value of M_p, a moment check is next made in step 6 by the "semi-graphical" method.

Step 6: It is found that the plastic moment values are not exceeded and therefore the solution is correct for this selection of k values.

The moment diagram (which constitutes the moment check) is shown in sketch (c). The construction (suitable for structures with this failure mechanism) is started by drawing out the base line 1–4. The determinate moment diagram 5–A–B–9, 11–C–D–15, and 17–E–F–21 is drawn next. The known moments for mechanism 14 are at sections 18 and 21, so these are laid off (downward from 21 and upward from E), the values being M_p. Next the pole point for the redundant moment diagram is located by the methods described in Table 7.3, Procedure 1, and line a–b can be drawn. Since the moment at section 18 is known, line b–c may be constructed. It is seen that the moment at sections 17 and 20 is also equal to M_p. Therefore the plastic moment condition is not violated in this span.

Now, the remaining portion of the structure is redundant, but the moment check may be continued assuming the magnitude of certain moments so long as the equilibrium condition at the joints is not violated, sway equilibrium is maintained and all $M \leq kM_p$. For example, the next step was to assume that M_{15} was the maximum possible value (k_2M_p). A new pole point can be found for the redundant diagram for span 2, the same formula applies for this failure mechanism (Eq. 7.8), and x_2 is found to be $5L$. Then line d–e may be drawn and it is seen that no moment is greater than k_2M_p. Line e–f will have a slope opposite to that of d–e.

The redundant diagram g–h–i is constructed in a similar manner, the moment at section 5 being assumed at the maximum possible value (k_1M_p). It is found that the moment at 9 is also equal to k_1M_p. Taking joint equilibrium, M_{10} must equal $k_2M_p - k_1M_p = M_p$. Thus the plastic moment condition is not violated for span 1.

As a check, the algebraic sum of the horizontal reactions should equal zero, and this is found to be the case.

Step 7: The analysis of case I could stop here and the design continued with an investigation of case II loading. In fact, since the M_p/P_uL values for mechanisms 13, 14, and 15 are reasonably close to one another, it is clear that effective use of the material has been made even though spans 1 and 2 remain redundant. (Span 1 is on the verge of failure—compare 1.95 with 2.0—and span 2 has somewhat more reserve strength.)

However, to make better use of the material, a reformulation of the M_p ratios will be made to obtain a more "complete" failure mechanism. For simultaneous failure of spans 1 and 3, the ultimate load computed for mechanism 13 (P_{13}) should equal the ultimate load for mechanism 14 (P_{14}). Similarly, P_{15} should equal P_{14}. Since k_3 is taken as unity, the resulting work equations for mechanisms 13, 14, and 15 enable us to find a relationship among k_1, k_2, and k_3, and this has been solved in this step. It is found that $k_1 = 1.74$ and $k_2 = 2.50$.

Step 8: A recalculation of mechanisms 13 and 15 gives $M_p = 2.00P_uL$, which is a check on the calculation, since the strength of these frames was made equivalent with that of the end span. A moment check is then completed for these mechanisms and the answer is found to be correct. The moment diagram is obtained by statics using Eqs. 7.10–7.12. The frame is overdeterminate at failure ($I = X - (M - 1) = 5 - (8 - 1) = -2$) because the failure of span 1 could occur quite separately from that of spans 2 and 3 acting together.

Another method of making the moment check would be by the graphical method. This is shown on the moment diagram of sketch (*c*) by the dotted lines. Both approaches show that the plastic moment condition is not violated. Since the basic construction of sketch (*c*) had already been completed, the graphical method is by far the most rapid in this instance. However, for illustration the moment check using the statical equilibrium equations is also shown.

Step 10: Pinned-base frames will ordinarily fail in the manner shown by mechanism 16 and therefore this case is investigated. $M_p = 2.17P_uL$ for this mechanism, and the moment check is made on this basis, since it is the maximum M_p value of all mechanisms.

Step 11: For the purpose of illustrating different techniques for making the moment check, the moment diagram for mechanism 16 is constructed, in part, with the aid of the equilibrium equations and in part by graphical construction using the known values of determinate moments M_{s1}, M_{s2}, M_{s3}, and T_uaL.

The determinate moments are first drawn to scale (line 1–*A*–*B*–*C*–*D*–*E*–*F*–*G*–*H*–*I*–21). The objective is to draw a possible fixing line which will not violate the plastic moment condition in any span. Next the known moments at plastic hinge locations are laid off to scale (sections 6, 9, 12, 15, 18, 21). V_4 and then M_{19} are determined from the equilibrium equations which makes it possible to construct *h*–*i* and *g*–*h*. M_{17} may then be obtained from the graph (230 kip-ft) or computed from the beam equilibrium equation (232 kip-ft).

At the left end of the frame, equilibrium equations are again used to determine that $H_1 = 13.5$ kips. This determines M_5 and hence line *a*–*b* can be drawn. Since M_9 is known, *b*–*c* may be drawn, and it is seen that $M \leq k_1M_p$ in span 1. M_{10} is next computed and this makes it possible to draw *c*–*e* and *e*–*f*, completing the moment check.

Step 12: The design for wind acting from the right (mechanism 17) shows that $(M_p)_{III} = 2.15\ P_uL$. A moment check is not made, but would follow a pattern similar to step 5.

Step 13: Case I controls the design, and the reactions are computed for this condition. "Symmetrical" failure mechanisms for each span facilitate the calculation of vertical reactions.

Step 15: In step 14 the sections were selected neglecting at first the effect of axial force. In only one case was the P/P_y ratio significantly in excess of 0.15. In the redesign of column 2–10 a 21WF68 was selected in lieu of the 21WF62 selected on the trial basis.

Step 16: The only remaining design guide that is illustrated in this example is an examination of the web at joint 15–17. According to Eq. 5.49 the web of the 24WF94 is found to be adequate.

9.5 SIMPLIFIED PROCEDURES FOR MULTI-SPAN FRAMES

When multi-span single story frames are considered, Ref. 7.4 makes possible an even greater saving in design time. Again, representation of the equilibrium equations in the form of charts may be used to facilitate the solution of these more complex problems.

Consider the two-span, flat-roof frame shown in Fig. 9.5*a*. It is the same structure, in fact, that was studied in Plate XI, p. 314. As the frame attains its ultimate load, the usual mode of failure will be that shown in sketch (*b*). The frame may be divided into two separate structures (or subassemblages) as shown in sketch (*c*) without changing the total internal and external work. The work done by the moments and forces as the two subassemblages move through a virtual displacement becomes zero when "continuity" is later restored at the cut section. The problem may be simplified still further by replacing all overturning forces and moments by imaginary moments acting about the column bases. The resulting separate structures (which are equivalent to the original structure) are shown in sketch (*d*).

Charts may then be prepared for the general case shown in sketch (*d*) of Fig. 9.5 just as described before. Panel A is given a virtual displacement and the corresponding work equation is written. It takes the following form:[7.4]

$$\frac{M_p}{wL^2} = \frac{1}{4}\left[\frac{\left(1 - \frac{x}{L}\right)\left(\frac{x}{L} + C - D\right) - 2DQ\frac{x}{L}}{1 + Q\frac{x}{L}}\right] \tag{9.6}$$

with

$$\left.\begin{aligned} x &= \frac{L}{Q}\{\sqrt{1 - Q[C(1+Q) - D(1-Q) - 1]} - 1\}, \quad (Q > 0) \\ x &= L\left(\frac{1 - C + D}{2}\right), \quad (Q = 0) \end{aligned}\right\} \quad (9.7)$$

Whereas D was zero in the single-span problem (Art. 9.3), for the multispan frame D becomes an additional parameter. Therefore it is neces-

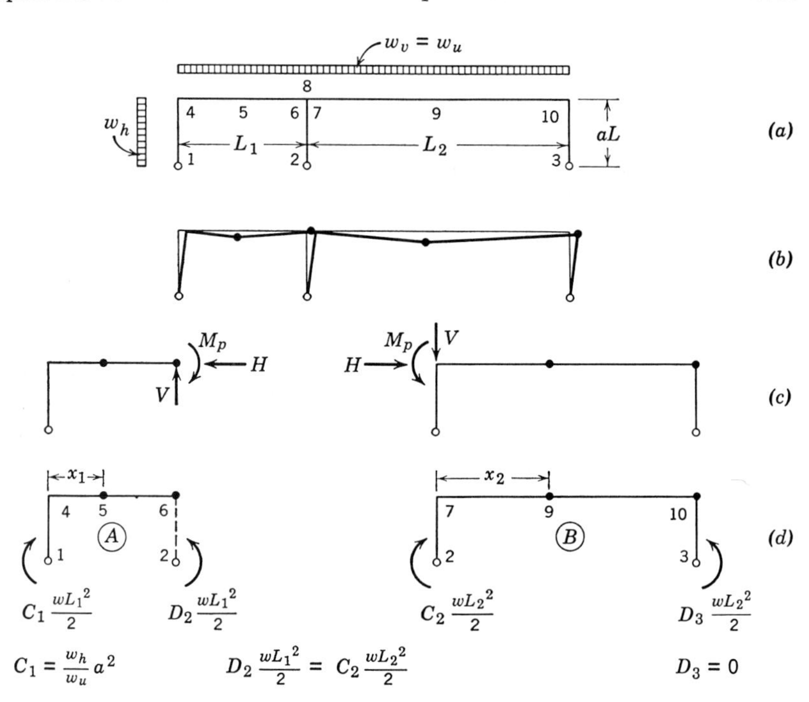

Fig. 9.5. Functions involved in the analysis of rectangular two-span frames.

sary to prepare one set of charts for each value of the ratio of roof rise to column height Q for which a solution is desired. Figure 9.6 represents the solution for the flat roof frame in chart form. The left hand portion represents Eq. 9.6 with $Q = 0$. The right is the second form of Eq. 9.7. Notice that the lower cut-off line on the chart is a beam mechanism in which $M_p/wL^2 = \frac{1}{16}$ (see Fig. 8.1, p. 265, case 5).

Now, to solve the problem it is noted from the loading (Fig. 9.5*d*) that $C_1 = (w_h/w_u)\, a^2$ and that $D_3 = 0$. The correct answer will be determined when the overturning moments at section 2 are equated.

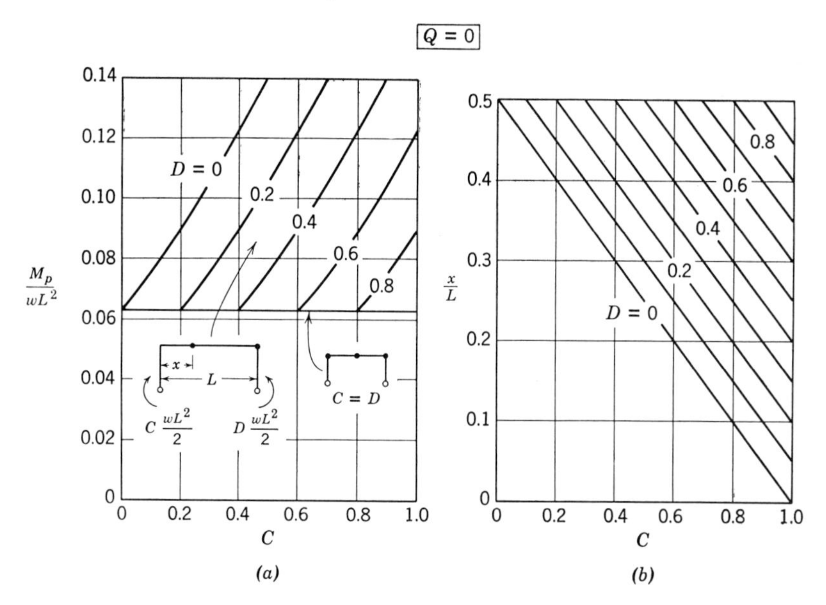

Fig. 9.6. Charts for the solution of multi-span, pinned base, rectangular frames.[7.4]

Thus, $D_2(wL_1^2/2) = C_2(wL_2^2/2)$. Consequently, the chart of Fig. 9.6*a* is used twice: once for structure A and once for structure B and an answer will be obtained in terms of M_p/wL^2 for which the overturning moments at section 2 will just cancel. An example should help illustrate this.

PLATE XIV Two-Span Rectangular Frame

The case for vertical load alone will result in beam mechanisms and need not be considered here to illustrate the use of the charts. (The problem is the same as Plate XI for which case I—without wind—was critical.)

In the first portion of Plate XIV the known quantities are indicated. The value of C_1 is found to be 0.125. D_3 is 0 since there is no external overturning moment applied to member 3–10. The only unknown values at this stage are D_2 and C_2, both of which may be found at the same time the value M_p/wL^2 is determined from the condition that $D_2 = 4C_2$.

Although the value of M_p/wL^2 that satisfies this condition cannot be selected at the outset by use of the chart of Fig. 9.6*a* one can determine *possible solutions* for each panel and find the *correct* answer graphically. Thus a table is prepared with the aid of the chart shown in Fig. 9.6 and this table is presented in Plate XIV. Panel A is first analyzed for $C_1 = 0.125$ and for various values of D. (Linear interpolation will be satisfactory if the range of C and D values is

PLATE XIV. TWO-SPAN RECTANGULAR FRAME

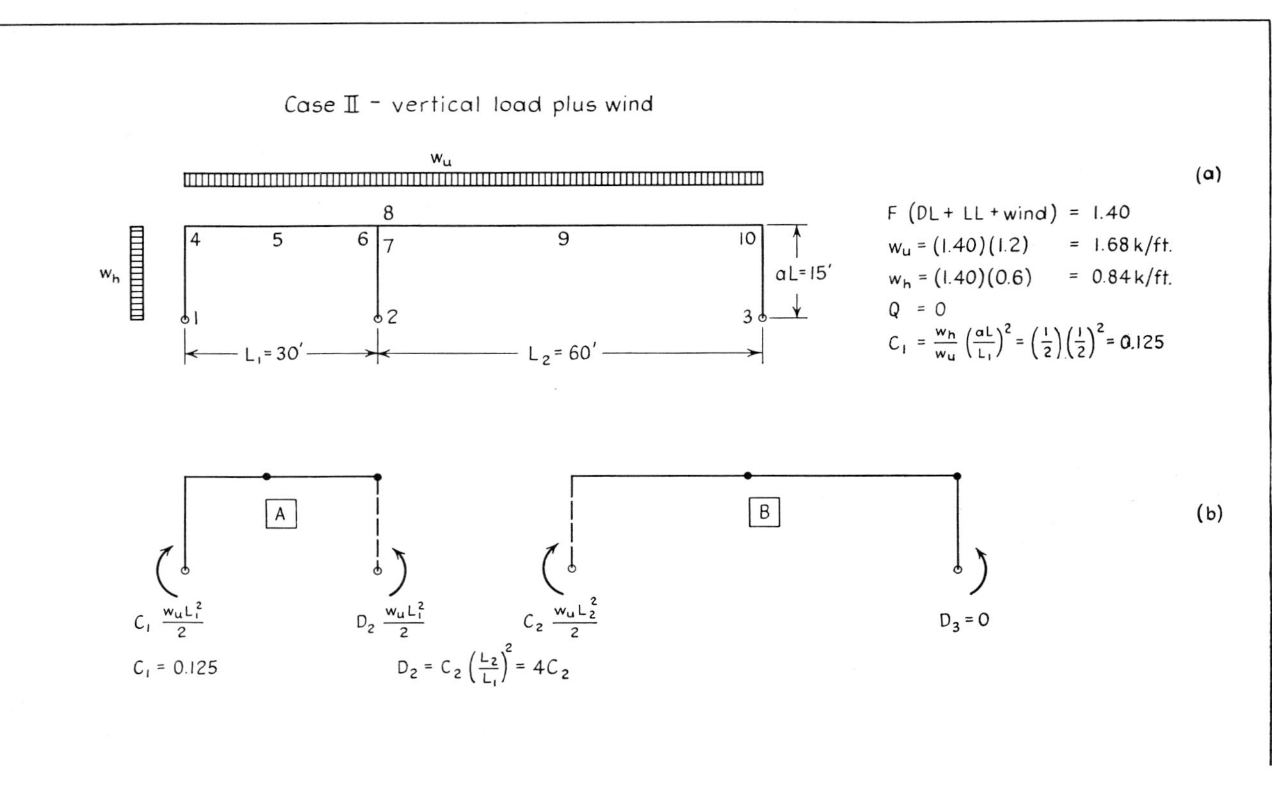

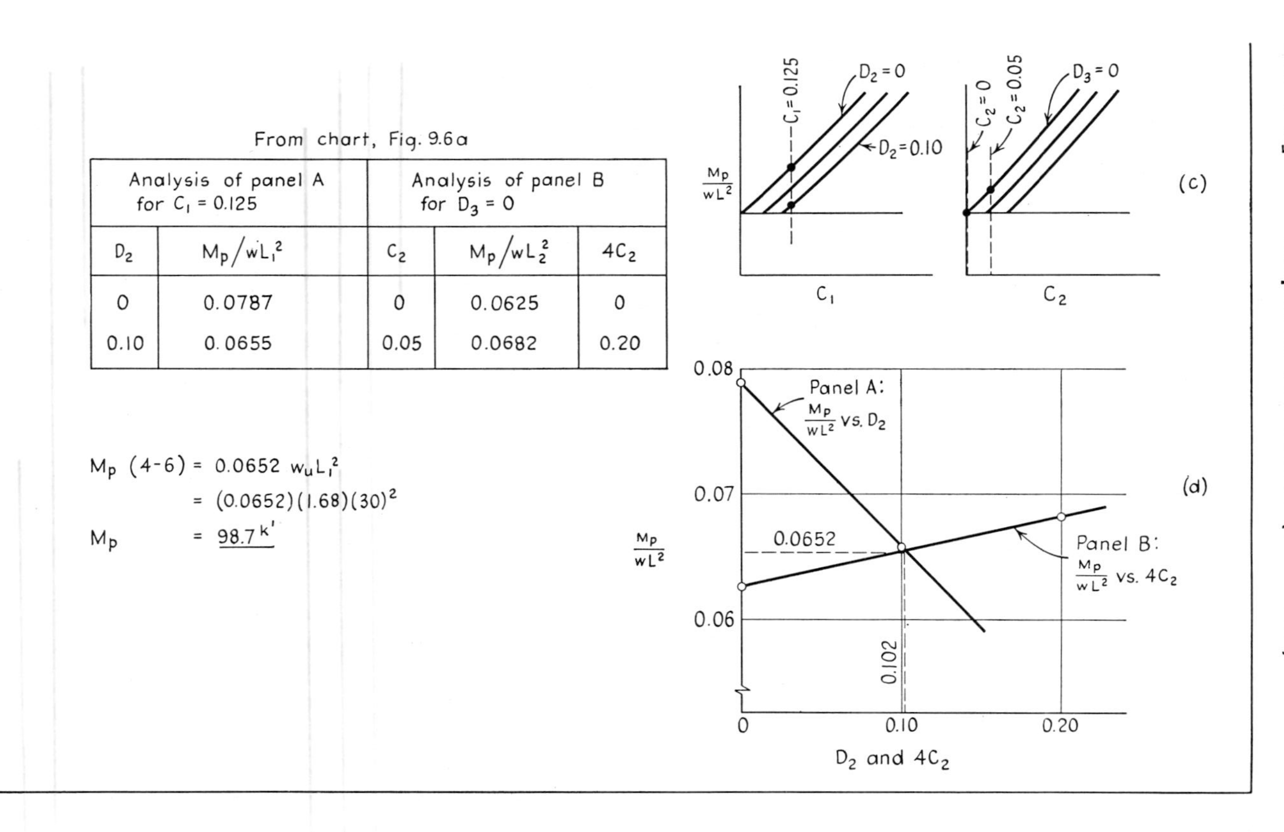
From chart, Fig. 9.6a

Analysis of panel A for $C_1 = 0.125$		Analysis of panel B for $D_3 = 0$		
D_2	M_p/wL_1^2	C_2	M_p/wL_2^2	$4C_2$
0	0.0787	0	0.0625	0
0.10	0.0655	0.05	0.0682	0.20

$$M_p\ (4\text{-}6) = 0.0652\ w_uL_1^2$$
$$= (0.0652)(1.68)(30)^2$$
$$M_p = \underline{98.7^{k'}}$$

small when compared with that of the chart. Therefore two points will be sufficient, and $D_2 = 0$ and $D_2 = 0.10$ were selected.) For $D_2 = 0$ and 0.10, the corresponding values of M_p/wL_1^2 are 0.0787 and 0.0655, respectively. The same thing is done for panel B except that, now, D_3 is known and C_2 is unknown. So, values of M_p/wL^2 are determined for two values of C_2 (0 and 0.05) and the corresponding values of M_p/wL_2^2 recorded (0.0625 and 0.0682). The sketches in (*c*) illustrate how the points are selected.

On a separate graph the information contained in the table may be plotted. It is shown in sketch (*d*). Where the curves for panel A and panel B intersect, $D_2 = 4C_2 = 0.102$, $M_p/wL^2 = 0.0652$, and the problem is solved.

Note that the value of M_p for member 4–6 ($M_p = 98.7$ kip-ft) agrees with the value determined for this same problem by direct use of the mechanism method. See Plate XI, p. 317 ($M_p = 99$ kip-ft).

In all of these procedures, the final step in the analysis would be to draw the moment diagram with the aid of charts similar to the one shown in Fig. 9.6*b*. Finally the secondary design considerations would be examined.

9.6 MULTI-STORY FRAMES

Up to the present time, attention has been restricted to one-story structures consisting of rectangular and gabled portal frames and to the multi-span frames that are typical of the industrial-type buildings to which plastic design can now be applied. This approach has come about for several reasons. First of all a considerable tonnage of steel goes into single-story structures, and therefore it is advantageous to document the necessary provisions which will enable the engineer to apply plastic design to the industrial building. Secondly (and perhaps more important), as the number of stories increases, the columns become more and more highly loaded. As already mentioned, the moment capacity of columns with relatively high axial load drops rapidly. The related problems are not completely solved and more research is needed before a "complete" plastic design can be applied to all classes of tier buildings. As will be noted below, however, the outlook is encouraging; partial application of plastic analysis may be made to the design of tier buildings.

The application of plastic design to multi-story structures of the tier building type has been discussed in Ref. 9.3. It is suggested there that the application of plastic design to such structures depends on the relative importance of horizontal forces. If horizontal forces are not a consideration (they may be so small that an ordinary masonry wall panel would carry any such small forces that might exist) then the regular connections are free of moments due to side sway and a large saving in steel is possible when comparing a plastic design of the beams to a conventional simple beam design. In fact, for uniformly distributed load the saving theoretically could be 50% if it weren't for other factors such as the cost of connections, etc. that tend to cancel out the potential saving due to economy in main material. When compared with rigidly connected elastic design there will, of course, be a saving through the plastic method.

The approach to design will be influenced largely by what is done about bracing against horizontal forces. Three situations may arise:

(1) No horizontal load must be resisted (any minor loads taken by wall panels).
(2) Horizontal forces carried by moment connections.
(3) Those cases in which the horizontal forces are carried by cross bracing around elevator shafts or elsewhere in walls.

The application of plastic design to situations 1 and 3 above will simply consist of a plastic analysis of continuous beams. For the second condition (horizontal forces resisted by moment connections) the area of possible application of plastic design is dependent to the greatest extent on further research because plastic hinges might form in the columns, and, as already mentioned, more needs to be known about the performance of columns under high axial load and as part of a framework.

When the horizontal forces are carried by cross-bracing, an approach using plastic design seems reasonable. If provision is made for wind bracing in wall panels, the beams and girders would be proportioned for full (plastic) continuity. The columns, on the other hand, would be proportioned according to present procedures. None of the plastic hinges would participate in the resistance to side load. All such load would be carried by the diagonal bracing. The only mechanisms are the beam mechanisms.

The top one or two stories might be designed by a "complete" plastic analysis, hinges forming both in the columns and in the beams. In those cases the vertical load in the columns would be relatively low and would be governed by considerations already described for the previous examples.

As far as the tall building is concerned, the column problem actually may not be as severe as first intimated. The most critical loading condition on a column is one which subjects it to equal end moments producing single curvature; the maximum moment then occurs at the mid-height of the member. On the other hand, in tall buildings the columns will usually be bent in double curvature with a point of inflection (zero moment) near the middle of the member. The critical sections in that case are at the ends. Such columns are better able to develop plastic hinges than columns loaded in single curvature.

The problem of the connection for tier buildings also relates to the ability of these components to form plastic hinges. In riveted work it is very difficult to design a connection of strength equal to that of the beams unless large brackets are used. Therefore, if riveting were to be used to achieve continuity at connections in a plastically designed structure and without the use of these large brackets, further studies would be needed. The use of high-strength bolts offers another method of achieving continuity.

As has already been emphasized, maximum continuity with minimum added connection material can often be achieved by the use of welding. Numerous design recommendations have been made in Chapter 5 that are directly applicable to multi-story buildings.

Naturally, no sharp dividing line exists between the form of structure that, on the one hand, may be designed by the plastic method and, on the other hand may not. An example will now be given of the plastic design of a two-story building in which cross-bracing is not used, but any possible side sway is resisted by moment connections. After the selection of member sizes, the design of some of the connections also will be examined.

PLATE XV Two-Span Two-Story Building

In Plate XV a two-span, two-story building frame is designed to support vertical floor loads and horizontal wind load acting on the vertical projection. The mechanism method of analysis will be used in the problem. Some short cuts will be taken, leading to a practical design, but these short cuts are such as to produce a design that is more conservative than would otherwise be the case. The M_p ratios chosen as the basis for proportioning the several members leave the frame as a whole partially indeterminate. Hence the last pound of economy of material presumably will not have been realized. However, any further saving in material through the use of a greater variety of shapes, would be small and would require a closer examination of more mechanisms. Since these would have involved the actual development of a number of hinges in the several columns, most of which would have to be furnished at a reduction from full tabulated value because of relatively high axial load, it is doubtful if any worth-while saving would actually be achieved. The increase in design computations, on the other hand, would have been considerable.

Step 1: Sketch (*a*) shows the loads acting on the two-bay frame. The load is uniformly distributed, and for this particular example the distributed loads are replaced by concentrated loads acting at the quarter points. The side loads T are derived on the basis of equivalent overturning moment about the bottom of the columns of the particular story. An alternate method of handling the distributed load would be to assume that the plastic hinge formed at the center and, after the subsequent analysis, to revise the design (upward) to suit the precise plastic moment requirements.

Step 3: Assuming that vertical load alone will control the design, the plastic moment ratios of the different members could be selected so that simultaneous failure would occur in beams A and B. For distributed load, the plastic moment ratios will vary as the square of span and directly as the magnitude of the load. An alternate method which gives the same answer for this problem is to write the virtual work equations for mechanisms 1 and 3 (see step 4) expressing the plastic moments as unknowns $k_A M_p$ and $k_B M_p$. For simultaneous failure, then, the load corresponding to mechanism 1 must be equal to that of mechanism 3.

In order to obtain simultaneous failure of beam C and beam A, the plastic moment ratio would be increased in proportion to the load, and therefore values of $k_C = 2.37$ and $k_D = 1.33$ are obtained. The assumed k values of the columns can be revised later in the problem if desired.

Step 4: The fourteen independent mechanisms are shown in sketch (*b*). Only two of the eight possible beam mechanisms are shown. The rest would all be similar.

Step 5: The solutions of the various mechanisms are worked out in tabular form in Plate XV. It is found that all of the beam mechanisms provide the same required plastic moment ($M_p = 1.69 P_u L$), and this is a check on the accuracy of step 3. As in the previous examples, the sequence of terms in the work equations follows the numbering sequence of sketch (*a*). Also, mechanism angles are shown on the tension side of the hinges.

With no side load acting there is no point in investigating mechanisms 9 and 10. Although we can be reasonably sure that the correct answer has been obtained at this stage, mechanism 15 is also investigated. It is a combination of every independent mechanism except Nos. 1, 3, 5, and 7. As expected, the required plastic moment is identical.

Step 6: The moment check for mechanism 15 follows: Since the failure loads are identical for all the mechanisms shown in the tabulation for step 4, plastic hinges will occur at all four possible points in each beam. Therefore the moment diagrams for these beams may be constructed without difficulty and quite evidently from sketch (*c*), the plastic moment condition is not violated in any beam.

A possible equilibrium moment diagram for the columns is shown in sketch (*d*). It is not precise since the frame is still indeterminate, but it shows that the plastic moment condition is not violated—and that is what was desired. The diagram is computed in one trial by assuming certain values and making use of Eqs. 7.11 and 7.12 that assure joint and sway equilibrium.

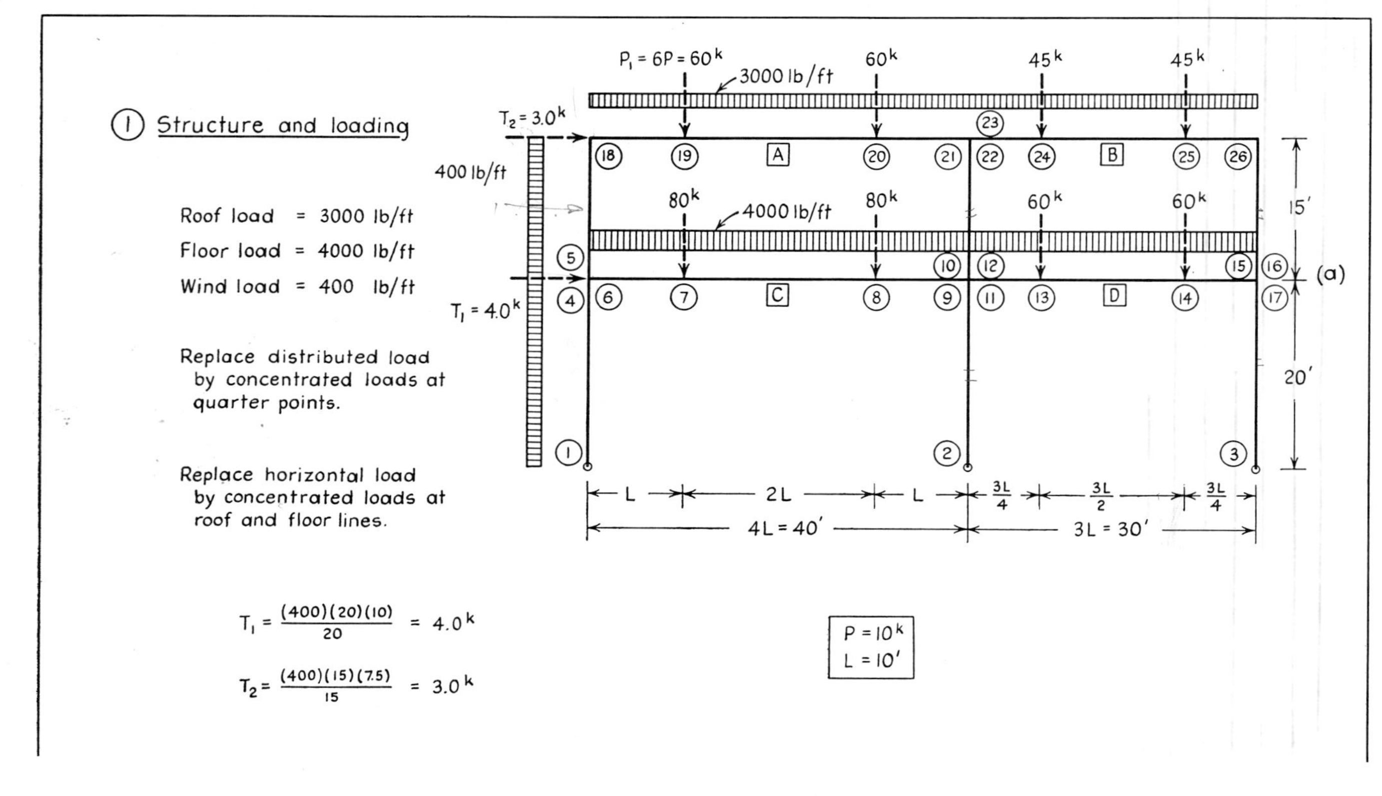
PLATE XV. TWO-SPAN TWO-STORY BUILDING
1 Structure and loading
Roof load = 3000 lb/ft
Floor load = 4000 lb/ft
Wind load = 400 lb/ft
Replace distributed load by concentrated loads at quarter points.
Replace horizontal load by concentrated loads at roof and floor lines.
T1 = (400)(20)(10)/20 = 4.0k
T2 = (400)(15)(7.5)/15 = 3.0k
P1 = 6P = 60k
3000 lb/ft
4000 lb/ft
60k
80k
45k
60k
400 lb/ft
T2 = 3.0k
T1 = 4.0k
A
B
C
D
15'
20'
(a)
L
2L
L
3L/4
3L/2
3L/4
4L = 40'
3L = 30'
P = 10k
L = 10'

(2) Loading conditions

(Table 7.1, Provision 6)

Case I:	DL	$F = 1.85$	$P_u = 18.5$
Case II:	DL + LL + wind from left	$F = 1.40$	$P_u = 14.0$
Case III:	DL + LL + wind from right	$F = 1.40$	$P_u = 14.0$

(3) Plastic moment ratios

(Table 7.2, Procedure 1)

For simultaneous failure of spans A & B under case I loading use plastic moments in ratio of square of spans. For equal spans, ratio is to vary as the load.

$$k_B = 1.00$$

$$k_A = \left(\frac{40}{30}\right)^2 k_B = 1.78$$

$$k_D = \frac{60}{45} k_B = 1.33$$

$$k_C = \left(\frac{40}{30}\right)^2 k_D = 2.37$$

Left column: Use $k = k_A = 1.78$

Center column: Use $k = k_B = 1.00$

Right column: Use $k = k_B = 1.00$

PLATE XV (*Continued*)

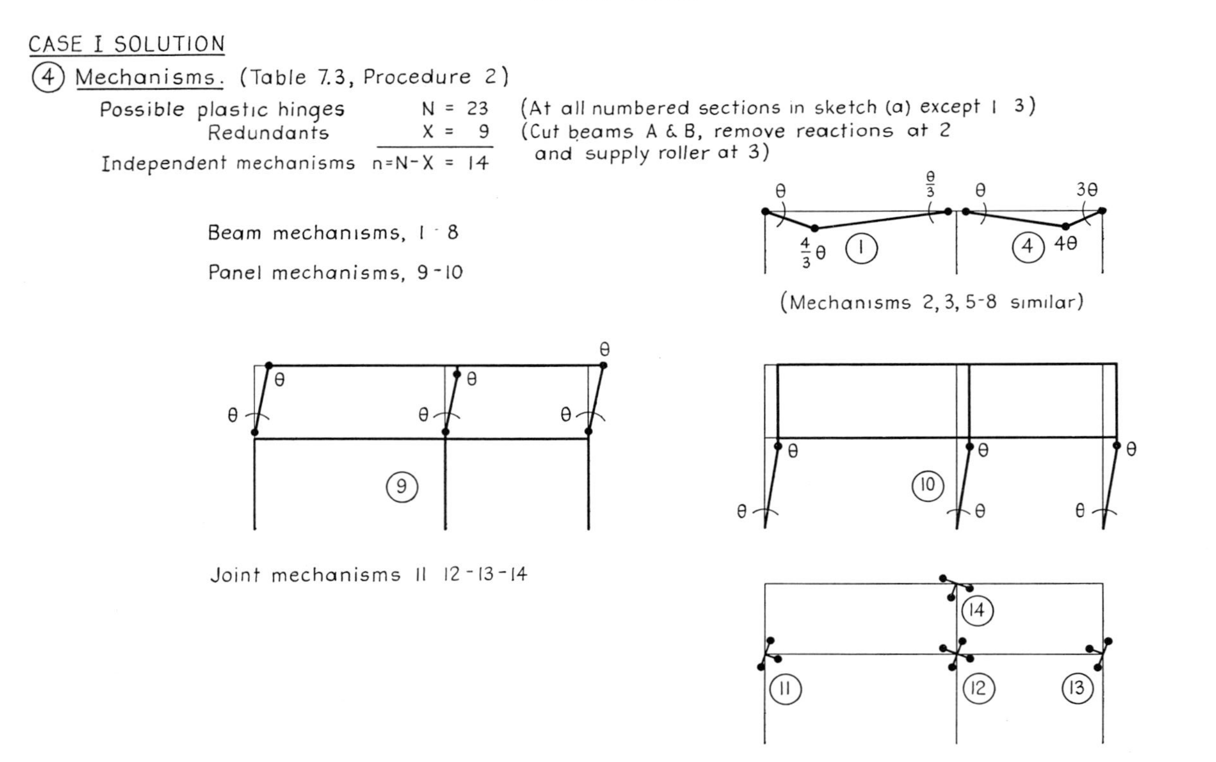

CASE I SOLUTION
4 Mechanisms. (Table 7.3, Procedure 2)
Possible plastic hinges N = 23
Redundants X = 9
Independent mechanisms n=N-X = 14
(At all numbered sections in sketch (a) except 1 3)
(Cut beams A & B, remove reactions at 2
and supply roller at 3)
Beam mechanisms, 1 8
Panel mechanisms, 9-10
(Mechanisms 2, 3, 5-8 similar)
Joint mechanisms 11 12-13-14
θ
θ/3
3θ
4/3 θ
4θ
1
4
9
10
11
12
13
14
(b)

⑤ Mechanism solutions:

No.	Mechanism	Internal work $\left(W_I/M_p\theta\right)$	External work $\left(W_E/PL\theta\right)$	$\frac{M_p}{P_uL}$
1 2	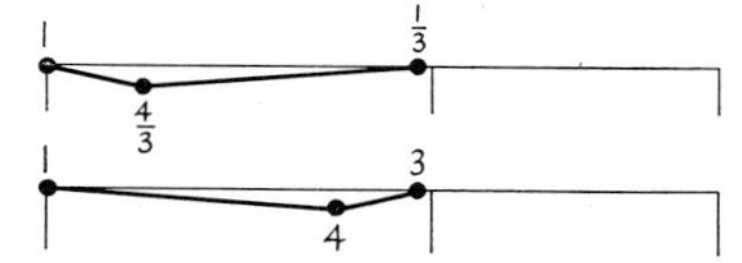	$k_A\left(1+\frac{4}{3}+\frac{1}{3}\right)=\frac{8}{3}(1.78)=4.74$ NOTE: 2 is identical with 1	$6(1)(1)+6\left(\frac{1}{3}\right)(1)=8$	1.69
3, 4		$k_B\left(1+\frac{4}{3}+\frac{1}{3}\right)=\frac{8}{3}(1.00)=2.67$	$4.5(1)\left(\frac{3}{4}\right)+4.5\left(\frac{1}{3}\right)\left(\frac{3}{4}\right)=4.5$	1.69
5, 8	NOTE: Mechanisms and virtual work equations are similar.			1.69
15 (2+4 +6+8 +9+10 +11+12 +13+14)	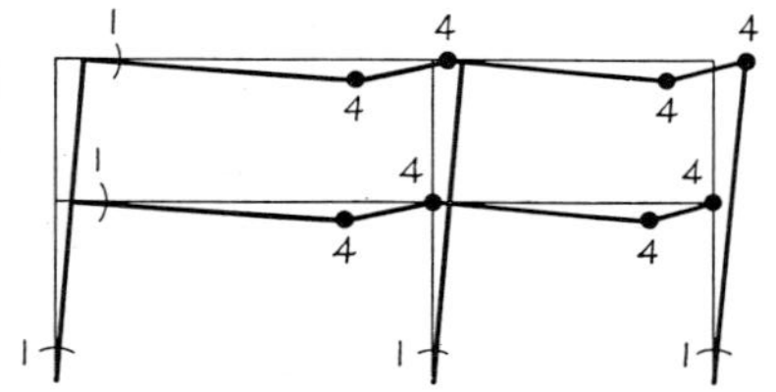	$k_C(4+4)+k_D(4+4)+k_A(4+4)$ $+k_B(4+4)=$ $=8(k_C+k_D+k_A+k_B)$ $=8(2.37+1.33+1.78+1)$ $=51.8$	$8(1+3)+6(1+3)\left(\frac{3}{4}\right)+$ $+6(1+3)+4.5(1+3)\left(\frac{3}{4}\right)=$ $=32+18+24+13.5$ $=87.5$	1.69

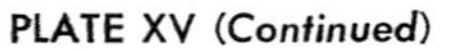

PLATE XV (*Continued*)

(6) Moment check for mechanism 15

$M_p = 1.69\ P_u L = (1.69)(18.5)(10) = 313^{k'}$
$M_{pA} = 1.78\ M_p = 557^{k'}$
$M_{pB} = 1.0\ M_p = 313^{k'}$
$M_{pC} = 2.37\ M_p = 741^{k'}$
$M_{pD} = 1.33\ M_p = 416^{k'}$

Joint 21-23 $M_{22} = -M_{21} - M_{23}$
$M_{22} = 557 - 313 = \underline{+244^{k'}}$

Assume $M_{17} = +M_p$ (in joint 15-17)

Joint 15-17 $M_{16} = -M_{15} - M_{17}$
$M_{16} = 416 - 313 = \underline{+103^{k'}}$

Assume $M_{10} = M_{22} = 244^{k'}$

Sway of top story

$M_5 + M_{10} + M_{16} + M_{18} + M_{22} + M_{26} = 0$
$M_5 = +244 + 103 - 557 + 244 + 313 = \underline{+347^{k'}}$

Joint 4-6 $M_4 = -M_5 - M_6 = +347 - 741 = -394^{k'}$

Sway of bottom story:

$M_4 + M_{11} + M_{17} = 0$
$+394 - 81 - 313 = 0$

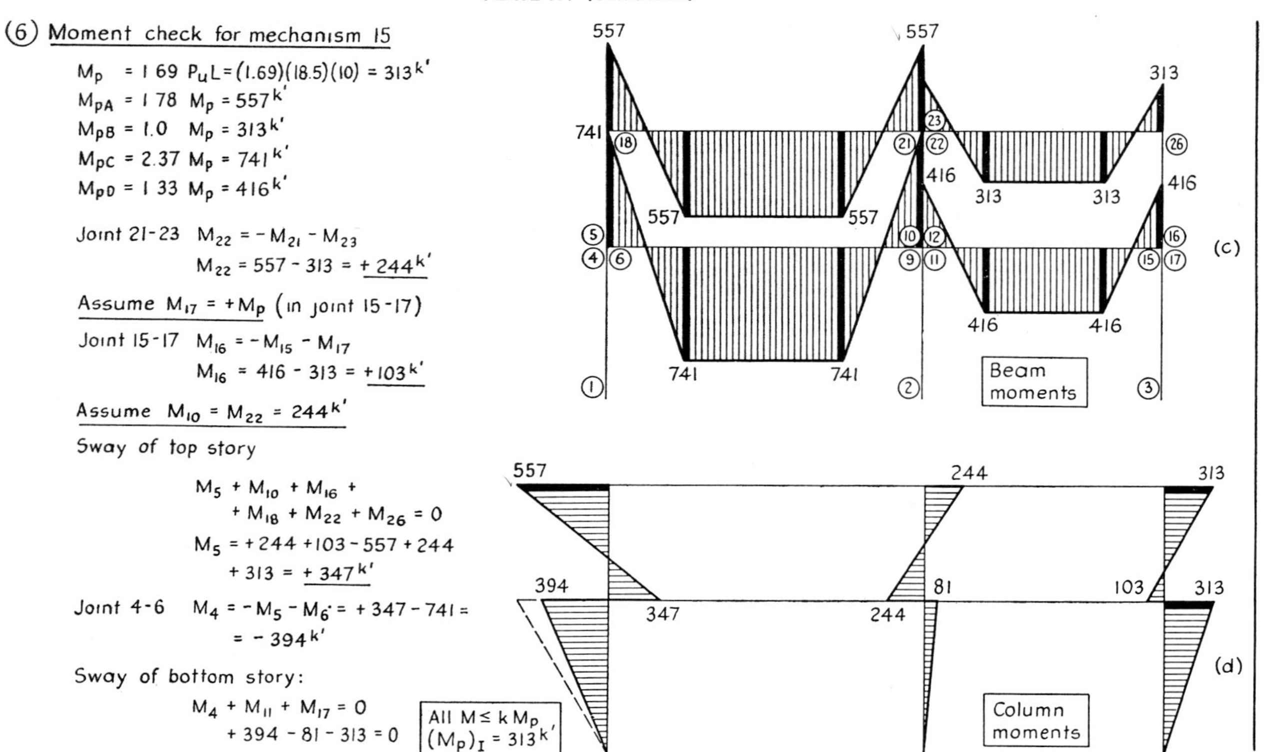

CASE II SOLUTION (Wind from left)

Independent mechanisms: See case I. sketch (b)

(7) Mechanism solutions

16	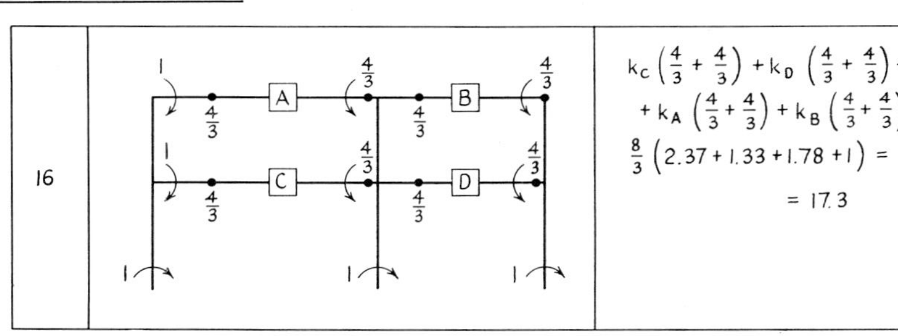	$k_C\left(\frac{4}{3}+\frac{4}{3}\right)+k_D\left(\frac{4}{3}+\frac{4}{3}\right)+$ $+k_A\left(\frac{4}{3}+\frac{4}{3}\right)+k_B\left(\frac{4}{3}+\frac{4}{3}\right)=$ $\frac{8}{3}\left(2.37+1.33+1.78+1\right)=$ $=17.3$	$(0.4)(1)(2)+8(1)(1)+8\left(\frac{1}{3}\right)(1)$ $+6(1)\left(\frac{3}{4}\right)+6\left(\frac{1}{3}\right)\left(\frac{3}{4}\right)$ $+(0.3)(1)(3.5)+6(1)\left(1+\frac{1}{3}\right)$ $+(4.5)\left(\frac{3}{4}+\frac{1}{4}\right)=$ $=0.8+8+\frac{8}{3}+6+1.05$ $+8+4.5=31.02$	1.80

(8) Moment check for mechanism 16 ($M_p = 1.80\ P_uL$)

$M_{pB} = 1.80\ P_uL = (1.80)(14.0)(10) = 252^{k'}$

$M_{pA} = 1.78\ M_p = 448^{k'}$

$M_{pC} = 2.37\ M_p = 597^{k'}$

$M_{pD} = 1.33\ M_p = 335^{k'}$

$M_{sB} = (4.5P)\left(\frac{3}{4}L\right) = (4.5)(14)\left(\frac{3}{4}\right)(10) = 472^{k'}$

$M_{sA} = (6.0P)(L) = 840^{k'}$

$M_{sC} = (8.0P)(L) = 1120^{k'}$

$M_{sD} = (6.0P)\left(\frac{3}{4}L\right) = 630^{k'}$

PLATE XV (*Continued*)

Joint 21-22-23: $M_{22} = -M_{21} - M_{23}$

$\underline{M_{22}} = 448 - 209 = \underline{+239^{k'}}$

Assume $M_{17} = M_p = \underline{252^{k'}}$

Assume $M_{11} = M_p = \underline{252^{k'}}$

Joint 15-17: $M_{16} = -M_{15} - M_{17}$

$\underline{M_{16}} = 335 - 252 = \underline{+83^{k'}}$

Sway of bottom span:

$(T_1 + T_2)(1.4)(20) + M_4 + M_{11} + M_{17} = 0$

$M_4 = -(7)(1.4)(20) + 252 + 252$

$\underline{M_4} = \underline{+308}^{k'}$

Joint 4-6: $M_5 = -M_4 - M_6$

$= 308 - 498$

$\underline{M_5 = -190^{k'}}$

Sway of top span:

$T_2\,(1.4)(15) + M_{18} + M_{22} + M_{26} +$

$+ M_5 + M_{10} + M_{16} = 0$

$M_{10} = -63 - 373 + 239 + 252 -$

$-190 + 83$

$\underline{M_{10} = -52^{k'}}$

Joint 9-12: $M_9 + M_{10} + M_{11} + M_{12} = 0$

$-597 + 52 + 252 + 282 = -597 +$

$+586 = -11$ (Adequate check)

$\boxed{(M_p)_{II} = 252^{k'}}$

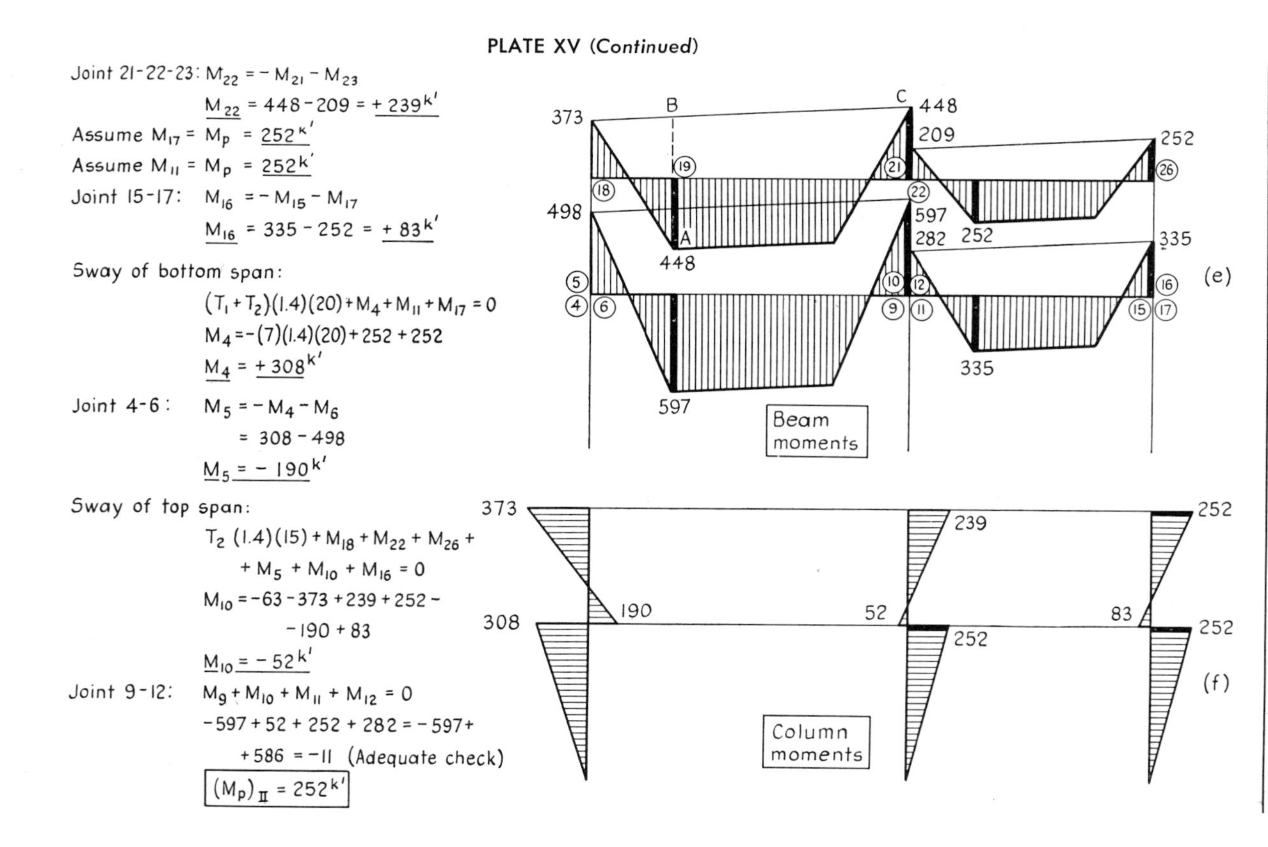

CASE III SOLUTION (Wind from right)

(9) Mechanism solutions

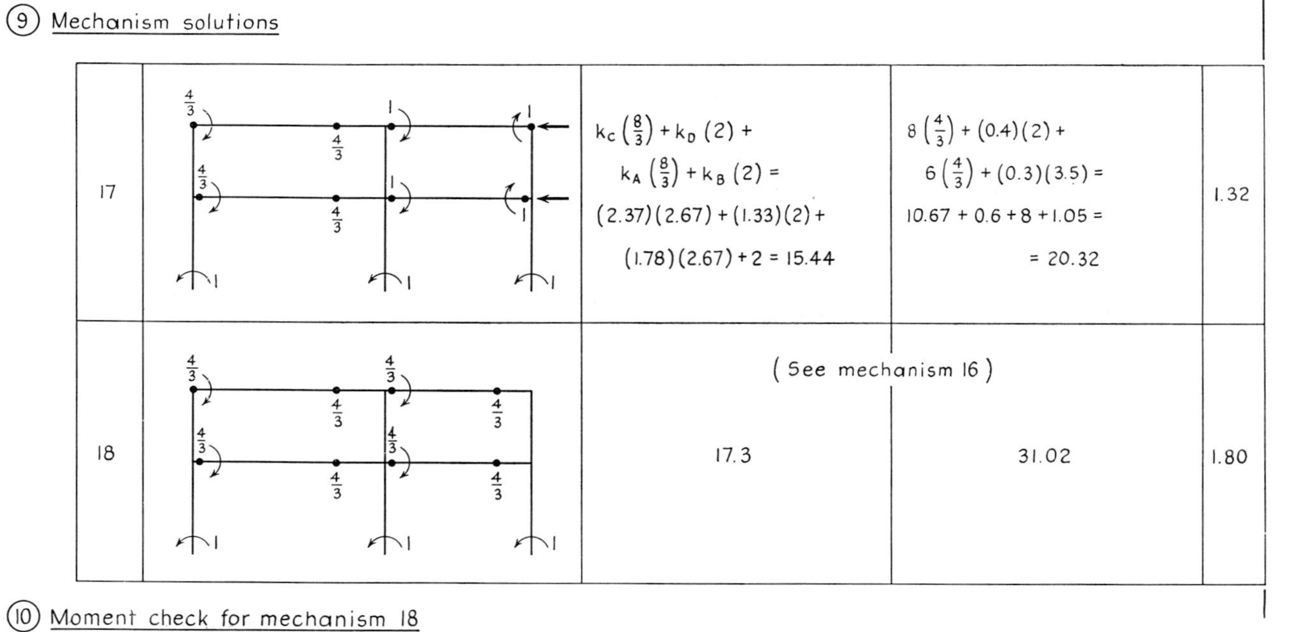

17		$k_C\left(\frac{8}{3}\right) + k_D(2) +$ $k_A\left(\frac{8}{3}\right) + k_B(2) =$ $(2.37)(2.67) + (1.33)(2) +$ $(1.78)(2.67) + 2 = 15.44$	$8\left(\frac{4}{3}\right) + (0.4)(2) +$ $6\left(\frac{4}{3}\right) + (0.3)(3.5) =$ $10.67 + 0.6 + 8 + 1.05 =$ $= 20.32$	1.32
18		(See mechanism 16) 17.3	31.02	1.80

(10) Moment check for mechanism 18

Procedure same as in step 8

Case I controls

PLATE XV (*Continued*)

(11) Reactions for case I

From sketch (d): $H_1 = \frac{394}{20} = 19.7^k$ (Maximum possible value $= \frac{557}{20} = 27.9^k$)

$H_2 = \frac{81}{20} = 4.0^k$ (Maximum possible value $= \frac{313}{20} = 15.7^k$)

$H_3 = \frac{313}{20} = 15.7^k$

From sketch (c) with symmetry:

$V_1 = 6P + 8P = (14)(10)(1.85) = 259^k$

$V_2 = 6P + 8P + 4.5P + 6P = 453^k$

$V_3 = 4.5P + 6P = 194^k$

(12) Selection of sections: (Table 7.1, Provision 5)

Member	Plastic moment (kip-ft)	$Z = Z_{trial}$ (in.^{3})	Section	Z furnished (in.^{3})
Beam 18-21	557	203	24 WF 84	224.0
Column 1-4 *			18 WF 96	206.0
Column 5-18			18 WF 96	206.0
Beam 23-26	313	114	18 WF 60	112.6
Column 2-11 *, 10-22 *			12 WF 79	119.3
Column 3-17 *, 16-26			12 WF 79	119.3
Beam 6-9	741	270	27 WF 94	277.7
Beam 12-15	416	152	21 WF 68	159.8

* Revised in step 14.

(13) Axial force: (Table 7.4, Guide 1)

Column	P in kips	Shape	P/P_y	Comparison with 0.15
1-4	259	18 WF 96	(259)/(33)(28.22) = 0.278	Requires revision
5-18	111		(111)/(33)(28.22) = 0.119	ok
2-11	453	12 WF 79	(453)/(33)(23.22) = 0.591	Requires revision
10-22	194		(194)/(33)(23.22) = 0.253	Requires revision
3-17	194	12 WF 79	(194)/(33)(23.22) = 0.253	Requires revision
16-26	83		(83)/(33)(23.22) = 0.108	ok

(14) Columns: (Table 7.6, Guide 2)

Column	Trial shape	P/P_y (step 13)	L/r_x	M_o/M_p (Fig. 4.23)	$Z = \frac{Z_t}{M_o/M_p}$ (in.^{3})	Revised section Shape	Z (in.^{3})
1-4	18WF96	0.278	(20)(12)/(7.79) = 31.0	0.822	$\frac{203}{0.822}$ = 247	18 WF 114	247.7
2-11	12WF79	0.591	18.4	0.478	= 238	18 WF 114 (Revised. See below)	247.7
10-22	12WF79	0.253	33.7	0.855	= 133	14 WF 78	134.0
3-17	12WF79	0.253	45.0	0.855	= 133	12 WF 92	140.2

PLATE XV (Continued)

Modification of column 2-11:

Eq 4.11: $Z = Z_t \left(\frac{P}{P_y} + 0.85 \right) = 114 \; (0.591 + 0.85) = 182 \text{ in.}^3$

Eq 4.9: $M_{pc} = 1.18 \left(1 - \frac{P}{P_y}\right) M_p$

$= 1.18 \; (1 - 0.42)(33)(196.0)$

$M_{pc} = 447^{k'} > 313^{k'}$ ok

Use 14WF111
$Z = 196.0 \text{ in.}^3$
$A = 32.65 \text{ in.}^2$

(15) Cross-section proportions (Table 7.8, Guide I)

Shape	b/t [< 17]	d/w [< 55]		Shape	b/t [< 17]	d/w [< 55]	
12WF79	16.4	26.4	ok	18WF96	14.1	35.5	ok
12WF92	14.2	23.2	ok	18WF114	11.9	31.1	ok
14WF78	16.7	32.9	ok	21WF68	12.1	49.2	ok
14WF111	16.8	26.6	ok	24WF84	11.68	51.3	ok
18WF60	10.9	43 9	ok	27WF94	13.4	55.0	ok

(16) Corner connection 26 (Table 7.7, Guide I)

$w \geq \frac{0.6M}{d_1 d_2} = \frac{(0.6)(313)}{(18.25)(12.38)} = 0.83 > 0.470$ [Stiffening req'd.]

$t_s = \frac{\sqrt{2}}{b} \left(\frac{s}{d} - \frac{wd}{\sqrt{3}} \right) = \frac{\sqrt{2}}{7.56} \left[\frac{107.1}{12.38} - \frac{(0.470)(12.38)}{1.732} \right] = 0.615''$

$\frac{b}{17} = \frac{7.56}{17} = 0.445$

Use diagonal plate stiffeners $3\frac{1}{2}'' \times \frac{5}{8}''$

18WF60
18.25"
(26)
(9)
12WF79
12.38"

(17) Side connection 15-16-17 (Table 7.7, Guide 10)

"Shear" stiffener

$$w = \frac{0.6\,\Delta M}{d_c\, d_b} \qquad (5.49)$$

$$\Delta M = M_{15} = 416^{k'}$$

$$d_c = d_{17} = 12.62''$$

$$d_b = d_{15} = 21.13''$$

$$w = \frac{(0.6)\quad(416)}{(12.62)(21.13)} = 0.925'' > 0.545'' \qquad \left[\text{Stiffening req'd.}\right]$$

Moment to be transmitted by stiffener

$$M = \left[\frac{0.925 - 0.545}{0.925}\right] 416 = 171^{k'}$$

Area of stiffener:

$$\frac{M}{\sigma_y\, d_{15} \cos 59^\circ} = \frac{(171)(12)}{(21.13)(33)(0.515)} = 5.70\ \square''$$

Use flange width of 21WF68 (b = 8.27")

$$t = \frac{5.70}{8.27} = 0.690''$$

$$\frac{b}{t} = \frac{8.27}{0.625} = 13.3 < 17 \quad \underline{ok}$$

Use ℞ stiffeners $3\frac{7}{8}'' \times \frac{5}{8}''$

Flange stiffeners

$$t_s = \frac{1}{2b}\left[A_b - w_c\,(d_b + 6\,k_c)\right] \qquad (5.42)$$

$$= \frac{1}{2\,(8.27)}\left[20.02 - 0.545\,(21.13 + (6)(1.438)\right] = 0.24''$$

$$\frac{b}{17} = \frac{8.27}{17} = 0.486''$$

Use ℞ stiffeners $3\frac{7}{8}'' \times \frac{1}{2}''$

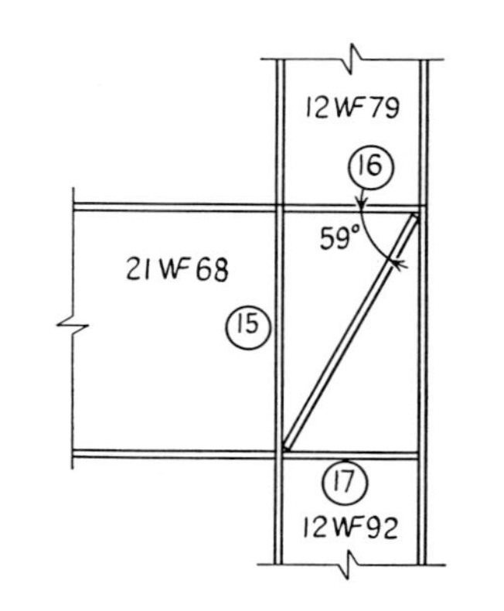

(h)

PLATE XV *(Continued)*

Frame layout:

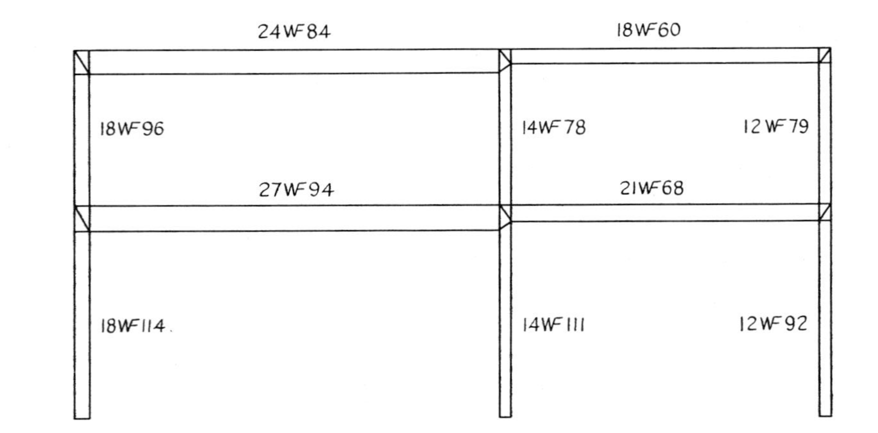

NOTE: Check remaining applicable design guides

The joint equilibrium equations are first used to obtain M_{18}, M_{22}, and M_{26}. The left and right columns are selected as having a moment strength equal to that of the beams which they restrain, and the same moment capacity is assumed for the full-column height at this stage. In order to obtain an estimate of the magnitude of moment at section 16, since the horizontal reaction at 3 would act to the left, it is assumed that M_{17} is equal to the full plastic value (313 kip-ft) in the direction indicated. The joint equilibrium equation gives the magnitude of M_{16}. M_{10} is assumed equal to M_{22}; this is a completely arbitrary assumption, but since there is no side load, any small value would be reasonable at this stage. M_5 is then obtained by the panel (sway) equilibrium equation (7.12) and is found to be 347 kip-ft. Since all moments are less than kM_p, the upper story is satisfactory thus far.

The moments at sections 4 and 11 can now be determined by joint equilibrium. Subsequently, the sway equation shows that the lower story is also in equilibrium, and therefore the moment check is completed. M_p for case I is thus equal to 313 kip-ft.

Step 7: For case II loading, the plastic hinges and independent mechanisms are the same as those in case I that are shown in sketch (*b*). Out of the large number of possible combinations that could be made, one can eliminate all of the beam mechanisms that would form without sidesway, since case I would automatically control. Therefore the failure mechanism must involve sway to the side of the upper panel, the lower panel or both panels. Mechanism 16 requires that all of the loads do external work and at the same time involves formation of a relatively small number of plastic hinges. Since this should lead to a maximum required M_p, mechanism 16 will be analyzed. It is found that $M_p = 1.80P_uL = 252$ kip-ft. The moment capacity of the other three beams are computed using the ratios determined for case I.

Step 8: The moment check is made for mechanism 16 by a combination of the "trial-and-error" and graphical methods. Beam moments are drawn in sketch (*e*); column moments in sketch (*f*). First, the known plastic moments are laid off in sketch (*e*) at sections 7, 9, 13, 15, 19, 21, 24, and 26. The rest of the beam moment diagrams may be constructed either graphically (for example in span 18-21, lay off $M_{sA} = 840$ kip-ft from point A to point B and then extend line CB to obtain $M_{18} = 373$ kip-ft), or beam equilibrium equations similar to 7.10 may be used to find M_{18}, M_{23}, M_6, and M_{12}. It is next assumed that $M_{17} = M_{11} = M_p$ in the direction shown in sketch (*f*), and with this information and with equations similar to 7.10, 7.11, and 7.12 the moment diagram may be completed.

The plastic moment condition is not violated at any section, and thus $(M_p)_{II} = 252$ kip-ft.

Step 9: Of the two mechanisms examined, number 18 requires the largest M_p. The procedure for making the moment check would be the same as in step 8.

Step 11: Since case I controls, the reactions are computed for case I with the aid of sketches (*c*) and (*d*). Although the *exact* values of the horizontal reactions have not been obtained in this problem, it is an easy matter to compute their maximum possible values and this has been done. The maximum reaction corresponds to formation of a plastic hinge at the opposite end of the column.

Step 12: The tabulation in this step shows the required plastic moments, the corresponding Z-values, and the sections selected to transmit the required plastic moments. Of course, the columns will need to be checked. In the third column the heading $Z = Z_t$ denotes the fact that the indicated plastic modulus is a "trial" value for some of the shapes. "Column" shapes are specified for the appropriate members.

Step 13: All columns except two (members 5–18 and 16–26) require an adjustment in plastic modulus to transmit the needed moment in the presence of axial force. The necessary correction will be made in one operation in the next step.

Step 14: Based upon the P/P_y ratios computed in step 13, and using Fig. 4.23 and Eq. 7.28, new values are computed for the required plastic modulus for each column that has an axial thrust greater than $0.15P_y$. L/r_x is very low—so low, in fact, that the effect of slenderness ratio could have been neglected altogether. The values of M_o/M_p shown in column 5 are obtained directly from Fig. 4.23. This figure applies for the condition wherein moment is applied at one end, the bottom end being pinned. It is clear from sketch (*c*) that this applies to the columns in the lower story and is a conservative approximation to the columns in the upper story. Actually all of the columns in the upper story are bent to a greater or lesser degree in double curvature.

In Eq. (7.28), solved in column 6 of the tabulation for this step, the value of the trial plastic modulus, Z_t, is the required value obtained in step 12, not the value actually furnished. It would be over-conservative to use the latter.

The shape originally selected for column 2–11 was a 12WF79 (step 12) with $P/P_y = 0.591$ (step 13). The shape selected in the redesign is an 18WF114 using a procedure that is overconservative for two reasons. In the first place, the moment at section 11 is considerably less than $M_p = 313$ kip-ft (the value shown in sketch (*d*) is 81 kip-ft). Secondly, the larger shape with greater area will mean a reduced P/P_y ratio, and this would increase the M_o/M_p ratio. In fact, the slenderness ratio in the strong direction is so low that this factor could be ignored and Eq. 4.11 could be used to compute the required value of Z. Thus,

$$Z = Z_t \left(\frac{P}{P_y} + 0.85 \right) = 114(0.591 + 0.85) = 182 \text{ in.}^3$$

A 14WF111 shape would be required ($Z = 196.0$ in.^{3}, $A = 32.65$ in.^{2}). Hence $P/P_y = 453/(33)(32.65) = 0.42$. Checking Eq. 4.9,

$$M_{pc} = 1.18 \left(1 - \frac{P}{P_y} \right) M_p = 1.18(1 - 0.42)(33)(196.0) = 447 \text{ kip-ft}$$

The maximum moment that must be transmitted is 313 kip-ft, so this refinement shows that the 14WF111 shape would be more than adequate.

Step 15: All cross-section proportions are adequate, although the 27WF94 shape used for beam 6–9 is on the borderline. Of course the P/P_y ratio is very small in the beams, so a d/w ratio of 55 would not be excessive for beam 6–9.

Step 16: For illustration, two of the connections are examined to see if the webs are adequate to resist the shear forces. Rather than use Eq. 5.7 for corner connection 26 as was done in previous examples, in this problem Eq. 5.49a is employed. It is found that the actual web thickness is considerably less than the required value. To determine the required thickness of diagonal stiffener, Eq. 5.12 is used even though the two members are not of the same depth. Minimum values of b, S, and d are used and a required thickness of 0.615 in. is obtained. A $\frac{5}{8}$-in. plate would therefore be adequate. An alternate method of proportioning this diagonal stiffener is to use procedure employed for side connection 15–16–17 as shown in step 17.

Step 17: Connection 15–16–17 is deficient in a similar way and therefore a diagonal stiffener is proportioned to carry the necessary additional shear.

The problem would be completed with the checking of such additional design guides as necessary. The final layout is shown in sketch (i).

An example of a tier building design with diagonals as wind bracing has been carried out at the University of Michigan in which an elastically designed eight-story steel frame building six bays wide by three bays deep was redesigned by use of plastic analysis.

The framing plan is shown in Fig. 9.7. The loading consists of 20 psf wind load on all of the vertical projected surfaces. Roof snow load was 40 psf and the floors were 4-in. concrete slabs designed for 60-psf live load. In the original elastic design, which relied upon the bending strength of the columns and beams to resist wind shear without the assistance of diagonal bracing, the columns were not affected by wind, but the three-span beams in the first four floors of the column bents were determined by wind moment requirements.

In the redesign, diagonal bracing was introduced as shown in Fig. 9.7 to take all of the wind load. The columns were designed for vertical load only. They were not counted upon at all in the plastic design. In other words, each beam was designed by the plastic method as an ordinary continuous beam using methods similar to those used in Art. 8.2. Connections to exterior columns used standard connections with zero design moment capacity and in the plastic analysis these were assumed as simple supports. For the 20-ft spans in both intermediate and column bents, 14WF30 beams were required for simple beam design. In the plastic design it was possible to change these to 10B17's in the center bay and 12B22's in the exterior bays.

Since the columns were the same for all designs they are not included in the following weight comparison which is for roof and floor framing only.

Original Design	Conventional Simple Beam Design with Wind Carried by Cross-Bracing Rather than Bending Moment	Plastic Redesign with Wind Carried by Cross-Bracing
262 tons	242 tons	189 tons

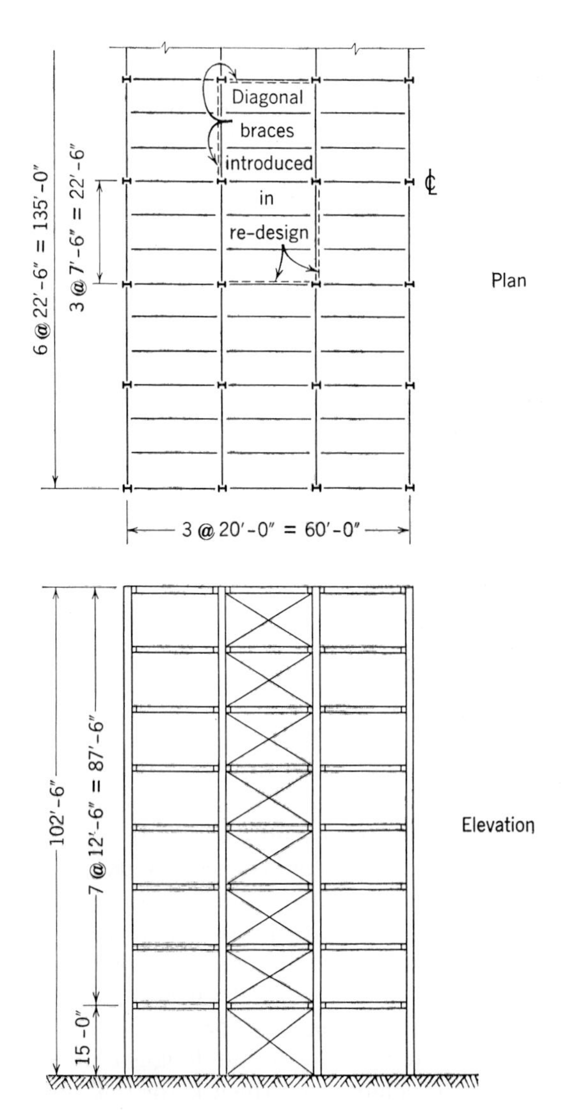

Fig. 9.7. Multi-story building.

Thus by plastic redesign of the beams a 28% saving in weight of the floor framing is possible when compared with the original design, the percentage being 22% when based on the conventional cross-bracing design.

Connection cost would be increased at the interior supports of the intermediate floor beams but would be decreased at all the exterior columns where simple supports would be substituted in lieu of wind-bracing connections. Connections in the column bents would be little changed in cost since, in the original design, they were necessarily continuous for wind load resistance. Bending moment in the exterior columns would be reduced in comparison with conventional design.

The use of diagonal bracing to resist wind is practicable when a bay can be found in the structure in which clearance or access would not otherwise suffer. The above weight comparison between two conventional designs using wind moment connections and using cross bracing confirms that the latter is more efficient. References 9.4 and 9.5 are recent examples.

In Refs. 1.5, 3.4, and 4.19, special problems of minimum weight design are treated and other techniques for analyzing multi-story structures are given. Reference is made to these works for the interested student. In summary, the outlook for the immediate future would involve application of the same principles as described earlier for structures in which the columns are designed primarily for bending and with relatively low axial force. Two-story and, in certain cases, three-story structures would probably fall in this category. Otherwise for tall multi-story buildings with diagonal cross-bracing, plastic analysis may be applied to the proportioning of the beams. Further research will more specifically define applications to multi-story frames.

References

CHAPTER 1 INTRODUCTION

1.1 Gabor Kazinczy, "Kiserletek Befalazott Tartokkal" ("Experiments with Clamped Girders"), *Betonszemle,* **2**(4), p. 68 (1914); **2**(5), p. 83 (1914); **2**(6), p. 101 (1914).

1.2 H. Maier-Leibnitz, "Contribution to the Problem of Ultimate Carrying Capacity of Simple and Continuous Beams of Structural Steel and Timber," *Die Bautechnik,* **1**(6) (1927).

1.3 J. A. Van den Broek, *Theory of Limit Design,* John Wiley and Sons, New York, 1948.

1.4 J. F. Baker, "A Review of Recent Investigations into the Behavior of Steel Frames in the Plastic Range," *J. Inst. Civil Eng.*, London, **31** No. 3, pp. 185–240, (1949).

1.5 J. F. Baker, M. R. Horne, and J. Heyman, *The Steel Skeleton, Vol. II: Plastic Behavior and Design,* Cambridge University Press, Cambridge, 1956.

1.6 P. S. Symonds and B. G. Neal, "Recent Progress in the Plastic Methods of Structural Analysis," *J. Franklin Inst.*, **252,** pp. 383–407, 469–492 (1951).

1.7 W. W. Luxion and B. G. Johnston, "Plastic Behavior of Wide Flange Beams," *Welding J.*, **27**(11), pp. 538-*s* to 554-*s* (1948).

1.8 B. G. Johnston, C. H. Yang, and L. S. Beedle, "An Evaluation of Plastic Analysis as Applied to Structural Design," *Welding J.*, **32**(5), p. 224-*s* (1953).

1.9 L. S. Beedle, B. Thurlimann, and R. L. Ketter, *Plastic Design in Structural Steel,* Lehigh University, Bethlehem, Pa., and American Institute of Steel Construction, New York, 1955.

1.10 W. S. Atkins and E. M. Lewis, "Developments of Design and Fabrication in Recent British Structures," *Trans. Inst. Welding,* **14,** pp. 74–84 (1951).

1.11 D. T. Wright, "Plastic Design Is Used Successfully," *Eng. News-Record,* **158**(15), p. 59 (April 11, 1957).

1.12 E. R. Estes, Jr., "Plastic Design of Warehouse Saves Steel," *Civil Engineering,* **27**(9), p. 608 (Sept. 1957).

1.13 J. Heyman, *Plastic Design of Portal Frames*, Cambridge University Press, Cambridge, 1957.

1.14 A. Huber and L. S. Beedle, "Residual Stress and the Compressive Strength of Steel," *Welding J.*, **33**(12), p. 589-*s* (1954).

1.15 J. F. Baker, "Shortcomings of Structural Analysis," *Trans. N. E. Coast Inst. Eng. Ship.*, **68**, pp. 31–50 (1951).

1.16 G. C. Driscoll, Jr., and L. S. Beedle, "The Plastic Behavior of Structural Members and Frames," *Welding J.*, **36**(6), p. 275-*s* (1957).

1.17 *Commentary on Plastic Design*, Fritz Laboratory Report No. 205.53, Lehigh University, 1958.

CHAPTER 2 FLEXURE OF BEAMS

2.1 *Steel Construction*, Am. Inst. Steel Constr., New York, 5th ed. (1950).

2.2 C. H. Yang, L. S. Beedle, and B. G. Johnston, "Residual Stress and the Yield Strength of Steel Beams," *Welding J.*, **31**(4), p. 205-*s* (1952).

2.3 L. S. Beedle and A. W. Huber, *Residual Stress and the Compressive Properties of Steel—a Summary Report*, Fritz Laboratory Report No. 220A.27, Lehigh University, 1957.

2.4 K. E. Knudsen, C. H. Yang, B. G. Johnston, and L. S. Beedle (with appendix prepared by W. H. Weiskopf), "Plastic Strength and Deflection of Continuous Beams," *Welding J.*, **32**(5), p. 240-*s* (1953).

2.5 C. Batho, "The Effect of Concrete Encasement on the Behavior of Beam and Stanchion Connections," *Struct. Eng.*, **16**, p. 427 (1938).

2.6 J. W. Roderick and I. H. Phillipps, "Carrying Capacity of Simply Supported Mild Steel Beams," *Engineering Structures*, p. 9, Academic Press, N. Y., 1950.

CHAPTER 3 ANALYSIS OF STRUCTURES FOR ULTIMATE LOAD

3.1 N. J. Hoff, *The Analysis of Structures*, John Wiley and Sons, New York, 1956.

3.2 H. J. Greenberg and W. Prager, "Limit Design of Beams and Frames," *Trans. ASCE*, **117**, p. 447 (1952).

3.3 B. G. Neal, and P. S. Symonds, "The Calculation of Collapse Loads for Framed Structures," *J. Inst. Civ. Engrs.*, **35**, 20 (1950).

3.4 M. R. Horne, "A Moment Distribution Method for the Analysis and Design of Structures by the Plastic Theory," *Proc. Inst. Civil Eng.*, **3** (Part 3) p. 51, April, 1954.

3.5 J. M. English, "Design of Frames by Relaxation of Yield Hinges," *Trans. ASCE*, **119**, p. 1143 (1954).

3.6 B. Thurlimann, *Analysis of Frames for Ultimate Strength*, Fritz Laboratory Report No. 205.29, Lehigh University, 1955.

3.7 W. J. Hall and N. M. Newmark, "Shear Deflection of Wide-Flange Steel Beams in the Plastic Range," *Trans. ASCE*, **122**, p. 666 (1957).

3.8 J. M. Ruzek, K. E. Knudsen, E. R. Johnston, and L. S. Beedle, "Welded Portal Frames Tested to Collapse," *Welding J.*, **33**(9), p. 469-*s* (1954).

3.9 C. G. Schilling, F. W. Schutz, Jr., and L. S. Beedle, "Behavior of Welded Single-Span Frames under Combined Loading," *Welding J.*, **35**(5), p. 234-*s* (1956).

3.10 J. F. Baker and J. W. Roderick, "Tests on Full-Scale Portal Frames," *Proc. Inst. Civil Eng.*, Part I, **1** (1952) p. 71.

3.11 J. F. Baker and K. G. Eickhoff, "The Behaviour of Saw-Tooth Portal Frames," Conference on the Correlation between Calculated and Observed Stresses and Displacements in Structures, Preliminary volume, *Inst. Civil Eng.*, p. 107, 1955.

3.12 J. F. Baker and K. G. Eickhoff, "A Test on a Pitched Roof Portal," *Br. Weld. Res. Assoc.*, Report No. FE 1/35 (Jan., 1954).

3.13 E. P. Popov and J. A. Willis, Plastic Design of Cover-Plated Continuous Beams, *J. ASCE*, **84**(EM 1), p. 1495-1 (1958).

CHAPTER 4 SECONDARY DESIGN PROBLEMS

4.1 M. R. Horne, "The Plastic Theory of Bending with Particular Reference to the Effect of Shear Forces," *Proc. Royal Soc.*, **207**, p. 216 (1951).

4.2 G. Haaijer, "Plate Buckling in the Strain-Hardening Range," *Proc. ASCE*, **83** (EM-2) p. 1212-1 (April 1957).

4.3 G. Haaijer and B. Thurlimann, "On Inelastic Buckling in Steel," *Proc. ASCE* **84** (EM-2) p. 1581 (April, 1958).

4.4 F. Bleich, *Buckling Strength of Metal Structures*, McGraw-Hill Book Co., New York, 1952.

4.5 G. C. Driscoll, Jr., *Rotation Capacity Requirements for Beams and Frames of Structural Steel* (Dissertation), Lehigh University, 1958.

4.6 M. W. White, *The Lateral Torsional Buckling of Yielded Structural Steel Members* (Dissertation), Lehigh University, 1956.

4.7 B. Thurlimann, T. Kusuda, R. G. Sarubbi, *Bracing of Beams in Plastically Designed Steel Structures*, Fritz Laboratory Report No. 205E.11, Lehigh University, 1958.

4.8 A. A. Topractsoglou, B. G. Johnston, and L. S. Beedle, "Connections for Welded Continuous Portal Frames," *Welding J.*, **30**(7), p. 359-*s* (1951); **30**(8), p. 297-*s* (1951); **31**(11), p. 543-*s* (1952).

4.9 *Guide to Design Criteria for Metal Compression Members*, Column Research Council, 1960.

4.10 Y. Fujita, *Built-Up Column Strength* (Dissertation), Lehigh University, 1956.

4.11 P. P. Bijlaard, G. P. Fisher, G. Winter, "Eccentrically Loaded, End Restrained Columns," *Trans. ASCE*, **120**, p. 1070 (1955).

4.12 T. V. Galambos and R. L. Ketter, *Further Studies of Columns under Combined Bending and Thrust*, Fritz Laboratory Report No. 205A.19, Lehigh University, June, 1957.

4.13 N. M. Newmark, "Numerical Procedure for Computing Deflections, Moments, and Buckling Loads," *Trans. ASCE*, **108**, p. 1161 (1943).

4.14 R. L. Ketter, E. L. Kaminsky, and L. S. Beedle, "Plastic Deformation of Wide-Flange Beam-Columns," *Trans. ASCE*, **120**, p. 1028 (1955).

4.15 B. Thurlimann, "Modifications to Simple Plastic Theory," *Proceedings, AISC National Engineering Conference*, p. 50 (1956).

4.16 Plasticity Committee (WRC), *Control of Steel Construction to Avoid Brittle Failure*," Welding Research Council, N. Y., 1957.

4.17 L. A. Harris and N. M. Newmark, *Effect of Fabricated Edge Conditions on Brittle Fracture of Structural Steels,* Univ. Illinois, Feb., 1957.

4.18 "Fatigue Tests of Beams in Flexure," *Welding J.,* **30**(3) p. 105-*s* (1951).

4.19 B. G. Neal, *The Plastic Methods of Structural Analysis,* John Wiley and Sons, N. Y., 1957.

4.20 A. T. Gozum and G. Haaijer, *Deflection Stability (Shakedown) of Continuous Beams,* Fritz Laboratory Report No. 205G.1, Lehigh University, 1955.

CHAPTER 5 CONNECTIONS

5.1 A. A. Toprac and L. S. Beedle, "Further Studies of Welded Corner Connections," *Welding J.,* **34**(7), p. 348-*s* (1955).

5.2 J. W. Fisher and G. C. Driscoll, Jr., *Corner Connections Loaded in Tension,* Fritz Laboratory Report No. 205C.23, Lehigh University, 1958.

5.3 D. T. Wright, "Design of Knee Joints for Steel Rigid Frames," *Br. Weld. J.,* (June 1957).

5.4 J. W. Fisher, G. C. Driscoll, Jr., and F. W. Schutz, Jr., *Behavior of Welded Corner Connections,* Fritz Laboratory Report No. 205C.21, Lehigh University, 1957.

5.5 J. E. Smith, *Behavior of Welded Haunched Corner Connections,* Fritz Laboratory Report No. 205C.20, Lehigh University, July 1956.

5.6 A. W. Hendry, "An Investigation of the Strength of Welded Portal Frame Connections," *The Structural Engr.,* **28**(10), p. 265 (1950).

5.7 A. H. Stang and M. Greenspan, "Strength of a Welded Steel Rigid Frame," *J. Research, Natl. Bur. Standards,* **23,** p. 145 (July 1939).

5.8 I. Lyse and W. E. Black, "An Investigation of Steel Rigid Frames," *Trans. ASCE,* **107,** p. 127 (1942).

5.9 F. Bleich, *Design of Rigid Frame Knees,* American Institute of Steel Construction, N. Y., 1943.

5.10 W. R. Osgood, "Theory of Flexure for Beams with Non-Parallel Extreme Fibers," *Trans. ASME,* **61** (1939).

5.11 H. C. Olander, "Stresses in the Corners of Rigid Frames," *Trans. ASCE,* **119,** p. 797 (1954).

5.12 J. D. Griffiths, *Single-Span Rigid Frames in Steel,* American Institute of Steel Construction, N. Y., 1948.

5.13 J. W. Fisher, *Plastic Analysis of Haunched Connections* (Thesis), Lehigh University, 1958.

5.14 A. N. Sherbourne and C. D. Jensen, *Direct Welded Beam-Column Connections,* Fritz Laboratory Report No. 233.12, Lehigh University, 1957.

5.15 R. N. Khabbaz and C. D. Jensen, *Four-Way Welded Interior Beam-Column Connections,* Fritz Laboratory Report No. 233.13, Lehigh University, 1957.

5.16 J. W. Fisher, G. C. Driscoll, Jr., and L. S. Beedle, "Plastic Design of Straight Rigid Frame Knees," *Welding Research Council Bulletin,* No. 39, 1958.

5.17 G. C. Driscoll, Jr., "Test of a Two-Span Gabled Portal Frame," *Proc. AISC National Engr. Conf.,* p. 74, 1956.

CHAPTER 6 DEFLECTIONS

6.1 C. H. Yang, L. S. Beedle, and B. G. Johnston, "Plastic Design and the Deformation of Structures," *Welding J.*, **30**(7), p. 348-*s* (1951).

6.2 P. S. Symonds and B. G. Neal, "The Interpretation of Failure Loads in the Plastic Theory of Continuous Beams and Frames," *J. Aero. Sci.*, **19,** p. 15 (1952).

6.3 R. J. Roark, *Formulas for Stresses and Strain*, 3rd ed., McGraw-Hill Book Co., N. Y., 1954.

CHAPTER 7 DESIGN GUIDES

7.1 *Plastic Design in Steel*, American Institute of Steel Construction, N. Y., 1959.

7.2 F. W. Schutz, "Plastic Design and the Steel Fabricator," *Proceedings, AISC Nat'l Engr. Conf.*, p. 69, 1956.

7.3 *The Collapse Method of Design*, British Constructional Steelwork Assn., Publication No. 5, 1952.

7.4 R. L. Ketter, *Plastic Design of Multi-Span Rigid Frames* (Dissertation), Lehigh University, 1956.

7.5 R. F. Pray and C. D. Jensen, "Welded Top-Plate Beam-Column Connections," *Welding J.*, **35**(7), p. 338-*s* (1956).

7.6 L. Schenker, C. G. Salmon, and B. G. Johnston, *Structural Steel Connections*, AFSWP Report No. 352, Univ. Michigan, June, 1954.

7.7 *Standard Code for Arc and Gas Welding in Building Construction*, American Welding Society, New York, 1946.

7.8 *Specifications for the Assembly of Structural Joints Using High-Strength Bolts*, Research Council on Riveted and Bolted Structural Joints, Feb., 1954.

7.9 *High Strength Bolting for Structural Joints*, Bethlehem Steel Company Booklet No. 414.

CHAPTER 9 STEEL FRAME DESIGN

9.1 *Procedure Handbook of Arc Welding Design and Practice*, 10th ed., Lincoln Electric Co., Cleveland, p. 904, 1955.

9.2 R. L. Ketter, "Solution of Multi-Span Frames," *Br. Welding J.*, **4**(1), p. 32 (1957).

9.3 W. H. Weiskopf, "Plastic Design and the Tier Building," *Proceedings, AISC Nat'l Engr. Conf.*, p. 66, 1956.

9.4 "Trussed Framing Stiffens Tall Slender Building," *Civ. Eng.* p. 54, Aug. 1956.

9.5 "A Skyscraper Crammed with Innovations," *Steel Construction Digest*, **14**(3), p. 8 (1957).

Appendix 1

SPACING OF LATERAL BRACING

The purpose of this appendix is to outline a procedure for checking the adequacy of the spacing of bracing to prevent lateral buckling and, in particular, when the selected spacing does not meet the requirements of Art. 4.5. The problem is a twofold one: First of all, what is the lateral buckling strength of an elastic-plastic segment of a member bent about the strong axis and which is required to sustain different amounts of rotation at a plastic hinge. Secondly, what is the necessary hinge rotation, namely the required rotation of a given plastic hinge, to assure that the total structure reaches the computed ultimate load? In view of the fact that research is currently (1958) being conducted actively on both phases of this problem, the procedure here described is considered as tentative. More work needs to be done to relax certain provisions that in some cases appear too restrictive.[1.17]

The procedure is as follows:

1. Assume a purlin spacing (usually dictated by available roofing materials). Compute L/r_y.
2. Examine the structure to see which segment (or segments) containing a plastic hinge will be the most critical. For equal spacing of bracing members it will be the segment at each hinge location with the largest moment ratio. Referring to the moment diagram in Fig. 4.18, this segment is called the "braced" span, with length L_B.

3. Compute the precise moment ratio for the laterally unsupported span being considered (length $= L_B$). This moment ratio is the ratio of the smaller moment to the plastic moment (M/M_p).

4. Compare the slenderness ratio existing in the structure with that which would be permitted for the particular moment ratio according to Eq. 4.28, neglecting for the time being the parameter $H_B/L_B\phi_p$ and the correction factor due to fixity C_f. The selected purlin spacing is adequate if its slenderness ratio is less than that permitted according to this equation. Otherwise, further refinements may be made as follows.

5. As a first refinement, evaluate the end fixity correction, C_f, which gives the "fixing" influence of the adjoining spans. If the distance between points of lateral support are equal,* then according to Ref. 4.7 the value of C_f is given by

$$C_f = 1.0 + 0.2\left\{1 - \left(\frac{L}{L_l}\right)^2\right\}\left\{0.9 + \frac{1 - \left(\frac{L}{L_s}\right)^2}{1 - \left(\frac{L}{L_l}\right)^2}\right\} \tag{A.1}$$

where C_f = correction factor for end fixity
L = distance between points of lateral support
L_l = critical length (with $C_f = 1.0$) of adjacent span which has the longer critical length
L_s = critical length (with $C_f = 1.0$) of adjacent span which has the shorter critical length.

The critical slenderness ratio will depend upon whether the member is elastic or plastic and upon the moment ratio of the braced span. Thus:

(a) For an elastic *adjacent* segment with M/M_p in the *braced* segment less than 0.9,

$$\frac{L}{r_y} = \frac{134}{\sqrt{|M/M_p|}} + 60(1 - g_A) \tag{A.2}$$

where g_A = moment ratio in the adjacent segment

(b) For an elastic adjacent segment with $M/M_p \geq 0.9$,

$$\frac{L}{r_y} = \frac{134}{\sqrt{|M/M_p|}} + 60(1 - g_A) - 1100\left(\frac{M}{M_p} - 0.9\right) \tag{A.3}$$

* When the lateral support spacing is not uniform, Ref. 4.7 suggests a procedure for determining C_f that is only slightly more involved than that which follows.

(c) For an elastic-plastic segment

$$\left.\begin{aligned} \frac{L}{r_y} &= 30 && (1.0 > g_A > 0.6) \\ \frac{L}{r_y} &= 48 - 30g_A && (0.6 > g_A > -1.0) \end{aligned}\right\} \quad \text{(A.4)}$$

If the segment is the critical segment, then $g_A = M/M_p$. The procedure to determine C_f, then, would be as follows: first, compute moment ratios from the moment diagram for the buckling segment and the adjacent segment. Next, compute the critical lengths for these segments, including the braced segment and assuming each of them to be torsionally unrestrained ($C_f = 1.0$)(Eqs. A.2, A.3, or A.4 as required). Then, C_f may be computed from Eq. A.1.

6. Multiply the allowable slenderness ratio obtained from Eq. 4.28 by C_f. This gives a value which can be compared to the ratio existing in the structure.

If the selected spacing is still too great and a closer spacing is undesirable, the rotation requirement might next be checked. The principles and general methods for computing hinge rotations are described in Chapter 6. However, the calculations are somewhat tedious and, if required, would tend to obviate one of the advantages of plastic design, namely, its simplicity. Alternatively, charts may be used which make possible a more rapid determination of the magnitude of hinge rotations and the sequence of formation of plastic hinges. If a hinge were the last to form, then a slenderness ratio of about 100 would be adequate (as shown, approximately, in Art. 4.5). A hinge that forms first requires more rotation capacity, thus decreasing the allowable L/r_y.*

Figure 4.19 presents some of the limiting values obtained as a result of a study of the rotation capacity problem for beams, single-span and multi-span frames with pinned bases.[4.5] It shows that the last hinge occurs in the rafter in most cases until the columns become unusually high with respect to the frame span.

As regards the hinge rotation, the value $H/L\phi_p$ in Fig. 4.19 is a nondimensional function in which H is the calculated total hinge angle, L is the frame or beam span, and ϕ_p is the curvature at $M_p(\phi_p = M_p/EI)$. Before this function can be used in Fig. 4.18 it must be corrected to L_B (the length of the braced segment) and H_B (the hinge angle *within the braced segment*). It was suggested in Ref. 4.6 that the value H_B may

* The British recommendation [7.3] for a solution to this problem is to use $L/r_y =$ 100.

be determined from the gradient of the moment diagram, in which case the following equation may be used to compute H_B:

$$H_B = \frac{H}{1 + \dfrac{M_p - M}{M_p - M_A}\left(\dfrac{L_A}{L_B}\right)} \tag{A.5}$$

where M = moment at the far end of the braced segment which has M_p at the near end

M_A = moment at far end of adjacent segment which has M_p at the near end

L_A = length of adjacent segment.

For the case shown in Fig. 4.18, $M_A = M_L$ and $L_A = L_L$. Thus, the final step in the procedure would be:

7. Determine the value $H/L\phi_p$ either from a deformation analysis or from charts (Fig. 4.19 summarizes a portion of the pertinent information). Compute $H_B/L_B\phi_p$ and revise the allowable slenderness ratio according to Fig. 4.18. In lieu of this step, of course, a larger shape could be specified which would supply the needed r_y value.

In applying the above procedure to members in a frame, two important assumptions are made in addition to those already noted: (1) It is assumed that the restraint provided by a girder to a column (or by a column to a girder) is the same as if one member were merely a continuation of the other. (2) It assumes that the effect of axial force on lateral buckling may be neglected in girders and columns as long as this force is of moderate intensity.

EXAMPLE

An illustration of the above procedure will now be given for the problem shown in Plate VI, p. 278.

1. The purlin and girt spacing is assumed.

Purlins	Girts
$L_B = 6.0$ ft	$L_B = 5.0$ ft from top
$\dfrac{L_B}{r_y} = \dfrac{(6.0)(12.0)}{1.63} = 43.2$	$\dfrac{L_B}{r_y} = \dfrac{(5.0)(12.0)}{1.63} = 36.8$

Since the procedure is developed for equal spacing of lateral support, the calculations will be made on the basis that the girt spacing is also 6 ft.

2. The critical segments with largest moment ratios are selected.

The critical segment in the rafter would be portion A–B (sketch d). However, the last hinge forms at the center of the rafter and since $L_B/r_y < 100$, the spacing there is satisfactory.

At the corner, segment D–E will be critical since it has a larger moment ratio than segment C–D.

3. Compute moment ratios for critical span ($L_B = D$–E) and for the adjacent spans ($L_L = C$–D, $L_R = E$–F).

$$\frac{M}{M_p} = \frac{140}{320} = +0.437$$

$$g_L = \frac{38}{320} = +0.12$$

$$g_R = \frac{-40}{140} = -0.29$$

4. Compare L_B/r_y with allowable value from Eq. 4.28:

$$\left(\frac{L_B}{r_y}\right)_{cr} = 48 - 30\frac{M}{M_p}$$

$$= 48 - 30(0.437) = 35.1 < 43.2 \quad \text{[More refined check is necessary]}.$$

5. Compute C_f, the correction due to fixity.

$$C_f = 1.0 + 0.2\left\{1 - \left(\frac{L}{L_l}\right)^2\right\}\left[0.9 + \frac{1 - \left(\frac{L}{L_s}\right)^2}{1 - \left(\frac{L}{L_l}\right)^2}\right]$$

Span L (segment C–D) is elastic-plastic, therefore use Eq. A.4 to compute critical slenderness ratio.

$$\left(\frac{L_L}{r_y}\right)_{cr} = 48 - 30g_L = 48 - 30(0.12) = \underline{44.4}$$

Span R (segment E–F) is elastic and M/M_p for the buckling span L_B is less than 0.9; therefore use Eq. A.2 to compute the critical length.

$$\left(\frac{L_R}{r_y}\right)_{cr} = \frac{134}{\sqrt{\left|\frac{M}{M_p}\right|}} + 60(1 - g_R) = \frac{134}{\sqrt{0.437}} + 60(1 + 0.29) = \underline{281}$$

$$(L_L)_{cr} = (44.4)(1.63) = 72.5 \text{ in.} = L_s$$

$$(L_R)_{cr} = (281)(1.63) = 457 \text{ in.} = L_l$$

$$C_f = 1.0 + 0.2\left\{1 - \left(\frac{72}{457}\right)^2\right\}\left\{0.9 + \frac{1 - (72/72.5)^2}{1 - (72/457)^2}\right\} = 1.18$$

6. Compute revised critical slenderness ratio on the basis of correction due to fixity.

$$\left(\frac{L_B}{r_y}\right)_{cr} = \frac{L_b}{r_y}C_f = (35.1)(1.18) = 41.5 < 43.2$$

Although the "allowable" slenderness ratio (41.5) is less than that of the assumed purlin spacing (43.2), it only differs by $3\frac{1}{2}\%$ and is considered adequate.

Appendix 2

PLASTIC MODULUS TABLE

Z	Shape	Z	Shape	Z	Shape
1255.0	**36 WF 300**	466.0	**33 WF 130**	253.0	**24 WF 94**
1167.0	**36 WF 280**	464.5	14 WF 246	247.9	18 WF 114
1076.0	**36 WF 260**	463.7	24 WF 160	242.7	14 WF 136
1008.0	**36 WF 245**	452.0	27 WF 145	238.8	24 I 100
942.7	**36 WF 230**	445.4	14 WF 237	226.5	18 WF 105
918.2	33 WF 240	436.7	30 WF 132	226.3	21 WF 96
869.3	14 WF 426	427.2	14 WF 228	225.9	14 WF 127
836.2	**33 WF 220**	416.0	24 WF 145	224.0	**24 WF 84**
803.0	14 WF 398	408.0	14 WF 219	220.5	24 I 90
767.2	**36 WF 194**	407.4	**30 WF 124**	210.9	14 WF 119
754.4	33 WF 200	391.7	14 WF 211	209.7	12 WF 133
737.3	14 WF 370	377.6	**30 WF 116**	206.0	18 WF 96
733.9	30 WF 210	373.6	14 WF 202	203.0	**24 I 79.9**
716.9	**36 WF 182**	369.2	24 WF 130	200.1	**24 WF 76**
666.7	**36 WF 170**	357.0	21 WF 142	196.0	14 WF 111
659.6	30 WF 190	355.1	14 WF 193	192.0	20 I 95
623.3	**36 WF 160**	345.5	**30 WF 108**	191.6	21 WF 82
611.5	14 WF 314	342.8	27 WF 114	186.4	12 WF 120
593.0	30 WF 172	337.5	14 WF 184	186.0	16 WF 96
592.2	14 WF 320	336.6	24 WF 120	181.0	14 WF 103
579.8	**36 WF 150**	321.3	14 WF 176	177.6	18 WF 85
558.3	33 WF 152	317.8	21 WF 127	177.3	20 I 85
556.9	27 WF 177	311.5	12 WF 190	172.1	**21 WF 73**
551.6	14 WF 287	307.7	24 WF 110	169.0	16 WF 88
513.2	**33 WF 141**	304.4	**27 WF 102**	166.6	14 WF 95
504.3	27 WF 160	302.9	14 WF 167	163.4	12 WF 106
502.4	14 WF 264	298.0	24 I 120	160.5	18 WF 77
		286.3	14 WF 158	159.8	**21 WF 68**
		278.0	21 WF 112	151.8	12 WF 99
		278.3	**24 WF 100**	151.5	20 I 75
		277.7	**27 WF 94**	151.3	14 WF 87
		273.0	24 I 105.9	147.5	10 WF 112
		270.2	14 WF 150	145.5	16 WF 78
		259.2	12 WF 161	145.4	14 WF 84
		254.8	14 WF 142	144.7	18 WF 70

Z	Shape
144.1	**21 WF 62**
140.2	12 WF 92
137.3	20 I 65.4
134.0	14 WF 78
131.8	18 WF 64
131.6	16 WF 71
130.1	10 WF 100
129.1	12 WF 85
125.6	14 WF 74
123.8	18 I 70
122.6	**18 WF 60**
119.3	12 WF 79
117.9	16 WF 64
114.8	14 WF 68
114.4	10 WF 89
111.6	**18 WF 55**
108.1	12 WF 72
106.2	16 WF 58
103.5	**18 I 54.7**
102.4	14 WF 61
100.8	**18 WF 50**
97.7	10 WF 77
97.0	12 WF 65
92.7	**16 WF 50**
90.7	10 WF 72
87.1	14 WF 53
86.5	12 WF 58
82.8	10 WF 66
82.0	**16 WF 45**
78.5	14 WF 48
78.2	12 WF 53
76.5	15 I 50
75.1	10 WF 60
72.7	**16 WF 40**
72.6	12 WF 50
70.1	8 WF 67
69.7	14 WF 43
68.6	15 I 42.9
67.0	10 WF 54
64.9	12 WF 45
63.9	**16 WF 36**
61.5	14 WF 38
60.7	12 I 50
60.3	10 WF 49
59.9	8 WF 58
57.6	12 WF 40
55.0	10 WF 45
54.5	**14 WF 34**
52.5	12 I 40.8
51.4	12 WF 36
49.0	8 WF 48
47.1	**14 WF 30**
47.0	10 WF 39
44.4	12 I 35
44.0	12 WF 31
41.6	12 I 31.8
39.9	8 WF 40
38.8	10 WF 33
38.0	**12 WF 27**
35.2	10 I 35
34.7	8 WF 35
34.7	10 WF 29
32.8	8 M 34.3
30.4	8 WF 31
29.5	**10 WF 25**
29.4	**12 B 22**
28.0	10 I 25.4
27.1	8 WF 28
24.8	**12 B 19**
24.1	10 WF 21
23.4	8 M 24
23.1	8 WF 24
21.6	**10 B 19**
20.6	**12 B 16.5**
19.2	8 I 23
19.1	8 WF 20
19.0	6 WF 25
18.6	10 B 17
17.9	6 M 25
17.5	8 M 20
17.4	**12 B 14**
16.3	8 I 18.4
16.0	10 B 15
15.8	8 WF 17
15.7	8 M 17
15.0	6 WF 20
14.6	6 M 20
14.4	7 I 20
14.2	**12 Jr 11.8**
13.6	8 B 15
12.1	**10 B 11.5**
11.9	7 I 15.3
11.6	6 B 16
11.4	8 B 13
11.4	5 WF 18.5
11.3	6 WF 15.5
11.1	5 M 18.9
10.5	6 I 17.25
9.6	5 WF 16
9.2	**10 Jr 9**
8.9	8 B 10
8.4	6 I 12.5
8.3	6 B 12
7.4	5 I 14.75
6.3	4 WF 13
6.1	4 M 13
5.7	**6 B 8.5**
5.6	5 I 10
5.4	**8 Jr 6.5**
4.0	**7 Jr 5.5**
4.0	4 I 9.5
3.5	4 I 7.7
2.8	**6 Jr 4.4**
2.3	3 I 7.5
1.9	3 I 5.7

Nomenclature

SYMBOLS

A	Area of cross-section
A_f	Area of two flanges of WF shape, $A_f = 2bt$
A_p	Area of plate
A_w	Area of web, $A_w = wd$
a	Column height ratio (column height $= aL$). Distance between centroids of areas above and below neutral axis. Distance from end of cantilever to critical section of beam.
b	Flange width. Breadth of rectangular cross section. Roof rise ratio (roof rise $= bL$).
C	Overturning moment parameter (windward side)
C_f	Correction factor due to end fixity (restraint)
c	Distance from neutral axis to the extreme fiber
D	Overturning moment parameter (leeward side)
d	Depth of section
d_p	Distance between two cover plates
E	Young's modulus of elasticity
E_{st}	Strain-hardening modulus
E_t	Tangent modulus
e	Eccentricity
F	Load factor of safety
f	Shape factor $= M_p/M_y = Z/S$
G	Modulus of elasticity in shear
G_{st}	Modulus of elasticity in shear at onset of strain-hardening

g_A	Moment ratio in adjacent segment
H	Hinge angle required at a plastic hinge. Horizontal reaction.
H_B	Portion of hinge angle that occurs in critical (buckling) segment of beam
h	Story height in multi-story frame
I	Moment of inertia (subscripts denote axis). Number of redundants remaining in a structure at ultimate load.
I_e	Moment of inertia of elastic part of cross section
I_p	Moment of inertia of plastic part of cross section
K	Torsion constant
KL	Effective (pin-end) length of column. K = Euler length factor.
k	Distance from flange face to end of fillet. Plastic moment ratio.
L	Span length. Actual column length. Length of bar.
L_B	Length of buckling (critical) segment
l	Length of segment (slope-deflection equation)
L_{cr}	Critical length for lateral buckling
L_l,L_s	Critical length (with C = 1.0) of adjacent spans; subscripts l and s denote larger and shorter critical lengths, respectively.
ΔL	Equivalent length of connection
M	Moment. Number of plastic hinges necessary to form a mechanism.
M_{cr}	Critical moment for lateral buckling of a beam
M_h	Moment at the haunch point
M_{max}	Maximum moment
M_o	Column end moment; a useful maximum moment
M_p	Plastic moment
M_{pc}	Plastic hinge moment modified to include the effect of axial compression
M_{ps}	Plastic hinge moment modified to include effect of shear force
M_s	Maximum moment of a simply supported beam
M_w	Moment at working (service) load
M_y	Moment at which yield point is reached in flexure
M_{yc}	Moment at which initial outer fiber yield occurs when axial thrust is present
N	Number of possible plastic hinges. Normal force.
n	Number of possible independent mechanisms. Shift of neutral axis.
P	Concentrated load
P_e	Euler buckling load
P_{max}	Maximum load

P_s	Stabilizing ("shakedown") load
P_t	Tangent modulus load
P_u	Ultimate load (theoretical)
P_w	Working (allowable) load
P_y	Axial load corresponding to yield stress level; $P = A\sigma_y$. Load on beam when yield point is reached in flexure.
Q	b/a = roof rise ratio/column height ratio
R	Rotation capacity. Radius of curved haunch.
r	Radius of gyration; subscripts denote flexure axis
S	Section modulus, I/c
S_e	Section modulus of elastic part of cross section
s	Length of compression flange of haunch
T	Force. Horizontal load applied at eaves which produces overturning moment equivalent to that of horizontal distributed load.
t	Flange thickness; subscripts c and t denote compression and tension
t_s	Stiffener thickness
t_{tr}	Transverse stiffener thickness
V	Shear force
$V_{\max}$	Maximum allowable shear force
u, v, w	Displacements in x, y, and z directions
W	Total distributed load
W_E	External work due to virtual displacement
W_I	Internal work due to virtual displacement
w	Distributed load per unit of length
w	Web thickness
X	Number of redundancies in original structure
x	Longitudinal coordinate. Distance to position of plastic hinge under distributed load.
y	Transverse coordinate
y_o	Ordinate to furthest still-elastic fiber. Distance from mid-height to neutral axis.
$\bar{y}$	Distance from neutral axis to centroid of half-area
Z	Plastic modulus, $Z = M_p/\sigma_y$
Z_e	Plastic modulus of elastic portion of cross section
Z_p	Plastic modulus of plastic portion of cross section
Z_t	Trial value of Z, neglecting axial force
z	Lateral coordinate
α	Central angle between points of tangency of curved connection
β	Angle between two nonparallel flanges

Δ	Virtual displacement
δ	Deflection. Subscripts u, w, and y denote deflection at ultimate, working, and yield load, respectively.
ϵ	Strain
ϵ_{max}	Strain at fracture
ϵ_{st}	Strain at strain-hardening
ϵ_y	Strain corresponding to theoretical onset of plastic yielding
θ	Measured angle change, rotation. Mechanism angle.
μ	Poisson's ratio
ρ	Radius of curvature
σ	Normal stress
σ_{ly}	Lower yield point
σ_p	Proportional limit
σ_r	Residual stress
σ_{ult}	Ultimate tensile strength of material
σ_{uy}	Upper yield point
σ_w	Working stress
σ_y	Yield stress level
τ	Shear stress
ϕ	Rotation per unit length, or average unit rotation; curvature
ϕ_p	M_p/EI
ϕ_{st}	Curvature at strain-hardening
ϕ_y	Curvature corresponding to first yield in flexure

ABBREVIATIONS

k	Kips
k′	Kip-ft
ksi	Kips per square inch
k/ft	Kips per foot
psf	Pounds per square foot
psi	Pounds per square inch
WF	Wide flange

GLOSSARY

Elastic Design: A design method which defines the limit of structural usefulness as the load at which a stress equal to the yield point of the material is first attained at any point.

Factor of Safety: As used in elastic design, it is a factor by which the yield point is divided to determine a working or allowable stress for the most highly stressed fiber.

Hinge Angle (H)*:* The angle of rotation through which a plastic hinge must sustain its plastic moment value.

Kip: 1000 lb.

Limit Design: A design based upon any chosen limit of structural usefulness.

Load Factor: As used in plastic design, a factor by which the working load is multiplied to determine the ultimate load.

Limit Load: The load under which a structure reaches a defined limit of structural usefulness.

Mechanism: A system of members containing a sufficient number of plastic (or real) hinges so that it is able to deform without a finite increase in load.

Mechanism Angle: The virtual angle of rotation at a plastic hinge corresponding to the virtual displacement of the mechanism.

Plastic Design: A design method which defines the limit of structural usefulness as the "ultimate load." The term "plastic" comes from the fact that the ultimate load is computed from a knowledge of the strength of steel in the plastic range.

Plastic Hinge: A yielded zone found in a beam when the plastic moment is applied. The beam rotates as if hinged, except for the constant restraining moment M_p.

Plastic Modulus: The resisting modulus of a completely yielded cross section. It is the combined statical moment about the neutral axis of the cross-sectional areas above and below that axis.

Plastic Moment: Maximum moment of resistence of a fully-yielded cross section.

Plastification: Gradual penetration of yield stress from the outer fiber towards the centroid of a section under increase of moment. Plastification is complete when the plastic moment M_p is reached.

Proportional Loading: A system of loading in which all loads increase in a constant ratio, one to the other.

Redistribution of Moment: A process which results in the successive formation of plastic hinges until the ultimate load is reached. Through transfer of moment which results from formation of "hinges," the less-highly stressed portions of a structure also may reach the M_p value.

Rotation Capacity: The angular rotation which a given cross-sectional shape can accept at the plastic moment value without prior local failure.

Shape Factor: The ratio M_p/M_y, or Z/S, for a cross section.

Ultimate Load: The largest load a structure will support, excluding such factors as instability and fracture. It is the load that is reached when a sufficient number of yield zones have formed to permit the structure to deform plastically without further increase in load.

Yield Moment: In a member subjected to bending, it is the moment at which an outer fiber attains yield-point stress.

Index

(2) The equilibrium condition is satisfied at ultimate load.
(3) There is "unrestricted" plastic flow at the ultimate load.

In principle, this simple example illustrates the essential features of the plastic method; what is required to complete the plastic analysis of an indeterminate beam or frame is to satisfy a plastic yield condition, an equilibrium condition, and an "unrestricted plastic flow" condition.

Later in this book more complete study will be given to the methods of analysis and design, and their simplicity as compared with the elastic design methods for the same indeterminate structures should become evident.

1.7 MARGIN OF SAFETY

In the opening article it was pointed out that the margin or factor of safety to be used in plastic design would be not less than that provided in usual past practice against failure of a simply supported beam. This may be demonstrated as follows:

In conventional elastic design, a member is selected in such a way that the maximum allowable stress is equal to 20,000 pounds per square inch at the working load. As shown in Fig. 1.1, p. 2, a simply supported beam has a reserve of elastic strength of 1.65 if the yield point stress is 33,000 pounds per square inch. Due to the ductility of steel there is an additional reserve against failure which amounts to about 12% of the yield load for a wide flange shape. Thus the total inherent overload factor of safety is equal to $1.65 \times 1.12 = 1.85$. The design basis is shown by the open arrow.

Now in plastic design the selection of member sizes is based upon the ultimate load. This load P_u is computed by multiplying the working load P_w by the same factor of safety that is *inherent* in simple beam design (in this case, 1.85), and a member is selected that will just support this factored load. In Fig. 1.1*b* is shown a fixed-ended beam designed plastically to support the same *working* load as the first beam, namely 1.0 k/ft. The corresponding ultimate load would therefore be 1.85 k/ft.

The load-versus-deflection curve for the restrained beam is also shown in Fig. 1.1*b*. The beam carries the same ultimate load as in the conventional design of the simple beam although its section modulus is reduced, and it is elastic at working load. At P_y is noted the increase of deflection rate which sets in when the yield stress is passed at the moment peaks at the beam ends. The deflection continues to be limited, however, until at P_u a final zone of yielding develops at the center of the beam.

The important thing to note is that the factor of safety is chosen to be the same in the plastic design of the indeterminate structure as is shown to be present in the elastic design of the simple beam. This inherent margin of 1.85 having been accepted for so many years for that most common structural element, the simple beam, it seems logical to adopt the same margin as adequate for any indeterminate structure similarly loaded.

1.8 THE CASE FOR PLASTIC DESIGN—SUMMARY

The goal in structural design is to provide a safe and enduring structure that incorporates maximum possible economy. If plastic analysis can be applied to design to realize these goals, it will be so applied, for the laws of evolution work as surely in the history of man-made structures as they do in a field such as biology.

The case for plastic design is supported by the following observations:

(1) Plastic design gives promise of economy in the use of steel, of saving in the design office by virtue of its simplicity, and of building frames more logically designed for greater over-all strength.

(2) The reserve in strength above the working loads computed by conventional elastic methods is considerable in indeterminate steel structures. Indeed, in some instances of elastic design, as much load-carrying capacity is disregarded as is used.

(3) Use of ultimate load as the design criterion provides at least the same margin of safety as is presently afforded in the elastic design of simple beams. (Fig. 1.1.)

(4) At working load the structure is normally in the so-called elastic range. (Fig. 1.1.)

(5) In most cases, a structure designed by the plastic method will deflect no more at working load than will a simply supported beam designed by elastic methods to support the same load. (Fig. 1.1.)

Plastic design is the realization of a goal that has been sought since the 1920's to see if some conscious design use could be made of the